Lignin-based Materials
Health Care and Medical Applications

Biomaterials Science Series

Editor-in-chief:
Julian Jones, *Imperial College London, UK*

Series editors:
Cole DeForest, *University of Washington, USA*
Changyou Gao, *Zhejiang University, China*

Titles in the Series:
1: Stimuli-responsive Drug Delivery Systems
2: Biodegradable Thermogels
3: Biofabrication and 3D Tissue Modeling
4: Biomaterial Control of Therapeutic Stem Cells
5: Antimicrobial Materials for Biomedical Applications
6: Decellularized Extracellular Matrix: Characterization, Fabrication and Applications
7: The Chemistry of Medical and Dental Materials, 2nd edition
8: Injectable Hydrogels for 3D Bioprinting
9: Extracellular Vesicles: Applications to Regenerative Medicine, Therapeutics and Diagnostics
10: Biomimetic Protein Based Elastomers: Emerging Materials for the Future
11: Phosphate and Borate Bioactive Glasses
12: Material-based Mechanobiology
13: Polysaccharide-based Biomaterials: Delivery of Therapeutics and Biomedical Applications
14: Biomaterial Based Approaches to Study the Tumour Microenvironment
15: Multicomponent Hydrogels: Smart Materials for Biomedical Applications
16: Lignin-based Materials: Health Care and Medical Applications

How to obtain future titles on publication:
A standing order plan is available for this series. A standing order will bring delivery of each new volume immediately on publication.

For further information please contact:
Book Sales Department, Royal Society of Chemistry, Thomas Graham House, Science Park, Milton Road, Cambridge, CB4 0WF, UK
Telephone: +44 (0)1223 420066, Fax: +44 (0)1223 420247,
Email: booksales@rsc.org
Visit our website at www.rsc.org/books

Lignin-based Materials
Health Care and Medical Applications

Edited by

Kuruvilla Joseph
Indian Institute of Space Science and Technology, India
Email: kjoseph.iist@gmail.com

Runcy Wilson
St. Cyrils College, India
Email: runcy21@gmail.com

Gejo George
St. Berchmans College (Autonomous), India
Email: gejochem@gmail.com

and

Saritha Appukuttan
Amrita Vishwa Vidyapeetham, India
Email: sarithatvla@gmail.com

ROYAL SOCIETY
OF **CHEMISTRY**

Biomaterials Science Series No. 16

Print ISBN: 978-1-83916-535-1
PDF ISBN: 978-1-83916-784-3
EPUB ISBN: 978-1-83916-785-0
Print ISSN: 2397-1401
Electronic ISSN: 2397-141X

A catalogue record for this book is available from the British Library

The Royal Society of Chemistry is a charity, registered in England and Wales, Number 207890, and a company incorporated in England by Royal Charter (Registered No. RC000524), registered office: Burlington House, Piccadilly, London W1J 0BA, UK, Telephone: +44 (0) 20 7437 8656.

For further information see our website at www.rsc.org

Printed in the United Kingdom by CPI Group (UK) Ltd, Croydon, CR0 4YY, UK

Preface

This book on lignin-based materials for biomedical applications is a one-point reference for lignin. This book opens up to the readers a beautiful description of the vast potential as well as sustainable utilization of lignin, which is currently discarded as a waste material. This book also brings to light the fact that lignin can be converted to various value-added products in the biomedical field, which might be an eye opener to emerging entrepreneurs and start-up aspirants. Since we are focused more on recycling and reducing waste materials, this book has great significance in the present scenario. We, the editors, have utilized the expertise of authors working in emergent areas of lignin research around the globe to make this book interesting as well as relevant to the present-day world. The chapters are arranged in such a way that beginners get an overview of the sources, extraction, chemistry, modifications, and applications of not only lignin but also its allied materials like its composites and blends. We have designed this book in such a way that it suits the requirements and satiate the curiosity of the students in universities as well as research scholars working in this field. We have also taken care to cater to the needs of the teachers and scientific experts in this area by careful consolidation of the future prospects of research and development that are anticipated in this area. We are grateful to all the contributors who have helped us in making this book a reality. The critical comments of the reviewers have helped us to make this book enticing. The vast expertise of the editorial team has made this book both appealing and up to date.

Kuruvilla Joseph
Runcy Wilson
Gejo George and Saritha Appukuttan

Biomaterials Science Series No. 16
Lignin-based Materials: Health Care and Medical Applications
Edited by Kuruvilla Joseph, Runcy Wilson, Gejo George and Saritha Appukuttan

Published by the Royal Society of Chemistry, www.rsc.org

Contents

Biomaterials Science Series No. 16
Lignin-based Materials: Health Care and Medical Applications
Edited by Kuruvilla Joseph, Runcy Wilson, Gejo George and Saritha Appukuttan

Published by the Royal Society of Chemistry, www.rsc.org

CHAPTER 1

Lignin and Its Composites: A Sustainable Tool for Health Care and Medical Applications – Introduction

A. S. SETHULEKSHMI,[a] SARITHA APPUKUTTAN,*[a] GEJO GEORGE,[b] RUNCY WILSON[c] AND KURUVILLA JOSEPH*[d]

[a] Department of Chemistry, School of Arts and Sciences, Amrita Vishwa Vidyapeetham, Amritapuri, Kollam, Kerala, India; [b] Research and Post Graduate Department of Chemistry, St. Berchmans College (Autonomous), Changanacherry, Kerala, India; [c] Department of Chemistry, St. Cyril's College, Adoor, Kerala, India; [d] Department of Chemistry, Indian Institute of Space Science and Technology, Valiyamala, Kerala, India
*Emails: sarithatvla@gmail.com; kjoseph.iist@gmail.com

1.1 Lignin

Cellulose, lignin and hemicellulose are the three important components present in the cell wall of plants and form an organic network using van der Waals forces, covalent bonds and intermolecular bridges.[1] The space between the cellulose and hemicellulose in plant cell walls is filled up with lignin where lignin acts as a binding component. Lignin is an important component in the cell wall of plants because it provides physical strength and durability to plants, protects plants from insects, microorganisms and

Biomaterials Science Series No. 16
Lignin-based Materials: Health Care and Medical Applications
Edited by Kuruvilla Joseph, Runcy Wilson, Gejo George and Saritha Appukuttan

Published by the Royal Society of Chemistry, www.rsc.org

physical disturbances and transports nutrients and water internally.[2] It is the second most abundant biopolymer with the highest aromatic moieties and constitutes 15–35% of lignocellulosic biomass.[3]

Lignin is a term derived from the Latin word "lignum", meaning wood.[4] Lignin has a glass transition temperature of about 90 °C and has a melting point of 170 °C.[5] It is easily oxidizable, soluble in hot alkalis, but insoluble in most solvents and is not hydrolyzable by acids. Lignin is highly hydrophobic and polymerized, is an amorphous complex organic biopolymer

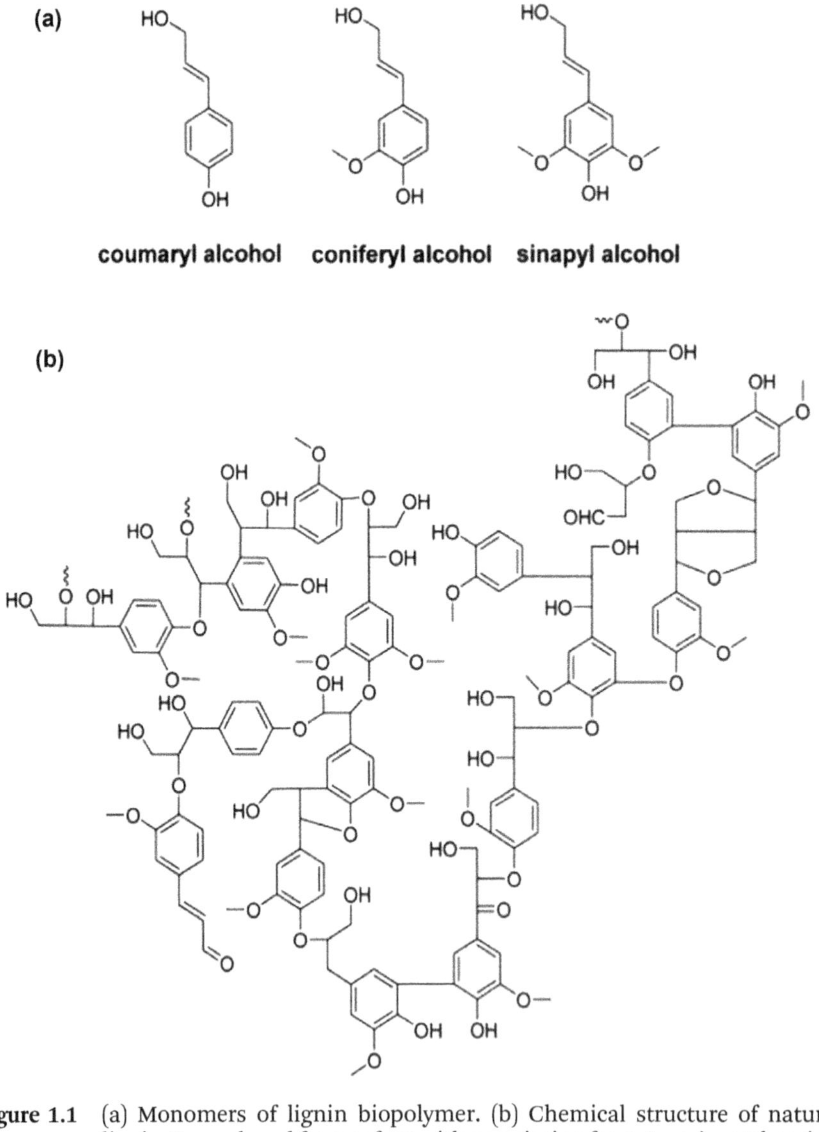

Figure 1.1 (a) Monomers of lignin biopolymer. (b) Chemical structure of natural lignin. Reproduced from ref. 20 with permission from American Chemical Society, Copyright 2017.

composed of phenylpropane groups and is one of the by-products in the paper and bioethanol industries.[6] Lignins with various compositions, structures and properties are obtained from various sources including trees, plants and crops.[7] They can be isolated from wood and agricultural residues and also from black liquor through a precipitation method. Usually, the lignin content is greater in wood than in grasses and the lignin amount changes with various plant sources. The diversity and complexity of lignin structures are mainly associated with their various types, sources, environmental conditions of their growth and extraction and purification processes.[8]

The structure of lignin has been intensively evaluated for several years, but the exact structure is uncertain. This three-dimensional (3D) macromolecule is made up of three monolignol monomers such as sinapyl alcohol, *p*-coumaryl alcohol and coniferyl alcohol and these three units are together known as phenylpropanoid units.[9] The most common linkages existing in high molecular weight heterogeneous lignin are β-O-4, α-O-4, β-5, 5-5, 4-O-5, β-1 and β-β. Various functional groups such as hydroxyl, carbonyl, carboxyl and methoxy groups are present on the lignin structure.[10] Figure 1.1 presents the structural units and structure of lignin. Numerous functional groups and linkages impart a complicated structure to lignin, and thus it exhibits biodegradability,[11] antioxidant properties,[12] antimicrobial properties,[13] adhesive properties,[14] biocompatibility,[15] UV absorption,[16] thermal stability,[17] reinforcing capability,[18] stabilizing effects,[19] *etc.* Depending on the source and isolation method, lignins are of various types such as organosolv lignin, technical kraft lignin, kraft lignin, lignosulfonate lignin (hardwood), lignosulfonate lignin (softwood) and dilute acid lignin.

1.2 Research Progress in Lignin

Around 70 million tons of lignin per year are produced by the paper and pulp industries, of which only 2% has been utilized to recover energy in the pulping industry and the rest is remaining as a waste material.[21] Heating of lignin to temperatures higher than 170 °C causes degradation of lignin, which limits its use in industrial purposes. It is very difficult to develop value added products from lignin due to the degradation of its phenolic structure and greater tar and char formation during the conversion process.[22] This in turn leads to an increase in the waste products from the paper and pulp industries. These lignin wastes have become an environmental problem when they are burned as low-grade fuel or dumped into earth. In recent decades, researchers' interests have been focused on the transformation of lignin from its use as an animal feed and low-grade fuel into advanced materials.[23]

The presence of numerous functional groups makes the functionalization of lignin easy. Usually chemical modifications of lignin can be categorized into three types: chemical modifications of the hydroxyl group, functionalization by introducing new chemically active sites and depolymerization or fragmentation. Reactions adopted for lignin chemical modifications are polycondensation, oxidation, graft copolymerization, sulfonation and deuteration. Surface functionalized lignin can be exploited in many fields than its unmodified form.[24]

Conversion of microsized lignin into nano-forms provides advanced performance and thus opens new applications for lignin-based materials. Like other biopolymers and materials, lignin can also show attractive properties when its size reaches the nanoscale. Synthesis of lignin nanoparticles has been extensively studied by various researchers, and this is still going on. Lignin nanoparticles with various morphologies like hollow, smooth colloidal, quasi-spherical and spherical shapes have been prepared by controlling reaction conditions such as temperature, pH, lignin concentration and solvent/antisolvent.[25] Self-assembly, anti-solvent precipitation and mechanical methods are the most common synthesis methods for lignin nanoparticles.[25] Researchers are trying to replace lengthy and organic solvent-based lignin nanoparticle synthesis by a green, safe and simple method.

Composite materials are advanced engineering materials having extensive applications in our daily life and also in various industries. Among them lignin-based composites and nanocomposites are gaining huge attention in the current society due to their bio-origin and their excellent applications in numerous fields.[26] The last few years have noticed a tremendous increase in the development of lignin-related composites with various applications like water purification, antioxidant, biomedical and energy storage applications. Lignin can also be used as an excellent reinforcing agent for various polymeric materials like epoxy,[27] polystyrene,[28] poly(vinyl alcohol) (PVA),[29] poly(methyl methacrylate),[30] poly(lactic acid),[31] polyurethane[32] and rubber[33] to fabricate composites with various applications such as coatings, lubricants, fire retardants, surfactants and waste water

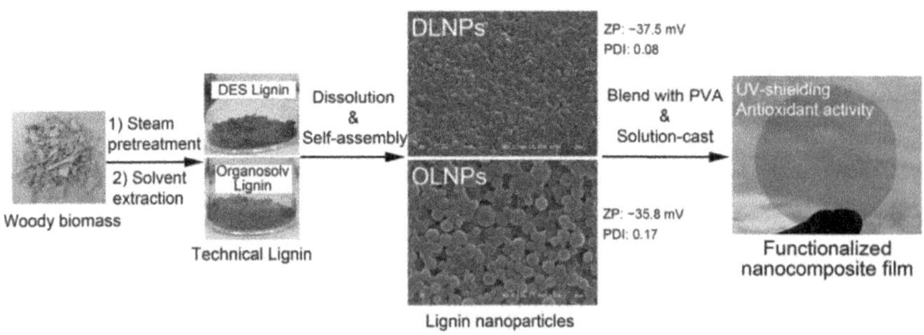

Figure 1.2 Synthetic procedure to fabricate lignin nanoparticles and the lignin nanoparticle/PVA composite film. ZP, zeta potential; PDI, polydispersity index. Reproduced from ref. 34, https://doi.org/10.1186/s13068-017-0876-z, under the terms of the CC BY 4.0 license https://creativecommons.org/licenses/by/4.0/.

treatment agents. Addition of lignin into polymer matrices has resulted in the formation of polymer composites with improved properties. Lignin can be incorporated into polymers with or without chemical modifications depending on their anticipated application. The high molecular weight and complex structure of the lignin molecule make the development of lignin nanocomposites a challenging area. Compared to micro-lignin, the reinforcing ability of nano-lignin in polymers has not been much studied. Figure 1.2 shows the synthetic procedure of polyvinyl alcohol–lignin nanocomposites.

1.3 Biomedical Applications of Lignin

Currently, researchers and industries concentrate more on the development of lignin-based biomaterials for various applications. Considering its pharmacological activities such as UV protection, anticancer, antioxidant and antibacterial activities, lignin plays a noticeable role in the biomedical field. Lignin proved its incredible role in biomedical fields especially in drug/gene delivery, tissue engineering,[35] 3D printing,[36] bio-sensors, bioimaging, dietary supplement additives, *etc.* Compounds isolated from lignin can be utilized to develop pharmaceuticals and drug delivery materials.[37] Among various lignins, kraft and sulfite lignins are the widely used lignins for the preparation of pharmaceuticals. Lignin has the ability to reduce cholesterol by binding with the bile acids in the intestine. Therefore, lignin can be utilized as an effective medicine against various diseases such as diabetes, viral infections, obesity, cancers and thrombosis.

Lignin nanoparticles have shown their incredible applications in the biomedical arena mainly in gene and drug delivery,[38] personal health care and medical device coating. Many studies have been done to combine lignin with other materials for constructing composites for medical applications. The high mechanical strength of lignin enables its use in hydrogels to protect wounds from further contamination.[39] Lignin is a major source of antimicrobial agents and can be used to replace fossil-based agents against many hazardous microbes.[40] The methoxyl and hydroxyl groups present in lignin are biologically active. The antimicrobial nature of lignin makes it a more interesting candidate than silver nanoparticles while considering its low environmental impact.[41] Cytotoxicity studies of lignin confirm that lignin does not disrupt cell viability and is thus safe to use.

Even though lignin is an outstanding candidate in biomedical applications, its applications are only limited to lab scale studies, and the *in vivo* evaluations are rarely conducted owing to the lack of clinical studies. By proper *in vivo* and *in vitro* assessments, lignin-based micro- and nanocomposites will open a new window in the biomedical field in the upcoming days. Figure 1.3 presents the biomedical applications of lignin and its composite materials. Figure 1.4 shows the wound healing applications of a chitosan/PVA hydrogel composite.

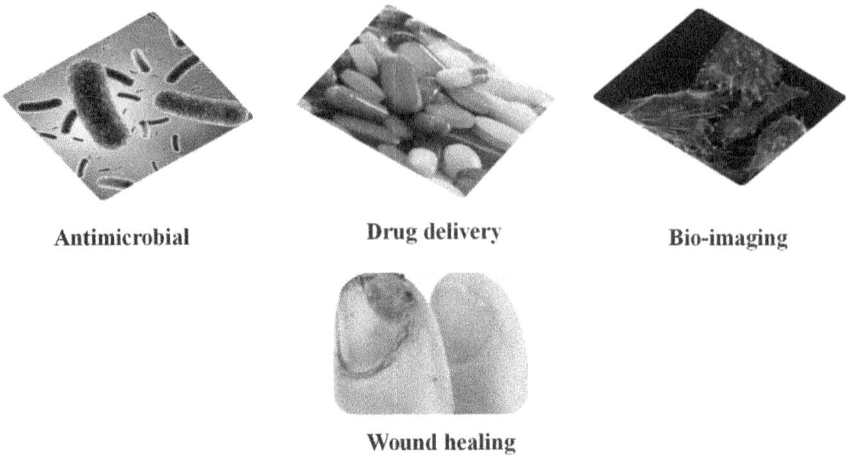

Figure 1.3 Biomedical applications of lignin and its composites.

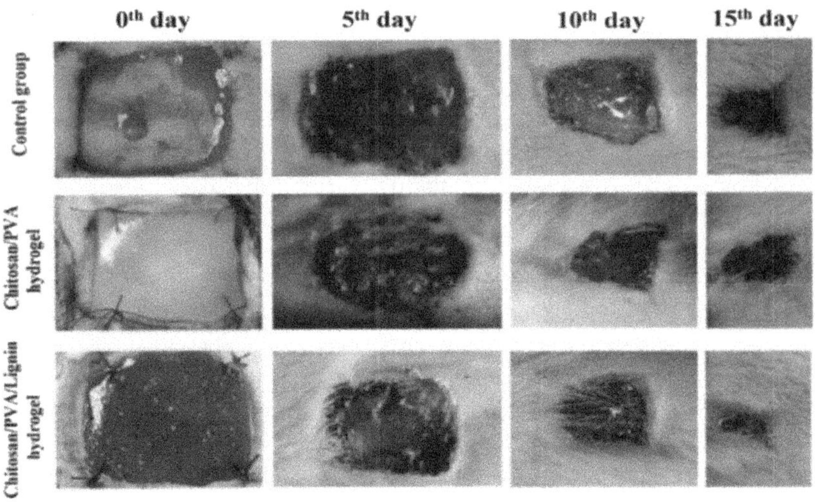

Figure 1.4 Photographs of wound healing results of the control, chitosan/PVA hydrogel and chitosan/PVA/lignin hydrogel. Reproduced from ref. 39 with permission from Elsevier, Copyright 2019.

1.4 Overview of the Chapters

This book entitled *Lignin-based Materials: Health Care and Medical Applications* encompasses 11 chapters written by prominent research groups and researchers from all over the world. This book will impart a detailed knowledge on the chemistry of lignin, its processing and the fabrication of lignin-based composites and nanocomposites with different matrices (thermosets, thermoplastics, elastomers, *etc.*) and their medical applications. It will provide almost all the details about lignin and its micro- and nanocomposites, thereby creating awareness about the sustainable utilization of waste lignin in the environment. This book will be of significance to scientists/researchers working in industries and the medical field and research scholars owing to the importance of lignin and its related materials for medical purposes.

Chapter 1 gives a general introduction to lignin and its composites and nanocomposites including their processing, properties and applications. Various methods for the synthesis of lignin nanoparticles and hydrogels are presented in Chapter 2. A brief overview of chemical modifications of lignin for biomedical applications is provided in Chapter 3. Various methods for lignin recovery and fractionation are discussed in Chapter 4. Chapter 5 focuses on the basic requirements and properties of lignin-based materials for biomedical applications. Chapter 6 deals with the antimicrobial properties of lignin and its composites. Moreover, this chapter describes the antioxidant properties and their applications in food packaging and plant protection and growth. Chapter 7 focuses on the drug delivery applications of lignin and its composites. Tissue engineering, wound dressing and bioimaging applications of lignin and its composites are described in Chapters 8–10, respectively. The UV-resistance properties and associated applications of lignin-based composites and their hybrid composites are described in Chapter 11.

Acknowledgements

The authors would like to thank Amrita Vishwa Vidyapeetham for providing the funding.

References

1. M. Zheng, *et al.*, Manipulation of lignin metabolism by plant densities and its relationship with lodging resistance in wheat, *Sci. Rep.*, 2017, 7, 1–12.
2. H. Chung and N. R. Washburn, Chemistry of lignin-based materials, *Green Mater.*, 2013, **1**, 137–160.
3. T. A. Khan, J.-H. Lee and H.-J. Kim, Lignin-based adhesives and coatings, in *Lignocellulose for future bioeconomy*, Elsevier, 2019, pp. 153–206.
4. A. Agrawal, N. Kaushik and S. Biswas, Derivatives and applications of lignin – an insight, *SciTech J.*, 2014, **1**, 30–36.
5. J. P. Patel and P. H. Parsania, Characterization, testing, and reinforcing materials of biodegradable composites, in *Biodegradable and Biocompatible Polymer Composites: Processing, Properties and Applications*, 2018, pp. 55–79.
6. K. Schröder, K. Matyjaszewski, K. J. T. Noonan and R. T. Mathers, Towards sustainable polymer chemistry with homogeneous metal-based catalysts, *Green Chem.*, 2014, **16**, 1673–1686.

7. A. Nasrullah, A. H. Bhat, A. S. Khan and H. Ajab, Comprehensive approach on the structure, production, processing, and application of lignin, in *Lignocellulosic Fibre and Biomass-Based Composite Materials*, Elsevier, 2017, pp. 165–178.
8. D. S. Bajwa, G. Pourhashem, A. H. Ullah and S. G. Bajwa, A concise review of current lignin production, applications, products and their environmental impact, *Ind. Crops Prod.*, 2019, **139**, 111526.
9. H. Hatakeyama and T. Hatakeyama, Lignin structure, properties, and applications, in *Biopolymers*, Springer, 2009, pp. 1–63.
10. D. Dakshinamoorthy, *et al.*, Streamlining the conversion of biomass to polyesters: bicyclic monomers with continuous flow, *Green Chem.*, 2014, **16**, 1774–1783.
11. A. P. Richter, *et al.*, Synthesis and characterization of biodegradable lignin nanoparticles with tunable surface properties, *Langmuir*, 2016, **32**, 6468–6477.
12. R. Kaur and S. K. Uppal, Structural characterization and antioxidant activity of lignin from sugarcane bagasse, *Colloid Polym. Sci.*, 2015, **293**, 2585–2592.
13. J. Sunthornvarabhas, S. Liengprayoon and T. Suwonsichon, Antimicrobial kinetic activities of lignin from sugarcane bagasse for textile product, *Ind. Crops Prod.*, 2017, **109**, 857–861.
14. M. Ghorbani, *et al.*, Lignin phenol formaldehyde resoles: the impact of lignin type on adhesive properties, *BioResources*, 2016, **11**, 6727–6741.
15. D. M. Rocca, *et al.*, Biocompatibility and photo-induced antibacterial activity of lignin-stabilized noble metal nanoparticles, *RSC Adv.*, 2018, **8**, 40454–40463.
16. H. Sadeghifar and A. Ragauskas, Lignin as a UV light blocker—a review, *Polymers*, 2020, **12**, 1134.
17. J. C. Domínguez, M. Oliet, M. V Alonso, M. A. Gilarranz and F. Rodríguez, Thermal stability and pyrolysis kinetics of organosolv lignins obtained from Eucalyptus globulus, *Ind. Crops Prod.*, 2008, **27**, 150–156.
18. C. Jiang, *et al.*, Converting waste lignin into nano-biochar as a renewable substitute of carbon black for reinforcing styrene-butadiene rubber, *Waste Manage.*, 2020, **102**, 732–742.
19. A. Gregorová, Z. Cibulková, B. Košíková and P. Šimon, Stabilization effect of lignin in polypropylene and recycled polypropylene, *Polym. Degrad. Stab.*, 2005, **89**, 553–558.
20. H. Liu and H. Chung, Visible-light induced thiol–ene reaction on natural lignin, *ACS Sustainable Chem. Eng.*, 2017, **5**, 9160–9168.
21. Y. Meng, J. Lu, Y. Cheng, Q. Li and H. Wang, Lignin-based hydrogels: a review of preparation, properties, and application, *Int. J. Biol. Macromol.*, 2019, **135**, 1006–1019.
22. L. Yang, K. Seshan and Y. Li, A review on thermal chemical reactions of lignin model compounds, *Catal. Today*, 2017, **298**, 276–297.

23. S. Sahoo, M. Ö. Seydibeyoğlu, A. K. Mohanty and M. Misra, Characterization of industrial lignins for their utilization in future value added applications, *Biomass Bioenergy*, 2011, **35**, 4230–4237.
24. X. Zhang, D. Jeremic, Y. Kim, J. Street and R. Shmulsky, Effects of surface functionalization of lignin on synthesis and properties of rigid bio-based polyurethanes foams, *Polymers*, 2018, **10**, 706.
25. Z. Zhang, V. Terrasson and E. Guénin, Lignin nanoparticles and their nanocomposites, *Nanomaterials*, 2021, **11**, 1336.
26. G. Parvathy, A. S. Sethulekshmi, J. S. Jayan, A. Raman and A. Saritha, Lignin based nano-composites: synthesis and applications, *Process Saf. Environ. Prot.*, 2021, **145**, 395–410.
27. Y. Zhang, H. Wang, T. L. Eberhardt, Q. Gu and H. Pan, Preparation of carboxylated lignin-based epoxy resin with excellent mechanical properties, *Eur. Polym. J.*, 2021, **150**, 110389.
28. N. Ghavidel and P. Fatehi, Synergistic effect of lignin incorporation into polystyrene for producing sustainable superadsorbent, *RSC Adv.*, 2019, **9**, 17639–17652.
29. X. Zhang, W. Liu, W. Liu and X. Qiu, High performance PVA/lignin nanocomposite films with excellent water vapor barrier and UV-shielding properties, *Int. J. Biol. Macromol.*, 2020, **142**, 551–558.
30. F. Avelino, D. R. de Oliveira, S. E. Mazzetto and D. Lomonaco, Poly(methyl methacrylate) films reinforced with coconut shell lignin fractions to enhance their UV-blocking, antioxidant and thermo-mechanical properties, *Int. J. Biol. Macromol.*, 2019, **125**, 171–180.
31. A. Chaubey, K. R. Aadil and H. Jha, Synthesis and characterization of lignin-poly lactic acid film as active food packaging material, *Mater. Technol.*, 2021, **36**, 585–593.
32. W. Hu, *et al.*, Synergism between lignin, functionalized carbon nanotubes and Fe_3O_4 nanoparticles for electromagnetic shielding effectiveness of tough lignin-based polyurethane, *Compos. Commun.*, 2021, **24**, 100616.
33. C. Jiang, H. He, H. Jiang, L. Ma and D. M. Jia, Nano-lignin filled natural rubber composites: Preparation and characterization, *eXPRESS Polym. Lett.*, 2013, 7, 480–493.
34. D. Tian, *et al.*, Lignin valorization: lignin nanoparticles as high-value bio-additive for multifunctional nanocomposites, *Biotechnol. Biofuels*, 2017, **10**, 1–11.
35. M. Witzler, *et al.*, Lignin-derived biomaterials for drug release and tissue engineering, *Molecules*, 2018, **23**, 1885.
36. M. Tanase-Opedal, E. Espinosa, A. Rodríguez and G. Chinga-Carrasco, Lignin: a biopolymer from forestry biomass for biocomposites and 3D printing, *Materials*, 2019, **12**, 3006.
37. O. Yu and K. H. Kim, Lignin to materials: a focused review on recent novel lignin applications, *Appl. Sci.*, 2020, **10**, 4626.
38. M. S. Alqahtani, A. Alqahtani, A. Al-Thabit, M. Roni and R. Syed, Novel lignin nanoparticles for oral drug delivery, *J. Mater. Chem. B*, 2019, 7, 4461–4473.

39. Y. Zhang, *et al.*, Novel lignin–chitosan–PVA composite hydrogel for wound dressing, *Mater. Sci. Eng., C*, 2019, **104**, 110002.
40. W. Yang, *et al.*, Effect of cellulose and lignin on disintegration, antimicrobial and antioxidant properties of PLA active films, *Int. J. Biol. Macromol.*, 2016, **89**, 360–368.
41. A. Alzagameem, *et al.*, Antimicrobial activity of lignin and lignin-derived cellulose and chitosan composites against selected pathogenic and spoilage microorganisms, *Polymers*, 2019, **11**, 670.

CHAPTER 2

Synthesis of Lignin Nanoparticles and Hydrogels

L. RUIZ-RUBIO*[a,b] AND J. L. VILAS VILELA[a,b]

[a] Macromolecular Chemistry Group (LQM), Physical Chemistry Department, Faculty of Science and Technology, University of the Basque Country, 48940 Leioa, Spain; [b] BCMaterials, Basque Center for Materials, Applications and Nanostructures, UPV/EHU Science Park, 48940 Leioa, Spain
*Email: leire.ruiz@ehu.eus

2.1 Introduction

The reduction of fossil resources together with the increase in plastic waste has generated great interest in finding clean and sustainable resources that could be available worldwide. Biomass, as the second most abundant material, can be an alternative source to replace many petroleum products.

Lignin is one of the three major constituents of lignocellulosic biomass, which includes cellulose (40–45%), hemicellulose (25–35%) and lignin (15–30%).[1] In addition, lignin is responsible for water transport and structural integrity in plants and the most abundant natural substance composed of an aromatic-based heteropolymer.[2] Lignin is present in 20–25% of the mass of hardwood (*i.e.* pine, fir, spruce and larch trees) and about 30% of softwood (*i.e.* eucalyptus, poplar (POP) and birch trees).[3] Furthermore, lignin is also an important by-product of the pulp and paper industry.

Biomaterials Science Series No. 16
Lignin-based Materials: Health Care and Medical Applications
Edited by Kuruvilla Joseph, Runcy Wilson, Gejo George and Saritha Appukuttan

Published by the Royal Society of Chemistry, www.rsc.org

2.1.1 Chemical Structure of Lignin

Lignin has a complex polymeric structure with methoxyl, phenolic hydroxyl and aldehyde groups on its terminal chains. Its network-like structure is composed of aromatic units derived from phenylpropanoid monomers, known as linked precursor hydroxycinnamic alcohols (Figure 2.1): *p*-coumaryl, coniferyl and sinapyl. Lignin is formed by a dehydrogenation polymerization carried out by enzymes in which the repeating units are the following: *p*-hydroxyphenyl (H), guaiacyl (G) and syringyl (S) are formed from their alcohols (Figure 2.1).[4] In addition, in the biosynthesis process, the lignin monolignols are generated by interunit bonds through *in situ* radical polymerizations: aryl ether (α-O-4′ and β-O-4′), resinol (β-β′), phenylcoumaran (β-5′), biphenyl (5–5′) and 1,2-diarylpropane (β-1).[5]

2.1.2 Extraction and Isolation of Lignin

There are numerous sources of lignin from which it can be isolated, such as woody biomass, trees, wheat straw and grass, among others. Lignin is usually obtained as a by-product of cellulose extraction in most paper industries. Currently, there are several lignin extraction and isolation processes industrially available. However, due to the heterogeneity of the sources, to date, there has been no method available for the quantitative isolation of lignin without some structural modification during the process.[6] The main chemical processes used for the extraction and purification of lignin are summarized in the following sections. However, it is important to notice that lignin could also be obtained by an enzymatic hydrolysis process.

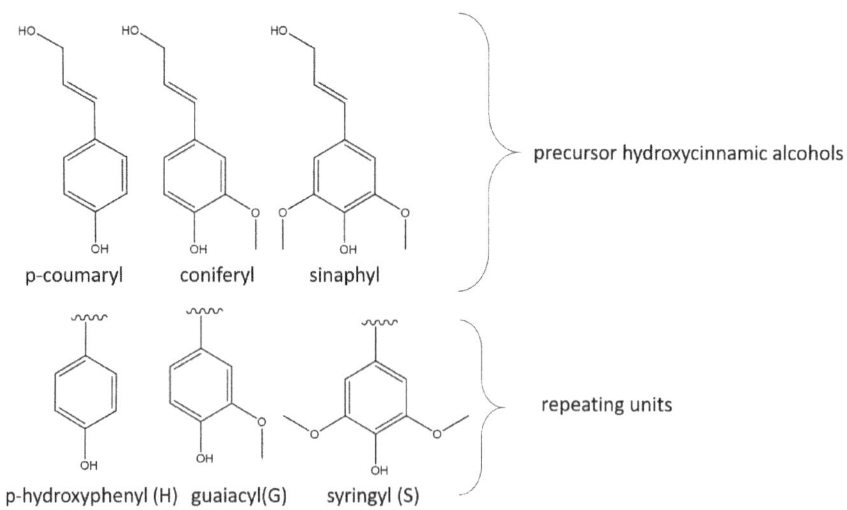

Figure 2.1 Lignin components: precursors and repeating units.

2.1.2.1 *Kraft Lignin*

The Kraft process is the main process for manufacturing paper pulp from wood. In the first stage, the wood is treated with a solution of sodium hydroxide and sodium sulfide in the temperature range of 155–175 °C, in which the cellulose and a black liquid in which the lignin is present can be separated.[7] In the second stage, the Kraft lignin is obtained through a delignification process, in which the black liquid is precipitated and is neutralized with an acidic solution (pH 1–2). Kraft lignin is highly modified and insoluble in water, but can be dissolved in alkaline solution.

2.1.2.2 *Lignosulfonates*

Another class of lignins is lignosulfonates. These are obtained from softwood residues, using various salts of sulfurous acid (sulfites or bisulfites). This process can be carried out under various conditions.[8,9] As main characteristics, lignosulfonates have hydrophobic and hydrophilic moieties. That is, these compounds are chemically modified with sulfonate groups, carbohydrates and small amounts of wood extracts and inorganic compounds, which makes them soluble in water. Basically, they are prepared as calcium, magnesium and ammonium salts.

2.1.2.3 *Organosolv Lignin*

In terms of chemical extraction methods, it should also be noted that many processes have been developed by using different organic solvents. Lignin obtained by any of these methods is called organosolv lignin (OSL).[10] In general, these methods are based on the extraction of lignin from biomass with an organic solvent/water mixture at high temperature/pressure. As for solvents, the most used solvents are alcohols. Currently, there are two types of commercial lignin obtained with these methods: Alcell and Organocell lignins. It should be noted that these lignins present fewer modifications than those obtained by the previous methods.

2.2 Synthesis of Lignin-based Hydrogels

The growing awareness of the current environmental problem together with the need for search for new raw materials to replace oil-based materials has pushed the development of lignin-based materials. Thus, hydrogels have not been alien to this revolution and each year the number of works based on these materials is growing.[11,12] In this section, the main syntheses for lignin-based hydrogels will be summarized. This section will be divided according to the main interactions of the synthesized hydrogels, similar to classical hydrogels: lignin hydrogels obtained by physical interactions and chemical interactions.[13,14]

2.2.1 Lignin Hydrogels Obtained by Physical Interaction

An interpenetrating polymeric network (IPN) can be defined as a mixture of two or more polymers in the form of a network. In this network, at least one of the components crosslinks in the presence of others.[15,16] Although crosslinking is not the main method of synthesis of lignin hydrogels, a few works based on lignin IPNs have been described in the literature. In one of the first works described on lignin IPNs, Raschip *et al.*[17,18] described the crosslinking of a mixture of xanthan and lignin in the presence of epichlorohydrin. The authors described that the semi-IPN of xanthan with lignin produces hydrogels with a very high degree of swelling; in addition, the nature of lignin strongly influences their swelling. Moreover, Peñaranda *et al.*[19] described an IPN obtained by polymerization of acrylamide (AAm) by free radical polymerization and crosslinking with *N*,*N*′-methylenebisacrylamide (NMBAm) in the presence of lignin. The authors reported a maximum water absorption at 100 hours and they demonstrated successful adsorption of Cu(ɪɪ) and Ni(ɪɪ) metal ions so these hydrogels could be potentially used in waste water remediation.

The lignin structure has abundant polar sites capable of forming hydrogen bonds, and taking advantage of these interactions, Oveissi and co-workers[20] introduced lignin in polyurethane hydrogels. The hydrogen bonds formed between the polyurethane and lignin units improved the mechanical properties and the processability of the final hydrogels. Similarly, Jiang *et al.*[21] introduced lignin in a PNIPAM hydrogel; as a result, hydrogels with enhanced thermal stability and increased hydrophobicity were obtained. The introduction of lignin in these gels leads to a decrease in lower critical solution temperature (LCST). In addition, the concentration of lignin present in the IPN varies the properties of the hydrogel, such as its compressive strength and swelling ratio, among others. On the other hand, the H-bonding with lignin was also described for physically crosslinked hydrogels by Huang *et al.*[22] In their study, lignin was incorporated into a physical hydrogel consisting of hydroxyethylcellulose (HEC) and polyvinyl alcohol (PVA) crosslinked with borax. The hydrogel was conductive, exhibited self-healing properties and was highly stretchable and thermosensitive.

More recently, some studies synthesized physical hydrogels based on lignin and chitosan (CHI). Ravishankar *et al.*[23] reported hydrogels formed from the interactions between CHI and alkali lignin by electrostatic interactions between phenoxide anions and amine groups of CHI. These hydrogels exhibited good biological properties and they provided a conductive surface for cell adhesion and growth. In another example, Zhang *et al.*[24] used lignin to improve the properties of a PVA-CHI hydrogel. Thus, the introduction of lignin improved the mechanical properties of the hydrogel.

2.2.2 Lignin Hydrogels Obtained by Chemical Interaction

Lignin has many reactive groups in its structure, such as phenolic and aliphatic hydroxyl groups in the side chain. For example, these groups can readily react with formaldehyde and epichlorohydrin to form crosslinked structures.[25]

Grishechko *et al.*[26] reported the synthesis of gels from lignin and phenol crosslinked with formaldehyde. In their work, they studied the influence of the composition on the gelation process as well as on its porous structure. In most cases, dark brown or black opaque hydrogels were obtained, depending on the composition. In fact, the formulations with too much lignin or little formaldehyde did not gel, with the maximum lignin content for gel production being 80%. This content is significantly higher than that reported by Chen and Li[27] for the lignin–resorcinol–formaldehyde system, for which the maximum was 50%. In addition, it should be noted that less formaldehyde was also needed for gelation. The authors concluded that lignin could replace phenol in this type of gel, due to its similar structure (with aromatic hydroxyl groups) and its similar crosslinking ability to phenol-formaldehyde resins. However, the steric hindrance of the lignin structure limits its reactivity with formaldehyde.

On the other hand, Cicolau and co-workers[28] prepared hydrogels from cellulose and lignin crosslinked with epichlorohydrin. The procedure was carried out in two steps: they dissolved the cellulose in a basic solution to mix it with lignin and finally crosslinked it with epichlorohydrin. The hydrogels had a high swelling capacity and proved to be suitable for the release of polyphenols. Furthermore, the increase in lignin content increased the percentage of polyphenols released. Similarly, Dai *et al.*[29] synthesized hydrogels of cellulose and lignin crosslinked with epichlorohydrin. In this work, a series of hydrogels with variable pore sizes were obtained that were evaluated for the controlled fractionation of lignin by means of specific absorption and its filtration.

Other polysaccharide (xanthan or agarose)/lignin-based hydrogels were also synthesized by using epichlorohydrin. Raschip *et al.*[30] reported xanthan/lignin hydrogels which were obtained by crosslinking xanthan and lignin with epichlorohydrin in alkaline medium. In this study, lignins obtained from five different sources, woody plants and annual fibre crops (softwood lignin, softwood lignin (Curran 100), pine wood lignin, aspen wood lignin and lignin from annual plants), were evaluated. The authors reported that the lignin type determines not only the morphology and porosity of the obtained hydrogels but also the interaction between the components of the hydrogels. These hydrogels have new hydrogen bonds with a bonding energy which varies with the nature of the used lignin. The intramolecular hydrogen bonds present bonding energy values between 16.9 and 17.3 kJ, whereas the bonding energy values of the intermolecular hydrogen bonds vary from 25.1 to 27.3 kJ. Considering the source of the lignin, the bonding energies of the xanthan/lignin hydrogels decrease in the following order: aspen lignin > softwood lignin > pine wood lignin > softwood lignin (Curran 100) > lignin from annual plants.

In another example of polysaccharide-based lignin hydrogels, Sathawong and co-workers prepared agarose/lignin hydrogels also crosslinked with epichlorohydrin.[31] The authors prepared different hydrogels starting from different formulations of agarose/lignin with epichlorohydrin as a crosslinking agent. Figure 2.2 shows the general scheme of the synthesis of lignin/agarose hydrogels. On the other hand, the hydrogels became browner as the amount of lignin present increased. It is important to highlight that both the lignin and epichlorohydrin contents have significant influences on the mechanical properties of the hydrogels. On the one hand, hydrogels with 1 wt% of agarose and lignin and a minimum epichlorohydrin (5 mL) present a hardness of 28.134 ± 1.88 gF, whereas a maximum hardness of 181.491 ± 5.59 gF is reached for formulations with 5 wt% of agarose and lignin and 10 mL of epichlorohydrin.

In addition, several hydrogels have been synthesized with lignin and PVA. Wu *et al.* developed a hydrogel with lignin from aspen wood and PVA crosslinked with epichlorohydrin.[32] In that work, the influence of the molecular weight and the concentration of PVA on the swelling capacity of the hydrogel was evaluated. The samples with the higher average molecular weight present higher viscosity that is capable of forming hydrogels at lower lignin concentrations. On the other hand, the effect of the epichlorohydrin concentration was evaluated; the swelling capacity of the hydrogel increases with increasing epichlorohydrin concentration until 3–4%: however, at higher epichlorohydrin concentration the swelling capacity decreases again. Finally, Wu and co-workers tested three different lignins (alkali lignin,

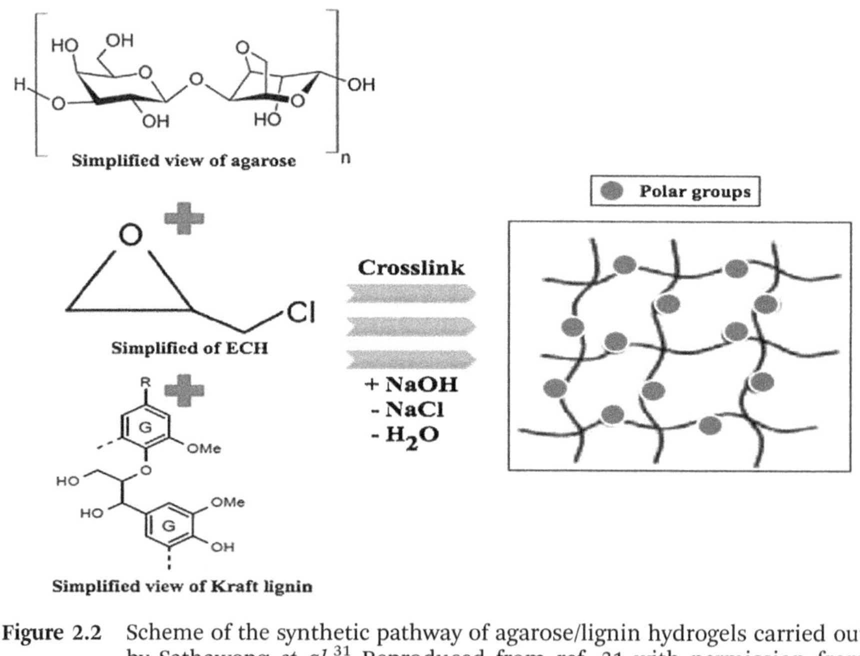

Figure 2.2 Scheme of the synthetic pathway of agarose/lignin hydrogels carried out by Sathawong *et al.*[31] Reproduced from ref. 31 with permission from Elsevier, Copyright 2018.

Kraft lignin and enzymatic hydrolysis lignin), with the maximum swelling ratio being around 550 $g\,g^{-1}$, for both Kraft lignin and enzymatic hydrolysis lignin.

Cortés-Treviño *et al.*[33] reported a direct approach to obtain a lignin-based oleogel. In their work, alkali lignin was modified with poly(ethylene glycol) diglycidyl ether (PEGDGE), a crosslinker, to directly form an oleogel in which lignin acted as a backbone. Similarly, Passauer *et al.*[34] prepared hydrogels based on oxidatively preactivated spruce OSL cross-linked with PEGDGE. These hydrogels had a high degree of swelling, which would allow them to be applied as superabsorbents. The authors reported that the swelling properties were related to the chemical structure and microstructure of the materials, with free and bound water in these hydrogels depending on the degree of modification of lignin with PEGDGE.

Another approach for the synthesis of lignin-based hydrogels was the esterification reaction with different copolymers containing maleic acid. Domínguez-Robles *et al.*[35] described the formation of lignin hydrogels by the esterification of the hydroxyl groups present in the lignin backbone with the acid groups formed by the opening of the anhydride group of the poly(methyl vinyl ether-*co*-maleic acid) in alkaline medium. A scheme of the reaction carried out in the synthesis of these hydrogels is depicted in Figure 2.3. These hydrogels had a high water absorption capacity, in some cases retaining up to 13 000%.

Larrañeta *et al.*[36] described a similar procedure using a GRANTEZ™, a commercial label for poly(methyl vinyl ether-*co*-maleic acid), and adding poly(ethylene glycol). However, in this study, the esterification reaction was carried out in a solid state by using microwave radiation, without using any solvents. In this case the water uptake of the hydrogels was smaller than those of the previously described materials, reaching 500%.

Finally, radical polymerization could also be used in the formation of lignin-based hydrogels. This approach turned out to be extensive due to the high amount of monomers available. This process is based on the formation of lignin-grafted copolymers within a polymerized network, forming IPN hydrogels. The radical polymerization starts with the formation of radicals on the phenolic hydroxyls in the presence of an initiator which consequently react with the monomers in the media.[37,38] As an example, Zerpa *et al.*[39] described the formation of Kraft lignin-based hydrogels by radical polymerization with *N*-isopropylacrylamide (NIPAM) and *N*,*N*′-methylenbisacrylamide (NMBA). The presence of NIPAM in the structure causes these hydrogels to have an LCST between 34 and 37 °C. The authors described that the hydrogels with lignin presented less swelling than those developed from NIPAM and NMBA. However, they presented greater thermal stability. In addition, the lignin content increased the rigidity and rheological stability of these lignin-based hydrogels.

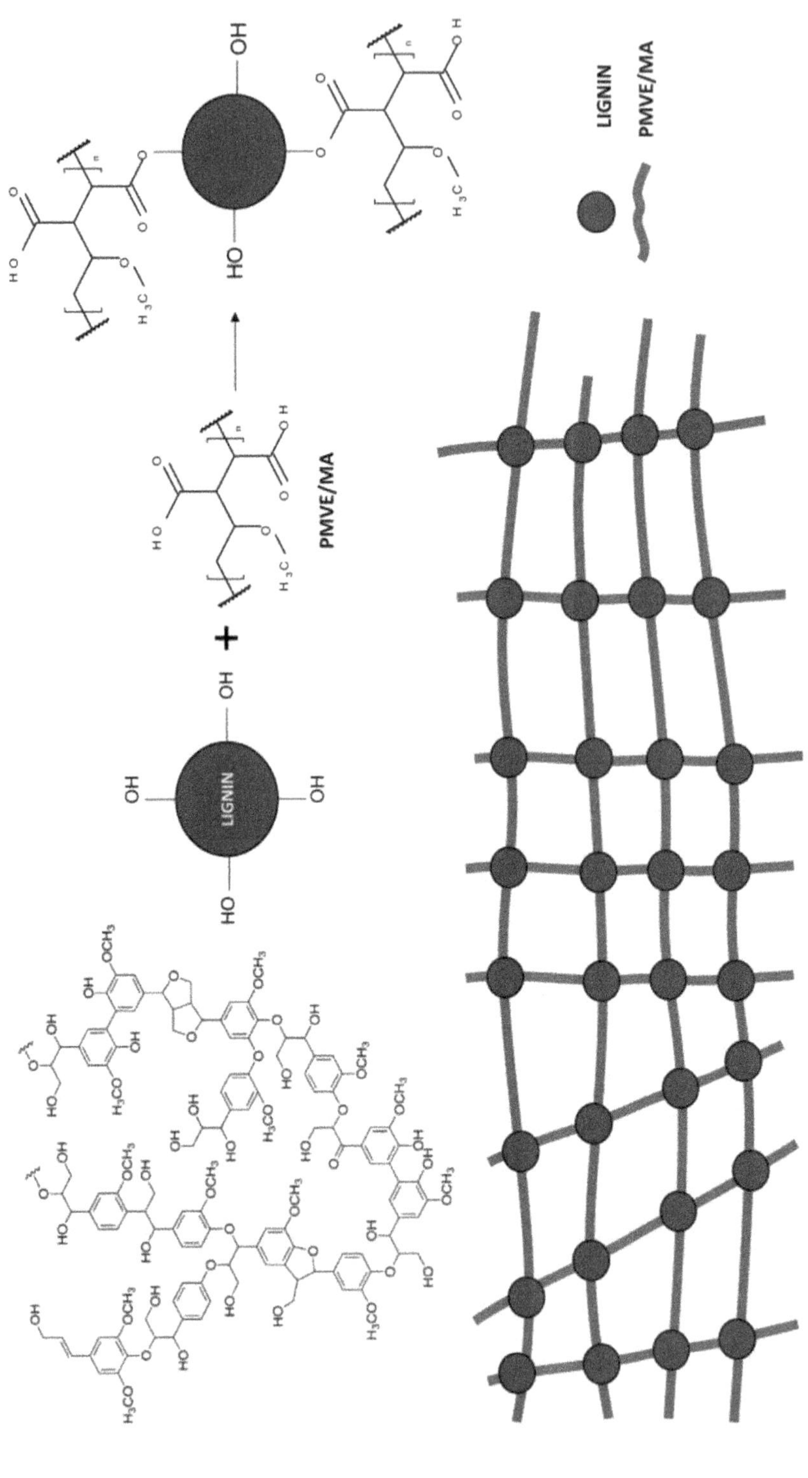

Figure 2.3 Scheme of the esterification process carried out to obtain lignin hydrogels. Reproduced from ref. 35 with permission from Elsevier, Copyright 2018.

2.3 Synthesis of Lignin-based Nanoparticles

The synthesis of lignin-based nanoparticles has gained the attention of many researchers recently due to their wide range of potential applications (Section 2.4). Lignin nanoparticles (LNPs) have been prepared by many different methods, such as acid precipitation, solvent exchange and ultrasonication. The main synthetic methods are summarized in the following sections.

Although there are different methods for the fabrication of LNPs, only a few attempts to increase their scale have been developed. In general, the large amount of energy and reagents added and very dilute solutions of nanoparticles obtained limit the viability of many of the methods.[40,41]

2.3.1 Acid-catalysed Precipitation

Acid-catalysed precipitation has been reported as a very adequate method to synthesize biocompatible, simple, cost-effective and stable LNPs. In addition, this method also allows control over the size of the LNPs.[42–46] It must be taken into account that lignins with low sulfonation degrees are almost insoluble in water. However, they dissolve easily in ethylene glycol or in highly alkaline solutions ($pH > 10$). Considering this, two different synthetic routes were described by Frangville *et al.*[43] So, in one of them, lignin is dissolved in organic solvents such as ethylene glycol, castor oil and acetone,[47–49] whereas in the other pathway lignin is dissolved in a highly basic aqueous solution ($pH > 10$). Considering the solubility of lignin, in the first part of the process the nucleation process and the growth of lignin in ethylene glycol or alkaline solution will take place. The gradual decrease in pH then produces LNPs. It should be noted that the gradual decrease in pH makes the particle size distribution narrower than when rapid mixing of solution occurs. However, the method using alkaline solution forms LNPs that are stable only below pH 5. The studies showed that the concentration and the pH can vary the particle size. Furthermore, the initial concentration of lignin in ethylene glycol solution also causes the particle size to vary. Thus, at low concentrations, approx. 0.5% wt, the particles had diameters of 100 nm. At concentrations around 1.5% wt, the average size was 200 nm; however, for lignin concentrations above 4% wt, stable NPs with an average size of <100 nm were produced.

Similarly, Gupta *et al.*[50] dissolved lignin in ethylene glycol, and once the solution was filtered, hydrochloric acid (HCl) was slowly added, allowing the precipitation of LNPs. In this work, 60% of the particles obtained have diameters around 150–250 nm. Although this method has good yields and promising particle size, its main disadvantage is that the LNPs obtained tend to aggregate.

2.3.2 Flash Precipitation and Nanoprecipitation

As mentioned earlier, the aggregation of the formed nanoparticles could be an important drawback for acid-catalysed precipitation; in order to overcome this problem, flash precipitation could be considered as an excellent alternative. Compared with acid precipitation, flash precipitation is capable of producing faster nucleation and precipitation, reducing significantly the aggregation of the LNPs. Richter *et al.* described two methods of manufacturing LNPs based on flash precipitation.[51,52] Thus, the first was based on the instantaneous precipitation induced by the pH of lignin in ethylene glycol, while the second was based on the instantaneous precipitation of a lignin solution in acetone by the action of an antisolvent. The authors obtained nanoparticles in the range between 45 and 250 nm. In addition, Richter *et al.*[51] reported the influence of concentration, temperature and solvent on the growth kinetics of LNPs and variation in particle size.

2.3.3 Dialysis

The dialysis process has been successfully used for LNP formation. Usually, for this method tubular dialysis membranes are used to remove the organic solvent. This method could be considered as low-cost and easy, considering the required components. However, the increased water and time consumption is considered as the main disadvantage. Lievonen *et al.*[53] reported a simple method in which Kraft lignin is dissolved in tetrahydrofuran (THF) and subsequently introduced into a dialysis tube, 6–8 kDa, and immersed in water. Similarly, Tian and co-workers[54] used dimethylsulfoxide (DMSO) as a solvent and placed the solution in a dialysis tube of 12–14 kDa. In this case, the lignin samples were obtained from corn stubble (CS), hardwood POP and softwood lodgepole pine (LPP). The obtained yields and nanoparticle sizes varied with the lignin source. The yields were 81.8%, 90.9% and 41.0%, and the average particle sizes were 218, 131 and 104 nm, for CS, hardwood POP and softwood LPP, respectively.

2.3.4 Solvent Exchange

The solvent exchange method is a widely used technique, similar to the previously described dialysis. This method has high yields and good nanoparticle size control. In this method, a solution of lignin, with a water miscible solvent, is mixed with an excess of water. When mixed, the nucleation and growth of the nanoparticles occurs, reducing the solubility of the LNPs in the mixture.[53,55–59] Despite the advantage of being a simple method, its low yield is its main disadvantage.[60] One of the first examples of this technique was described by Qian *et al.*[58] In their work, they obtained monodisperse nanoparticles of 110 nm by adding water dropwise to a solution of lignin in THF. Nanoparticles were formed by self-assembly of the lignin molecules due to hydrophobic interactions. In another example, Li *et al.* prepared LNPs from a solution of Kraft lignin in ethanol by adding water dropwise; in this case the particle sizes are in the range of 10 to 100 nm.

2.3.5 Supercritical Antisolvent Process

Supercritical CO_2 can be used in the formation of LNPs. This technology is based on supercritical antisolvent precipitation; that is, the solubility of lignin in a supercritical fluid is lower than the solubility in a solvent.[61] Thus, the solubility of lignin in the solvent is reduced upon contact with supercritical CO_2, increasing its saturation and precipitation in the form of nanoparticles. The main advantage of this method is the low dispersibility of the nanoparticles and their small size. Furthermore, supercritical CO_2 can be considered as a green solvent. Myint *et al.*[62] used supercritical CO_2 as an antisolvent for the manufacture of LNPs. They obtained highly monodisperse 38 nm LNPs with excellent stability, UV absorption and non-cytotoxicity. Furthermore, they reported that as the temperature increased and the pressure decreased, the size of the LNPs increased. Similarly, Lu *et al.*[63] formed LNPs from a solution in acetone. In Figure 2.4, a non-nanoscale lignin particles and LNPs obtained in their work are depicted. In this case, the reported nanoparticle size was higher than the previously described one, reaching 144 nm. This difference could be mainly attributed to the variation in the solvent used for the dissolution of LNPs. As in the previous case, the nanoparticles presented good stability.

2.3.6 W/O Microemulsion Methods

Microemulsions are widely used in the development of polymeric nanoparticles. Some authors have described the use of microemulsions for the formation of nanolignin.[64–66] For example, Nypelö *et al.*[65] dissolved lignin in sodium hydroxide, followed by NaOH solution, and then mixed with an oily phase, to study different sulfates. The nanoparticles were crosslinked by means of epichlorohydrin at 55 °C and for 40 minutes. From this method, LNPs whose size ranged from 90 to 600 nm were obtained.

Figure 2.4 SEM images of lignin and LNPs obtained by Lu *et al.* Reproduced from ref. 63 with permission from Elsevier, Copyright 2012.

2.3.7 Mechanical Treatment: Homogenization and Ultrasonication

Mechanical treatments can also be used in the development of LNPs. High shear homogenization[67–69] and ultrasonication[70,71] have been used to apply force to disintegrate lignin particles down to nanometer size. The main advantage of mechanical treatments is that they can be used with a wide range of lignin. It should be noted that they can also be used with insoluble lignin without chemical pre-treatment. However, these methods produce irregularly shaped LNPs with a wide size distribution.[72]

Nair *et al.*[67] presented a method to prepare nanolignin by means of a high shear homogenizer. Kraft lignin particles with a size distribution ranging from large to nanometric were homogenized, affording nanolignin with sizes less than 100 nm after 4 h of mechanical shearing. Rao *et al.*[73] obtained nanolignin by homogenizing OSL/ethanol/water mixtures for 45 min. Their work concluded that by increasing the ethanol content of the solution, the size distribution obtained was more uniform. In addition, the increase in homogenization time decreased the size.

Ultrasonication as a method for the development of LNPs offers a series of advantages such as its simplicity and the lack of organic solvents. This method consists of applying ultrasound waves, 20 kHz to 10 MHz, to a medium, causing microscopic bubbles that generate heat and pressure when they collapse during the cavitation process.[74,75] In this method, lignin is usually suspended in an aqueous solution, and the suspension is sonicated, generating a pressure capable of disintegrating the lignin particles, resulting in nanoparticles. The control over time, intensity and temperature could produce nanoparticles of different sizes.[76] Furthermore, mechanical methods can cause chemical modification of lignin. In their study, Gilca *et al.*[76] showed that during ultrasonication a process of depolymerization of lignin polymer chains and the oxidative coupling of their phenolic groups also took place.

The main advantages of ultrasonication are its easy operation, low equipment requirements and harmlessness to the environment.[71] However, similar to those described for the homogenization method, one of the disadvantages is poor size uniformity. Thus, in recent years combined methodologies have been developed to improve the resulting size distribution. For example, Wang *et al.*[56] combined solvent exchange with ultrasonication. Moreover, some authors also reported a method based on the combination of acid precipitation and ultrasound.[44] Agustín *et al.*[77] studied the optimization of acid-catalysed precipitation conditions using different acids (HNO_3, H_2SO_4 and HCl) previously to ultrasonication of samples. All the samples obtained in their study were monodisperse. After two minutes of sonication, samples treated with HCl and HNO_3 produced LNPs with an average size of 93 nm. On the other hand, for H_2SO_4 the nanoparticle size was slightly higher, 100 nm. Regarding the yield, the samples obtained with HCl were the ones that showed the highest yield.

2.4 Applications of Lignin-based Hydrogels and Nanoparticles

The abundance of lignin together with the environmental concern has exponentially increased the development of lignin-based materials for a plethora of applications. In this section some of the latest applications of nanolignin and lignin-based hydrogels will be summarized.

2.4.1 Antibacterial and Antioxidant Activity of LNPs

Some studies have described the antibacterial and antioxidant effects of some lignin extracts. Moreover, some authors have described relationships between the antioxidant capacity and its antibacterial properties.[78,79] In this context, Dong *et al.*[80] described that the antibacterial activity of the lignin extract from corn straw was related to its antioxidant activity.

The antibacterial activity of LNPs was used by Yang *et al.*[81] in the development of PVA/CHI hydrogels with antibacterial activity induced by LNPs. In this study, PVA/CHI hydrogels were prepared with 1 and 3 wt% of LNPs. The authors observed that the presence of LNPs produced a different swelling than pure PVA/CHI hydrogels, forming honeycomb-like structures due to the presence of LNPs. In addition, a synergistic antioxidant effect between the CHI of the hydrogels and the LNPs was described. Finally, antimicrobial tests confirmed that LNPs are effective against Gram-negative bacteria (*Escherichia coli*) compared to Gram-positive strains (*Staphylococcus aureus* and *Staphylococcus epidermidis*).

In another example, Kim *et al.* reported the fabrication of chitosan–lignosulfonate nanoparticles by a sonochemical method.[82] In this work, the lignosulfonates act as counterpolyions stabilizing the chitosan particles, generated by high intensity ultrasound. These particles showed greater stability to lysozyme degradation, biocompatibility with human cells and antimicrobial activity by incorporating lignosulfonates in the chitosan nanoparticles.

As already indicated, lignin has antioxidant characteristics.[83] Moreover, a study carried out by Lu *et al.*[63] demonstrates that the antioxidant activity increases when lignin is in the form of nanoparticles compared to bulk lignin. The antioxidant properties of LNPs were also successfully studied in PVA/CHI hydrogels by Yang and co-workers.[81] In their study they described an increase of over 78% compared to the control hydrogel without LNPs.

2.4.2 Lignin-based Drug Delivery Systems

One of the most promising applications of both LNPs and lignin-based hydrogels is their use as drug delivery systems. Considering that LNPs are considered to be cost-effective, non-toxic and highly biodegradable, they could be an excellent candidate for nanocarriers.[84] On the other hand, the hydrogels are known for their excellent properties for drug delivery applications, so hydrogels fabricated with sustainable materials such as lignin could be an interesting development.

As an example of nanolignin-based system development, Figueiredo *et al.* studied three types of LNP-based nanocarriers: pure LNPs, iron(III)-complexed LNPs (Fe-LNPs) and Fe_3O_4-infused LNPs (Fe_3O_4-LNPs).[85] All the developed systems presented low cytotoxicity. In addition, they evaluated the loading and release capacity of drugs that are poorly soluble in water and cytotoxic, such as sorafenib and benzazulene (BZL). Pure LNPs presented the best sustained release profiles in the pH range close to physiological pH. In addition, Fe_3O_4-LNPs exhibited superparamagnetic behaviour, which makes them promising for cancer therapy such as magnetic targeting. Recently, Pourmoazzen *et al.*[86] (2020) developed cholesterol-modified LNPs. The authors modified two fractions of Kraft lignin, low and high molecular weight lignin, with cholesterol, increasing the hydrophobicity of the LNPs. The synthesized nanoparticles had diameters between 200 and 500 nm. In addition, they were able to adequately load and release hydrophobic molecules such as folic acid.

On the other hand, several examples of lignin-based hydrogels for delivery of active agents have been reported in the literature.[87,88] Lignin hydrogels have been used not only for the release of drugs but also for the release of other agents such as essential oils and pesticides, among others. For example, Raschip *et al.*[89] synthesized lignin/xanthan gum hydrogels for vanillin release. The authors evaluated different formulations for these hydrogels, concluding that the 70/30 xanthan/lignin hydrogel released vanillin more slowly than the 90/10 hydrogel. They also suggested that this variation was due to the different intermolecular and intramolecular interactions between the matrix and the active substance. In another example, hydrogels based on lignin and acrylic acid were synthesized. The loading and release capacities of these hydrogels were evaluated for three different pesticides, obtaining the following loading order: cyfluthrin > paraquat > cyhalofop-butyl.[90]

2.4.3 Stimuli-responsive and Smart Lignin-based Materials

Some materials are capable of responding to certain stimuli (pH, light, temperature) by modifying their properties. This ability to respond to stimuli makes the materials expand their potential applications such as sensors, actuators and drug delivery systems, among others. The structural complexity of lignin, in many cases, limits the preparation of stimuli-sensitive materials, so there are fewer lignin-based materials compared to other synthetic materials.[91] However, several examples have been reported in this field by using lignin-based materials. Although most examples of these types of materials are based on hydrogels, examples of stimuli-responsive LNPs have also been described. Some of the most relevant examples for both types of materials, hydrogels and nanoparticles, are described in this section.

Among the possible stimuli-responsive systems that have been reported in the literature, those triggered by a temperature variation are the most developed ones. The most common synthesis method is based on developing materials that incorporate temperature-sensitive monomers, such as NIPAM.

Dual-responsive lignin-based nanoparticles were synthesized by Dinari and co-workers[92] by combining temperature- and pH-sensitive monomers with lignin. In this study, a lignin nanogel was developed by graft copolymerization of NIPAM and *N*,*N*-dimethylaminoethylmethacrylate (DMAEMA) by atom transfer radical polymerization (ATRP) by using fully brominated lignin as a macroinitiator (Figure 2.5). The lignin-*g*-P(NIPAM-*co*-DMAEMA) nanogels (LNDNGs) presented different LCSTs of 32, 34, 37 and 42 °C depending on their composition. These nanogels presented a diameter of 150 nm and they were pH sensitive. Moreover, the curcumin loading capacity and the encapsulation efficiency of the LNPs were 49.69% and 92.62%, respectively. In addition, a release capacity of 65.36% after 72 h was reported for these nanogels.

Parvathy *et al.*[93] followed a similar approach to the previous one to synthesize hydrogels. Temperature- and pH-sensitive monomers, NIPAM and acrylic acid, were grafted onto methacrylated lignin by redox initiated radical polymerization by using potassium persulfate (KPS) and sodium bisulfite (SBS). At room temperature and in basic medium, a maximum swelling was observed, while at 40 °C a significant reduction in swelling was observed at different pH values. The increase in lignin content improved the storage modulus of the gels. It is important to note that the hydrogels presented adequate cell viability.

In addition, acetic acid modified lignin was also copolymerized with thermosensitive NIPAM to synthesize responsive hydrogels.[94] The acetic acid lignin was subsequently copolymerized with NIPAM using NMBAm as a crosslinker and H_2O_2 as an initiator. These hydrogels presented an LCST of around 31 °C and the lignin content had a significant influence on several properties, increasing the pore size and the swelling ratio (at 22 °C) with the increase in lignin content.

Similar to the examples described for nanolignin, macroscopic hydrogels with dual responses (to temperature and pH) have also been developed. Thus, for example, Jin *et al.*[95] synthesized lignin hydrogels modified with NIPAM and itaconic acid (IA). In this case, the hydrogels were formed from methacrylate lignosulfonate (MLS). On the one hand, the hydrogels presented an LCST of around 35 °C, that is, close to body temperature. On the other hand, the hydrogels showed pH sensitivity in the range of pH 3.0 to 9.1. In addition, some of the properties of the hydrogels also varied with the lignin ratio; that is, the increase in the lignin content increased the mechanical resistance of the material.

In addition to temperature-sensitive and dual-sensitive materials, pH-responsive lignin-based nanoparticles and hydrogels have been developed. In the development of thermoresponsive materials, NIPAM is the

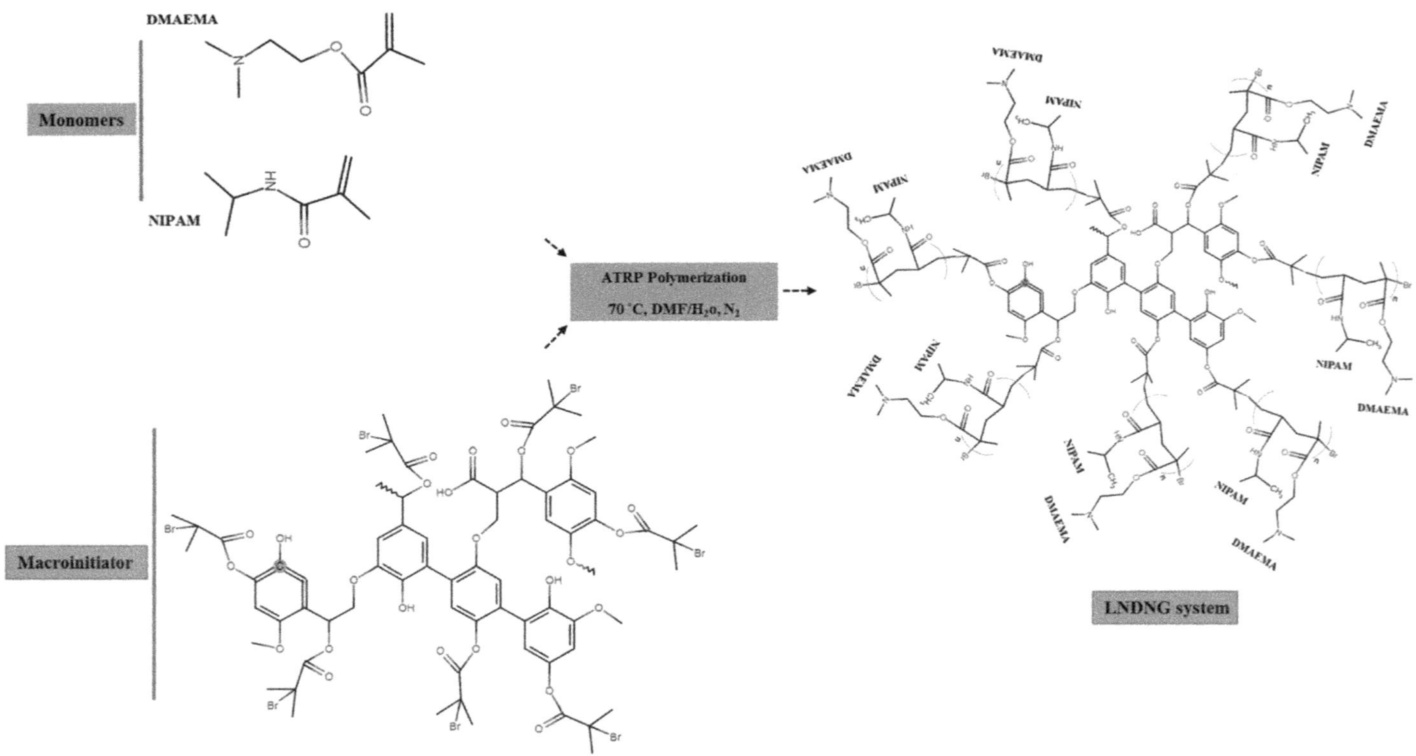

Figure 2.5 Scheme of the synthetic pathway used for obtaining a lignin-*g*-P(NIPAM-*co*-DMAEMA) nanogel (LNDNG). Reproduced from ref. 92, https://doi.org/10.1038/s41598-021-81393-3, under the terms of the CC BY 4.0 license https://creativecommons.org/licenses/by/4.0/.

most widely used monomer, while for pH-sensitive materials derivatives with carboxylic acid groups such as acrylic acid are mainly used. Besides, basic functional groups, such as amines, could be used for the synthesis of these kinds of pH-sensitive materials.

Figueiredo *et al.*[96] developed carboxylated nanolignin as nanocarriers in anticancer therapies. In their work, carboxylated lignin was prepared by modification of Kraft lignin with succinic anhydride and 4-dimethylaminopyridine. In a subsequent reaction, carboxylated LNPs were modified by conjugating them with poly(ethylene glycol)-*block*-poly(L-histidine) (NH_2-PEG-PHIS) and further functionalized them with cell penetrating peptides (CPP) to prepare CLNP-CPP and CLNP-PEG-PHIS-CPP. In addition, the releasing capacity of the nanoparticles was tested with a poorly water-soluble cytotoxic compound, BZL, used in cancer therapies. The authors evaluated the release profiles at different pH values, describing a greater release at acidic pH. Overall, the developed nanoparticles showed a good capacity for the controlled release of BZL. These nanoparticles presented good biocompatibility and stability in physiological media, being suitable for the transport of hydrophobic drugs such as BZL.

In order to improve the response to pH changes, recently Yan *et al.*[97] proposed to adjust the hydroxyl group content of the synthesized lignin hydrogels. In this study, lignins with different hydroxyl contents were developed. The increase in functional groups made the gels more compact and with a higher tensile strength (4.10 MPa). Furthermore, the authors described that the bending behaviour of these hydrogels varied as a function of pH. In Figure 2.6, the application test performed on the hydrogels is illustrated. In this experiment, a wire was hooked (1400 mg) from the hydrogel (140 mg, 5.0 cm $\times$ 0.25 cm $\times$ 0.2 cm), observing that it could accomodate 10 times more weight than itself under both acidic (pH 5) and alkaline (pH 9) conditions. While at pH 7, the hydrogel maintained its shape.

Among the different stimuli-sensitive hydrogels, it is worth highlighting those that present a rapid response. The fast response capacity improves the potential application of these materials as sensors and actuators.[99] In this context, the lignin-based hydrogels developed by Dai *et al.*[100] presented an interesting response time. The authors prepared Kraft lignin-based hydrogels by crosslinking Kraft lignin with PEGDGE. To test the responsiveness and the potential actuating performance of the hydrogel, strip-shaped samples were prepared. First, the sample was immersed in 0.1 M KOH solution for 1.0 min and then in 0.1 M HCl. When the sample was immersed in acidic solution, it bent spontaneously in less than 30 s. In addition, the sample recovered the original shape (30 s) in a subsequent immersion in basic solution, presenting only a slight deformation (Figure 2.7a and b). Moreover, the authors evaluated the deformation reproducibility, reporting an excellent repeatability between the bent and straight shapes 10 times (Figure 2.7c and d). The observed responsiveness added to the repeatability could improve the use of these materials as sensors or actuators.

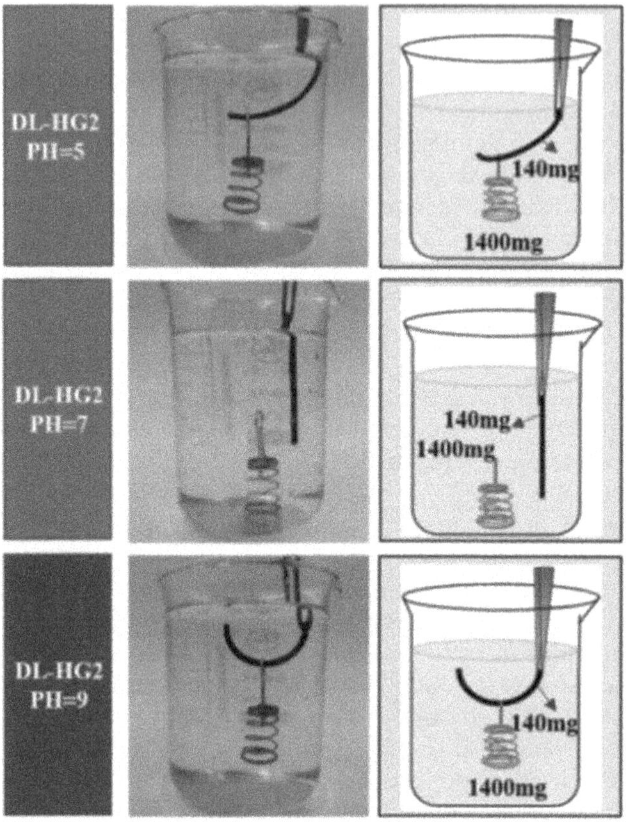

Figure 2.6 Performance of the hydrogels synthesized by Yan *et al.* at different pHs when a wire of 1400 mg was hooked onto them. Reproduced from ref. 98 with permission from Elsevier, Copyright 2022.

2.4.4 Lignin as a UV Protective Agent

The skin is the largest and most visible organ of the body and is the most affected part of the human body by the incidence of UV rays from the sun. UV rays cause inflammation, immunological changes, physical changes, impaired wound healing and DNA damage that promote cellular senescence and carcinogenesis, with melanoma being the deadliest form of skin cancer.[101] Thus, in order to protect the skin against solar radiation, there is great interest in the development of new and improved sunscreens. Gutiérrez-Hernández *et al.*[102] developed and evaluated the effectiveness of lignin and ZnO nanoparticles as active additives for sunscreens. In this study, *Agave tequilana* was used as a source of lignin. The developed formulations gave sun protection factor (SPF) values in the range of 4–13. It is relevant to highlight that LNPs showed absorption in the UVB and UVC regions, improving the SPF value of sunscreens composed only of zinc oxide nanoparticles. In addition, LNPs present other advantages such as their

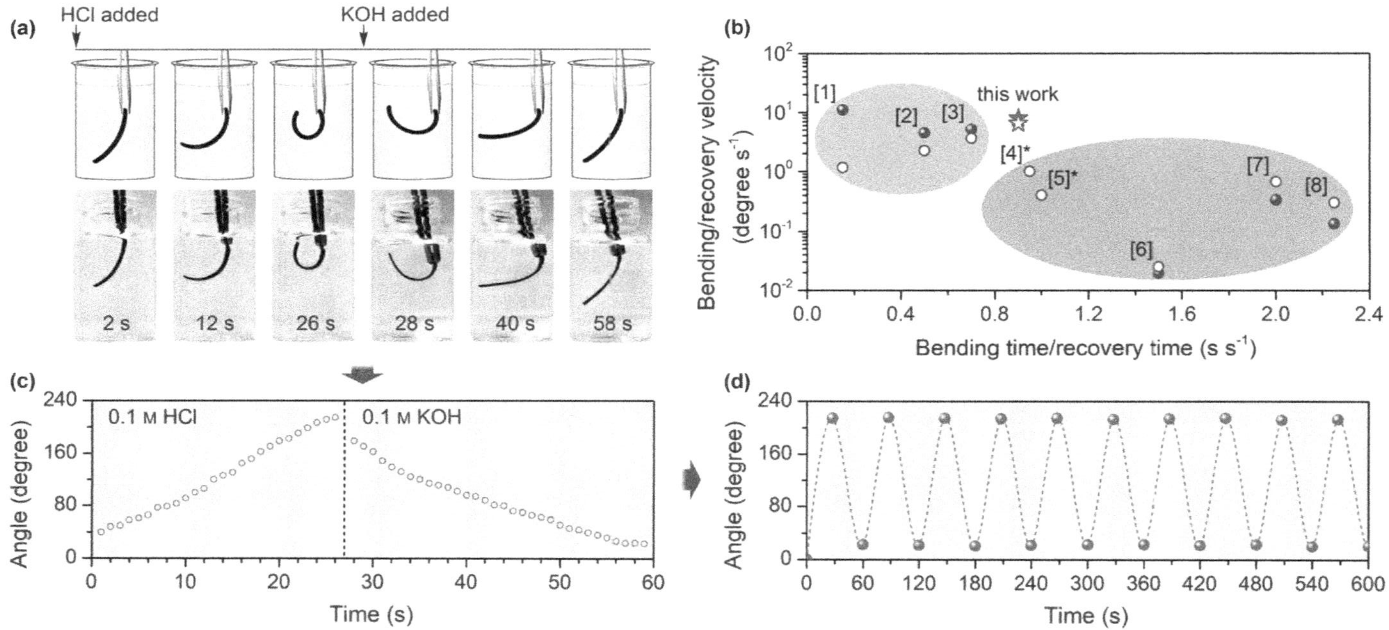

Figure 2.7 Study of the pH responsiveness and evaluation of the repeatability of the hydrogels developed by Dai *et al.*: (a) deformation and recovery with pH variation, (b) velocity and response ratios of bending (solid points)/recovery (hollow points), (c) bending angle and (d) bending and recovering cycles. Reproduced from ref. 100 with permission from American Chemical Society, Copyright 2020.

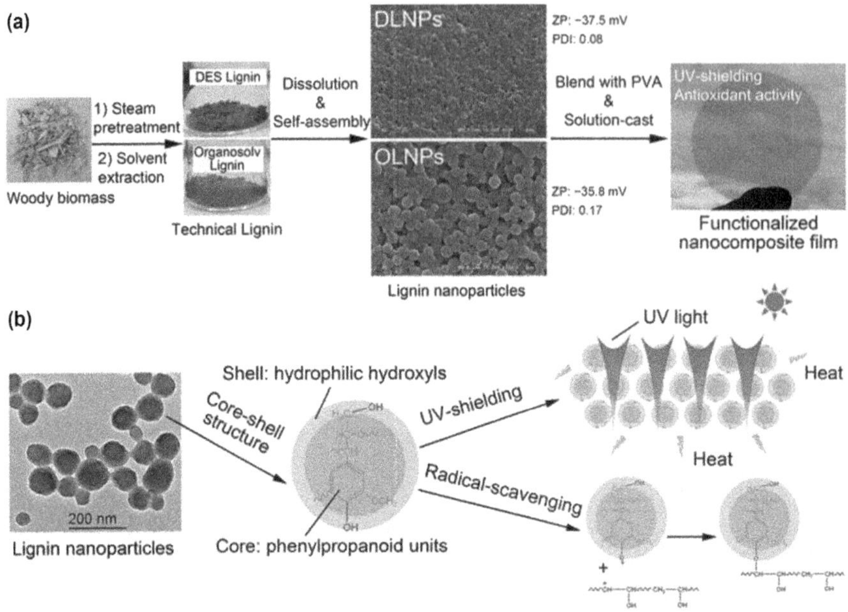

Figure 2.8 (a) Scheme of the synthetic procedure used by Tian *et al.*[103] to fabricate LNPs. (b) UV absorption mechanism proposed in the study. Reproduced from ref. 103, https://doi.org/10.1186/s13068-017-0876-z, under the terms of the CC BY 4.0 license https://creativecommons.org/licenses/by/4.0/.

organic nature and their brown colour that could be similar to the skin tone of the wearer.

In another example, Tian and co-workers introduced LNPs into a PVA polymeric matrix, forming a transparent film[103] (Figure 2.8). The results showed a UV protection efficiency of ∼80% at 400 nm with an amount of 4% by weight of LNPs.

2.4.5 Environmental Applications

The presence of heavy metal ions in the environment, both in soil and in water, is an important concern due to their high toxicity and carcinogenicity. So, the development of materials capable of removing these metals from aquatic media for their further treatment or degradation still represents a significant percentage of research carried out in this area. In this context, lignin derivatives could be considered as suitable alternatives. The sustainable origin of lignin and the presence of polyphenols in its structure capable of interacting with heavy metals make them materials of interest for water remediation technologies.[104–106] To date, lignin-based materials have been used in the removal of Ni^{2+}, Cu^{2+}, Co^{2+}, Zn^{2+}, Cr^{3+}, As^{2+}, Hg^{2+} and Pb^{2+}.[105–111] In addition to metal ions, lignin-based hydrogels have also been tested for organic pollutant adsorption.

As an example, Yao *et al.*[107] developed absorbent hydrogels based on bentonite/sodium lignosulfonate graft polymerized with AAm and maleic anhydride (BLPAMA). These hydrogels were capable of removing high amounts of Pb^{2+}, reaching up to 322.70 mg g^{-1} from aqueous solution. The studies showed that the Pb^{2+} adsorption presented a pseudo-second-order kinetics and, on the other hand, the adsorption isotherms followed the Langmuir model, monolayer chemical adsorption. In addition, the authors described the preferential absorption of Pb^{2+} when it was in competition with other ions such as Cu^{2+}, Zn^{2+} and Cd^{2+}. In another example, Le *et al.* synthesized a lignosulfonate-modified graphene hydrogel.[108] The hydrogel had a high adsorption capacity reaching up to 1210 mg g^{-1}, being significantly higher than those of the previous example and other examples described in the literature such as lignin/acrylic acid hydrogels (235 mg g^{-1}).

Another example of hydrogels to highlight is those developed by Parajuli *et al.*[109] The authors developed hydrogels based on lignocatechol. Ion adsorption selectivity studies performed with those hydrogels resulted in the following order: $Pb^{2+} \sim La^{3+} > Fe^{3+} > Al^{3+} > Ni^{2+} \sim Zn^{2+} \sim Cd^{2+} \sim Co^{2+}$.

In addition to being used successfully to the capture of heavy metals, numerous examples of lignin-based hydrogels for the capture of organic contaminants have also been reported. Wang *et al.*[110] developed lignin hydrogels with AAm and NIPAM grafts incorporating montmorillonite. The studies carried out showed a good capacity for removing methylene blue from aqueous media. These lignin-based hydrogels had a higher contaminant adsorption capacity than the AAm and NIPAM hydrogels without lignin. It is worth highlighting the adsorption value of methylene blue reported in this study, which is 9646.92 mg g^{-1}. This adsorption capacity is significantly higher than those of other reported lignin-based hydrogels. Lignosulfonate-*g*-acrylic acid hydrogels fabricated by Yu *et al.* also presented a high methylene blue adsorption capacity of 2013 mg g^{-1}, but lower than that of the previous example. On the other hand, Albadarin and co-workers demonstrated a maximum methylene blue adsorption capability of 36.25 mg g^{-1} for lignin/chitosan hydrogels.[111]

2.5 Future Perspectives and Conclusions

The search for new environmentally sustainable resources has activated the interest not only of the scientific community but also of industry. Considering the abundance of lignin both in its natural state and as the main by-product of the pulp and paper industry, it is considered as a promising sustainable resource. In addition, the current United Nations 2030 agenda in its Sustainable Development Goals (SDGs) underlines the need to develop new sustainable resources. In this context and considering the current environmental problems, the revaluation of a residue such as lignin for its use as a raw material in the development of new materials represents a very interesting field of business.

In the last decade, many studies have been carried out on the development of thermosets and polyurethanes derived from lignin. Even if studies on these materials continue to be numerous, the number of articles focused on the development of hydrogels and lignin-based nanoparticles is increasing.

In recent years, the development of lignin-based hydrogels and nanoparticles has increased significantly. In addition, it is worth highlighting the diversity of applications studied and the interesting results obtained. Specifically, lignin-based hydrogels can have applications as biomaterials and drug delivery systems and in water remediation. On the other hand, in the case of LNPs whether isolated or within nanocomposites, the results suggested that they can be widely used in the search for new sustainable materials capable of totally or partially replacing materials based on petroleum derivatives. Recently, nanolignin has been described in applications ranging from its use as an antioxidant, antibacterial, UV protective and nanocarrier agents in controlled release systems to its advanced use as a nano-reinforcement.

Thus, it is expected that in the future the quantity of materials fabricated from both lignin-based hydrogels and LNPs will increase, as well as the applications in which these materials can be used.

Acknowledgements

The authors acknowledge MINECOG within the project "Polímeros autorreparables con memoria de forma basados en interacciones orgánicas–inorgánicas reversibles" (MAT2017-89553-P) and the Basque Government within ELKARTEK projects.

References

1. W. Boerjan, J. Ralph and M. Baucher, *Annu. Rev. Plant Biol.*, 2003, **54**, 519–546.
2. B. M. Upton and A. M. Kasko, *Chem. Rev.*, 2016, **116**, 2275–2306.
3. F. G. Calvo-Flores and J. A. Dobado, *ChemSusChem*, 2010, **3**, 1227–1235.
4. W. D. H. Schneider, A. J. P. Dillon and M. Camassola, *Biotechnol. Adv.*, 2021, **47**, 107685.
5. E. Dorrestijn, L. J. J. Laarhoven, I. W. C. E. Arends and P. Mulder, *J. Anal. Appl. Pyrolysis*, 2000, **54**, 153–192.
6. G. Jiang, D. J. Nowakowski and A. V Bridgwater, *Thermochim. Acta*, 2010, **498**, 61–66.
7. J. Hu, Q. Zhang and D. Lee, *Bioresour. Technol.*, 2018, **247**, 1181–1183.
8. A. A. Komissarenkov and T. L. Lukanina, *Russ. J. Gen. Chem.*, 2012, **82**, 985–990.
9. T. Aro and P. Fatehi, *ChemSusChem*, 2017, **10**, 1861–1877.
10. F. G. Calvo-Flores and J. A. Dobado, *ChemSusChem*, 2010, **3**, 1227–1235.
11. S. Sugiarto, Y. Leow, C. L. Tan, G. Wang and D. Kai, *Bioact. Mater.*, 2022, **8**, 71–94.

12. W.-J. Chen, C.-X. Zhao, B.-Q. Li, T.-Q. Yuan and Q. Zhang, *Green Chem.*, 2022, **24**, 565–584.
13. N. A. Peppas and A. S. Hoffman, in *Biomaterials Science*, Elsevier, 2020, pp. 153–166.
14. A. S. Hoffman, *Adv. Drug Delivery Rev.*, 2002, **54**, 3–12.
15. V. Mishra, F. E. Du Prez, E. Gosen, E. J. Goethals and L. H. Sperling, *J. Appl. Polym. Sci.*, 1995, **58**, 331–346.
16. L. H. Sperling and R. Hu, in *Polymer Blends Handbook*, Springer Netherlands, Dordrecht, 2003, pp. 417–447.
17. I. E. Raschip, C. Vasile, D. Ciolacu and G. Cazacu, *High Perform. Polym.*, 2007, **19**, 603–620.
18. I. E. Raschip, E. G. Hitruc and C. Vasile, *High Perform. Polym.*, 2011, **23**, 219–229.
19. J. E. A. Peñaranda and M. A. Sabino, *Polym. Bull.*, 2010, **65**, 495–508.
20. F. Oveissi, S. Naficy, T. Y. L. Le, D. F. Fletcher and F. Dehghani, *ACS Appl. Bio Mater.*, 2018, **1**, 2073–2081.
21. P. Jiang, Y. Cheng, S. Yu, J. Lu and H. Wang, *Polymers*, 2018, **10**, 1109.
22. S. Huang, S. Shuyi, H. Gan, W. Linjun, C. Lin, X. Danyuan, H. Zhou, X. Lin and Y. Qin, *Carbohydr. Polym.*, 2019, **223**, 115080.
23. K. Ravishankar, M. Venkatesan, R. P. Desingh, A. Mahalingam, B. Sadhasivam, R. Subramaniyam and R. Dhamodharan, *Mater. Sci. Eng., C*, 2019, **102**, 447–457.
24. Y. Zhang, M. Jiang, Y. Zhang, Q. Cao, X. Wang, Y. Han, G. Sun, Y. Li and J. Zhou, *Mater. Sci. Eng., C*, 2019, **104**, 110002.
25. F. Chen, M. Xu, L. Wang and J. Li, *BioResources*, 2011, **6**, 1262–1272.
26. L. I. Grishechko, G. Amaral-labat, A. Szczurek, V. Fierro, B. N. Kuznetsov and A. Celzard, *Microporous Mesoporous Mater.*, 2013, **168**, 19–29.
27. F. Chen and J. Li, *Adv. Mater. Res.*, 2010, **113–116**, 1837–1840.
28. D. Ciolacu, A. M. Oprea, N. Anghel, G. Cazacu and M. Cazacu, *Mater. Sci. Eng., C*, 2012, **32**, 452–463.
29. L. Dai, W. Zhu, J. Lu, F. Kong, C. Si and Y. Ni, *Green Chem.*, 2019, **21**, 5222–5230.
30. I. E. Raschip, G. E. Hitruc, C. Vasile and M.-C. Popescu, *Int. J. Biol. Macromol.*, 2013, **54**, 230–237.
31. S. Sathawong, W. Sridach and K. Techato, *J. Environ. Chem. Eng.*, 2018, **6**, 5879–5888.
32. L. Wu, S. Huang, J. Zheng, Z. Qiu, X. Lin and Y. Qin, *Int. J. Biol. Macromol.*, 2019, **140**, 538–545.
33. E. Cortés-Triviño, C. Valencia, M. Delgado and J. Franco, *Polymers*, 2018, **10**, 670.
34. L. Passauer, M. Struch, S. Schuldt, J. Appelt, Y. Schneider, D. Jaros and H. Rohm, *ACS Appl. Mater. Interfaces*, 2012, **4**(11), 5852–5862.
35. J. Domínguez-Robles, M. S. Peresin, T. Tamminen, A. Rodríguez, E. Larrañeta and A. Jääskeläinen, *Int. J. Biol. Macromol.*, 2018, **115**, 1249–1259.

36. E. Larrañeta, M. Imízcoz, J. X. Toh, N. J. Irwin, A. Ripolin, A. Perminova, J. Dom, A. Rodr and R. F. Donnelly, *ACS Sustainable Chem. Eng.*, 2018, **6**, 9037–9046.
37. Y. Meng, J. Lu, Y. Cheng, Q. Li and H. Wang, *Int. J. Biol. Macromol.*, 2019, **135**, 1006–1019.
38. X. Wang, Z. Zhou, X. Guo and Q. He, *RSC Adv.*, 2016, **6**, 35550–35558.
39. A. Zerpa, L. Pakzad and P. Fatehi, *ACS Omega*, 2018, **3**, 8233–8242.
40. R. P. Bangalore Ashok, P. Oinas, K. Lintinen, G. Sarwar, M. A. Kostiainen and M. Österberg, *Green Chem.*, 2018, **20**, 4911–4919.
41. C. Abbati de Assis, L. G. Greca, M. Ago, M. Y. Balakshin, H. Jameel, R. Gonzalez and O. J. Rojas, *ACS Sustainable Chem. Eng.*, 2018, **6**, 11853–11868.
42. O. U. Rahman, S. Shi, J. Ding, D. Wang, S. Ahmad and H. Yu, *New J. Chem.*, 2018, **42**, 3415–3425.
43. C. Frangville, M. Rutkevičius, A. P. Richter, O. D. Velev, S. D. Stoyanov and V. N. Paunov, *ChemPhysChem*, 2012, **13**, 4235–4243.
44. P. Figueiredo, M. H. Lahtinen, M. B. Agustin, D. M. Carvalho, S. Hirvonen, P. A. Penttilä and K. S. Mikkonen, *ChemSusChem*, 2021, **14**, 4718–4730.
45. W. Yang, E. Fortunati, D. Gao, G. M. Balestra, G. Giovanale, X. He, L. Torre, J. M. Kenny and D. Puglia, *ACS Sustainable Chem. Eng.*, 2018, **6**, 3502–3514.
46. Y. Pang, S. Wang, X. Qiu, Y. Luo, H. Lou and J. Huang, *J. Agric. Food Chem.*, 2017, **65**, 11011–11019.
47. A. K. Gupta, S. Mohanty and S. K. Nayak, *Mater. Focus*, 2015, **3**, 444–454.
48. X. He, F. Luzi, X. Hao, W. Yang, L. Torre, Z. Xiao, Y. Xie and D. Puglia, *Int. J. Biol. Macromol.*, 2019, **127**, 665–676.
49. M. Yang, W. Zhao, S. Singh, B. Simmons and G. Cheng, *Nanoscale Adv.*, 2019, **1**, 299–304.
50. A. K. Gupta, S. Mohanty and S. K. Nayak, *RSC Adv.*, 2015, **5**, 56028–56036.
51. A. P. Richter, B. Bharti, H. B. Armstrong, J. S. Brown, D. Plemmons, V. N. Paunov, S. D. Stoyanov and O. D. Velev, *Langmuir*, 2016, **32**, 6468–6477.
52. A. P. Richter, J. S. Brown, B. Bharti, A. Wang, S. Gangwal, K. Houck, E. A. Cohen Hubal, V. N. Paunov, S. D. Stoyanov and O. D. Velev, *Nat. Nanotechnol.*, 2015, **10**, 817–823.
53. M. Lievonen, J. J. Valle-Delgado, M. L. Mattinen, E. L. Hult, K. Lintinen, M. A. Kostiainen, A. Paananen, G. R. Szilvay, H. Setälä and M. Österberg, *Green Chem.*, 2016, **18**, 1416–1422.
54. D. Tian, J. Hu, R. P. Chandra, J. N. Saddler and C. Lu, *ACS Sustainable Chem. Eng.*, 2017, **5**, 2702–2710.
55. M. H. Sipponen, M. Farooq, J. Koivisto, A. Pellis, J. Seitsonen and M. Österberg, *Nat. Commun.*, 2018, **9**, 2300.
56. B. Wang, D. Sun, H.-M. Wang, T.-Q. Yuan and R.-C. Sun, *ACS Sustainable Chem. Eng.*, 2019, 7, 2658–2666.

57. F. Zikeli, V. Vinciguerra, A. D'Annibale, D. Capitani, M. Romagnoli and G. S. Mugnozza, *Nanomaterials*, 2019, **9**(2), 281.
58. Y. Qian, Y. Deng, X. Qiu, H. Li and D. Yang, *Green Chem.*, 2014, **16**, 2156–2163.
59. F. M. C. Freitas, M. A. Cerqueira, C. Gonçalves, S. Azinheiro, A. Garrido-Maestu, A. A. Vicente, L. M. Pastrana, J. A. Teixeira and M. Michelin, *Int. J. Biol. Macromol.*, 2020, **163**, 1798–1809.
60. S. Beisl, A. Miltner and A. Friedl, *Int. J. Mol. Sci.*, 2017, **18**, 1244.
61. Z. Zhang, V. Terrasson and E. Guénin, *Nanomaterials*, 2021, **11**(5), 1336.
62. A. A. Myint, H. W. Lee, B. Seo, W.-S. Son, J. Yoon, T. J. Yoon, H. J. Park, J. Yu, J. Yoon and Y.-W. Lee, *Green Chem.*, 2016, **18**, 2129–2146.
63. Q. Lu, M. Zhu, Y. Zu, W. Liu, L. Yang, Y. Zhang, X. Zhao, X. Zhang, X. Zhang and W. Li, *Food Chem.*, 2012, **135**, 63–67.
64. S. Lam, K. P. Velikov and O. D. Velev, *Curr. Opin. Colloid Interface Sci.*, 2014, **19**, 490–500.
65. T. E. Nypelö, C. A. Carrillo and O. J. Rojas, *Soft Matter*, 2015, **11**, 2046–2054.
66. N. Chen, L. A. Dempere and Z. Tong, *ACS Sustainable Chem. Eng.*, 2016, **4**, 5204–5211.
67. S. S. Nair, S. Sharma, Y. Pu, Q. Sun, S. Pan, J. Y. Zhu, Y. Deng and A. J. Ragauskas, *ChemSusChem*, 2014, **7**, 3513–3520.
68. L. Matsakas, A. Karnaouri, A. Cwirzen, U. Rova and P. Christakopoulos, *Molecules*, 2018, **23**, 1822.
69. R. Xu, H. Du, H. Wang, M. Zhang, M. Wu, C. Liu, G. Yu, X. Zhang, C. Si, S.-E. Choi and B. Li, *Front. Bioeng. Biotechnol.*, 2021, **9**, 677963.
70. M. N. Garcia Gonzalez, M. Levi, S. Turri and G. Griffini, *J. Appl. Polym. Sci.*, 2017, **134**, 45318.
71. H. Yin, L. Liu, X. Wang, T. Wang, Y. Zhou, B. Liu, Y. Shan, L. Wang and X. Lü, *Colloids Surf., A*, 2018, **545**, 51–59.
72. J. Y. Zhu, U. P. Agarwal, P. N. Ciesielski, M. E. Himmel, R. Gao, Y. Deng, M. Morits and M. Österberg, *Biotechnol. Biofuels*, 2021, **14**, 114.
73. X. Rao, Y. Liu, Q. Zhang, W. Chen, Y. Liu and H. Yu, *ACS Omega*, 2017, **2**, 2858–2865.
74. K. S. Suslick and G. J. Price, *Annu. Rev. Mater. Sci.*, 1999, **29**, 295–326.
75. D. Peters, *J. Mater. Chem.*, 1996, **6**, 1605.
76. I. A. Gilca, V. I. Popa and C. Crestini, *Ultrason. Sonochem.*, 2015, **23**, 369–375.
77. M. B. Agustin, P. A. Penttilä, M. Lahtinen and K. S. Mikkonen, *ACS Sustainable Chem. Eng.*, 2019, **7**, 19925–19934.
78. T. Dizhbite, *Bioresour. Technol.*, 2004, **95**, 309–317.
79. W. Yang, E. Fortunati, F. Dominici, G. Giovanale, A. Mazzaglia, G. M. Balestra, J. M. Kenny and D. Puglia, *Int. J. Biol. Macromol.*, 2016, **89**, 360–368.
80. X. Dong, M. Dong, Y. Lu, A. Turley, T. Jin and C. Wu, *Ind. Crops Prod.*, 2011, **34**, 1629–1634.

81. W. Yang, E. Fortunati, F. Bertoglio, J. S. Owczarek, G. Bruni, M. Kozanecki, J. M. Kenny, L. Torre, L. Visai and D. Puglia, *Carbohydr. Polym.*, 2018, **181**, 275–284.
82. S. Kim, M. M. Fernandes, T. Matamá, A. Loureiro, A. C. Gomes and A. Cavaco-Paulo, *Colloids Surf., B*, 2013, **103**, 1–8.
83. A. García, A. Toledano, M. Á. Andrés and J. Labidi, *Process Biochem.*, 2010, **45**, 935–940.
84. S. Iravani and R. S. Varma, *Green Chem.*, 2020, **22**, 612–636.
85. P. Figueiredo, K. Lintinen, A. Kiriazis, V. Hynninen, Z. Liu, T. Bauleth-Ramos, A. Rahikkala, A. Correia, T. Kohout, B. Sarmento, J. Yli-Kauhaluoma, J. Hirvonen, O. Ikkala, M. A. Kostiainen and H. A. Santos, *Biomaterials*, 2017, **121**, 97–108.
86. Z. Pourmoazzen, H. Sadeghifar, G. Yang and L. Lucia, *Colloids Surf., B*, 2020, **186**, 110685.
87. S. Park, S. H. Kim, K. Won, J. W. Choi, Y. H. Kim, H. J. Kim, Y.-H. Yang and S. H. Lee, *Carbohydr. Polym.*, 2015, **115**, 223–229.
88. M. Witzler, A. Alzagameem, M. Bergs, B. El Khaldi-Hansen, S. E. Klein, D. Hielscher, B. Kamm, J. Kreyenschmidt, E. Tobiasch and M. Schulze, *Molecules*, 2018, **23**, 1885.
89. I. E. Raschip, E. G. Hitruc, A. M. Oprea, M.-C. Popescu and C. Vasile, *J. Mol. Struct.*, 2011, **1003**, 67–74.
90. Y. Sun, Y. Ma, G. Fang, S. Ren and Y. Fu, *BioResources*, 2016, **11**, 2361–2371.
91. D. Rico-García, L. Ruiz-Rubio, L. Pérez-Alvarez, S. L. Hernández-Olmos, G. L. Guerrero-Ramírez and J. L. Vilas-Vilela, *Polymers*, 2020, **12**, 81.
92. A. Dinari, M. Abdollahi and M. Sadeghizadeh, *Sci. Rep.*, 2021, **11**, 1962.
93. P. A. Parvathy, A. V. Ayobami, A. M. Raichur and S. K. Sahoo, *Int. J. Biol. Macromol.*, 2021, **192**, 180–196.
94. Q. Feng, F. Chen and H. Wu, *BioResources*, 2011, **6**, 4942–4952.
95. C. Jin, W. Song, T. Liu, J. Xin, W. C. Hiscox, J. Zhang, G. Liu and Z. Kong, *ACS Sustainable Chem. Eng.*, 2018, **6**, 1763–1771.
96. P. Figueiredo, C. Ferro, M. Kemell, Z. Liu, A. Kiriazis, K. Lintinen, H. F. Florindo, J. Yli-Kauhaluoma, J. Hirvonen, M. A. Kostiainen and H. A. Santos, *Nanomedicine*, 2017, **12**, 2581–2596.
97. S. Yan, L. Chai, W. Li, L.-P. Xiao, X. Chen and R.-C. Sun, *Colloids Surf., A*, 2022, **643**, 128815.
98. S. Yan, L. Chai, W. Li, L.-P. Xiao, X. Chen and R.-C. Sun, *Colloids Surf., A*, 2022, **643**, 128815.
99. A. Bin Imran, T. Seki and Y. Takeoka, *Polym. J.*, 2010, **42**, 839–851.
100. L. Dai, M. Ma, J. Xu, C. Si, X. Wang, Z. Liu and Y. Ni, *Chem. Mater.*, 2020, **32**, 4324–4330.
101. A. Amaro-Ortiz, B. Yan and J. D'Orazio, *Molecules*, 2014, **19**, 6202–6219.
102. J. M. Gutiérrez-Hernández, A. Escalante, R. N. Murillo-Vázquez, E. Delgado, F. J. González and G. Toríz, *J. Photochem. Photobiol., B*, 2016, **163**, 156–161.

103. D. Tian, J. Hu, J. Bao, R. P. Chandra, J. N. Saddler and C. Lu, *Biotechnol. Biofuels*, 2017, **10**, 192.
104. Y. Ge and Z. Li, *ACS Sustainable Chem. Eng.*, 2018, **6**, 7181–7192.
105. S. Thakur, P. P. Govender, M. A. Mamo, S. Tamulevicius, Y. K. Mishra and V. K. Thakur, *Vacuum*, 2017, **146**, 342–355.
106. S. Wei, W. Chen, Z. Li, Z. Liu and A. Xu, *Environ. Technol.*, 2022, 1–14.
107. Q. Yao, J. Xie, J. Liu, H. Kang and Y. Liu, *J. Polym. Res.*, 2014, **21**, 465.
108. F. Li, X. Wang, T. Yuan and R. Sun, *J. Mater. Chem. A*, 2016, **4**, 11888–11896.
109. D. Parajuli, K. Inoue, K. Ohto, T. Oshima, A. Murota, M. Funaoka and K. Makino, *React. Funct. Polym.*, 2005, **62**, 129–139.
110. Y. Wang, Y. Xiong, J. Wang and X. Zhang, *Colloids Surf., A*, 2017, **520**, 903–913.
111. A. B. Albadarin, M. N. Collins, M. Naushad, S. Shirazian, G. Walker and C. Mangwandi, *Chem. Eng. J.*, 2017, **307**, 264–272.

CHAPTER 3

Chemical Modifications of Lignin for Biomedical Applications

IULIANA SPIRIDON AND NARCIS CATALIN ANGHEL*

"Petru Poni" Institute of Macromolecular Chemistry, Grigore Ghica-Voda Alley no. 41A, 700487 Iasi, Romania
*Email: anghel.narcis@icmpp.ro

3.1 Introduction

Lignin plays a vital role in plant development. It functions as a barrier to pathogen invasion.[1] It also serves as a natural adhesive, binding plant carbohydrates together. Despite its availability and structural potential, lignin is still an underutilized natural resource, and the interest in making use of it in the design of nanomaterials is only beginning to grow.

Being the second most abundant natural substance in the world, lignin represents between 15 and 35% of the biomass used by the paper industry. It is made up of cross-linked phenylpropanoid structures that promote the structural integrity of plants and are crucial for the development of cell walls.

Sadly, lignin is still seldom used in biomedicine. It is quite challenging to generate lignin-based materials for biomedical applications owing to the heterogeneity of natural lignin. The many resources and extraction techniques used to obtain lignin are the causes of its heterogeneity. Additionally, the *in vivo* biocompatibility and biodegradability of lignin have not been dealt with sufficiently in recent research or application work on lignin-based

Biomaterials Science Series No. 16
Lignin-based Materials: Health Care and Medical Applications
Edited by Kuruvilla Joseph, Runcy Wilson, Gejo George and Saritha Appukuttan

Published by the Royal Society of Chemistry, www.rsc.org

biomaterials. One of the biggest obstacles to using lignin as a biomaterial today is the absence of fundamental knowledge of the mechanism by which the human body breaks it down.

Despite these disadvantages, lignin has numerous special qualities and advantages over other natural polymers, including antioxidant protection and effective antibacterial activity. With these benefits, lignin has the potential to be employed as a bioactive additive to support widely used biomaterials, such as xanthan, chitosan, alginates, and cellulose. There has been an upsurge of interest in using lignin and its derivatives as functional materials in biomedical applications, which highlights the potential of lignin as a leading contender for the creation of novel, long-lasting biomaterials.

The structure, reactivity, and physicochemical properties of lignin play important roles in the development of novel products. At the same time, it is worth mentioning that extraction processes, as well as the source of lignin, influence the physicochemical properties of the isolated product. For example, Ten *et al.*[2] reported that lignin nanotubes can be used as delivery vehicles of DNA (deoxyribonucleic acid). The researchers indicated that the efficacy of gene delivery depended on the physicochemical properties of lignin, which varied substantially based on the lignin source and the isolation method. The authors also highlighted that the derivatization of lignin nanotubes offers numerous ways to develop a smart delivery system for DNA in living organisms.

Many exhaustive studies[3–6] have pointed out the substantial benefits as well as considerable bottlenecks of delignification processes. We have chosen some examples to highlight the impact of the delignification method on the properties of lignin. Thus, kraft delignification is the main process to separate biomass components. It accounts for about 85% of all lignin generated. Sulfur groups are present in moderate amounts in kraft lignin (KL, 1–3% by weight), and it is only soluble[7] at $pH > 10$. Alternatively, sulfite delignification gives lignosulfonates, which have sulfur contents between 4 and 8% and are water-soluble under most conditions.[8] Soda delignification is usually used for the treatment of non-wood materials, the resulting lignin being a good candidate for high-value products due to the absence of sulfur.[9] Organosolv delignification is a promising alternative process to the conventional delignification processes. It uses organic solvents to isolate lignin. Organosolv lignins are sulfur-free and have high purity and low ash content due to their lower carbohydrate content.

The emergence of the biorefinery theory for biomass treatment has drawn the attention of academic and industrial researchers towards lignin.

Chemical processes can easily modify lignin owing to its functional groups such as carbon–carbon double bonds and phenolic and alcoholic hydroxyl groups. Thus, several strategies, including chemical modification and the use of compatibilizers, have been proposed to prepare homogeneous blends/composites comprising lignin. Packaging plastics, biomedical devices, adhesives, and stimuli-responsive materials could all benefit from the use of lignin-based products.[10]

The renewability and biodegradability properties make lignin an interesting precursor for developing environmentally friendly nanomaterials. Lignin and its derivatives are promising candidates for the development of materials with medical applications.[11–15]

3.2 Biomedical Applications of Lignin

3.2.1 Lignin-containing Nanoparticles

It is well known that nanoparticles improve the stability of encapsulated drug molecules by protecting them from the effects of the surrounding pH value and premature metabolic degradation. At the same time, the sustained release and adsorption of drug molecules are improved,[16] while the risk of systemic toxicity is reduced. Antimicrobial nanoparticles with biodegradable cores were developed from lignin laden with silver cations and coated with a cationic polyelectrolyte.[17] These lignin-based nanoparticles were found to have biocidal activity against Gram-negative and -positive pathogens, while lignin capped with silver nanoparticles[18] exhibited antifungal, antiplatelet, and antioxidant properties. Lignin nanoparticles did not exhibit hemolytic activity, suggesting their non-toxic nature. Fe_3O_4-lignin nanoparticles may be used in cancer therapy and diagnosis, according to Figueiredo *et al.*[19] Fe_3O_4-lignin nanoparticles were demonstrated to serve well as a vehicle for delivering therapeutic molecules to a specific target in the body. In addition, alkaline lignin self-assembling with resveratrol and Fe_3O_4 magnetic particles led to the generation of a stable nanodrug carrier.[20] These magnetic nanoparticles proved their anticancer activity.

The use of lignin nanoparticles as a vaccination adjuvant was recently demonstrated in a study. It was found that these nanoparticles enhanced the uptake of antigen into dendritic cells, and subsequently caused higher immune cell stimulation.[21] Other authors[22] reported the use of lignin in the preparation of green biosensors for the detection of glucose or prion protein.[23]

3.2.2 Prime Applications of Esterified Lignins

Through processes such as esterification, phenolation, and etherification, modifications of lignin open ways to introduce new chemical sites onto the compound. A major emphasis is placed on functionalizing hydroxyl groups, with lignin modification playing an important role in the creation of novel biomedical materials.

One of the simplest methods for changing the hydroxyl groups in lignins is esterification.

Here, we provide a review of the latest data related to the biomedical applications of lignin derivatives.

Esterification is one of the most straightforward paths to alter the hydroxyl groups of lignin. For instance, it was discovered that the porphyrin complex

5,10,15,20-tetrakis(4-hydroxyphenyl)-21*H*,23*H*-porphyrin (THPP) could be encapsulated, in aqueous medium, into acetylated lignin nanoparticles[24] (see Figure 3.1).

These nanoparticles were tested against Gram-positive (*Enterococcus faecalis*, *Staphylococcus epidermidis*, and *Staphylococcus aureus*) strains. The system was able to produce reactive oxygen species (ROS) and perform photodynamic eradication, possibly due to the synergistic work between the flocculant characteristics of lignin nanoparticles and the light-driven ROS production of THPP. According to another study,[25] acetylated lignin nanoparticles are a good carrier for a variety of porphyrins, while retaining their photochemical and antibacterial capabilities.

Lignin was combined with poly(ethylene glycol) (PEG) and poly(methyl vinyl ether-*co*-maleic acid) (GANTREZ™) by an esterification reaction[26] (see Figure 3.2) that took place in the solid phase. According to the mechanism postulated by Singh *et al.*,[27] some OH groups contained in PEG/lignin align with the COOH groups found in poly(methyl vinyl ether-*co*-maleic acid) strings before the cross-linking reaction, generating hydrogen bonds. The thus-obtained hydrogels provided sustained curcumin delivery for 4 days. In addition, the materials cross-linked with PEG 10 000 displayed antibacterial activity against two microorganisms associated with medical device-related infections (*S. aureus* and *Proteus mirabilis*).

KL was modified through esterification with 2-bromo-isobutyric bromide and by atom transfer radical polymerization (ATRP)[28] of 2-(dimethylamino)ethyl methacrylate (DMAEMA) of esterified lignin. The technique depicted in Figure 3.3 was used to make several cationic amphiphilic lignin-based graft copolymers.

MDA-MB-231, Cos-7, and HeLa cell lines were used to assess the gene transfection efficiency. The cytotoxicity and gene transfection efficiency depended on the chain length of PDMAEMA, *i.e.* poly(2-(dimethylamino)ethyl methacrylate). As a result, it was shown that lignin-PDMAEMA copolymers with short PDMAEMA grafts had very high gene transfection effectiveness, which is likely due to the lignin core structure. The aforementioned examples demonstrate the potential of KL derivatives to be used as non-viral gene vectors for prospective gene delivery applications.

3.2.3 Bioavailability of Intracellular Drugs Through Carboxylated Lignin

In order to promote the interactions between lignin and polar functional groups of delivered compounds to the sites of action, lignin was carboxylated with a view to increasing the carboxyl group content on the lignin surface, hence regulating the bioavailability of the embedded bioactive substances.

Figuerido *et al.*[29] synthesized carboxylated lignin to obtain different dentin phosphophoryn-derived peptides. The dentin phosphophoryn-derived peptide and iRGD peptides (a 9-amino acid cyclic peptide and a molecular

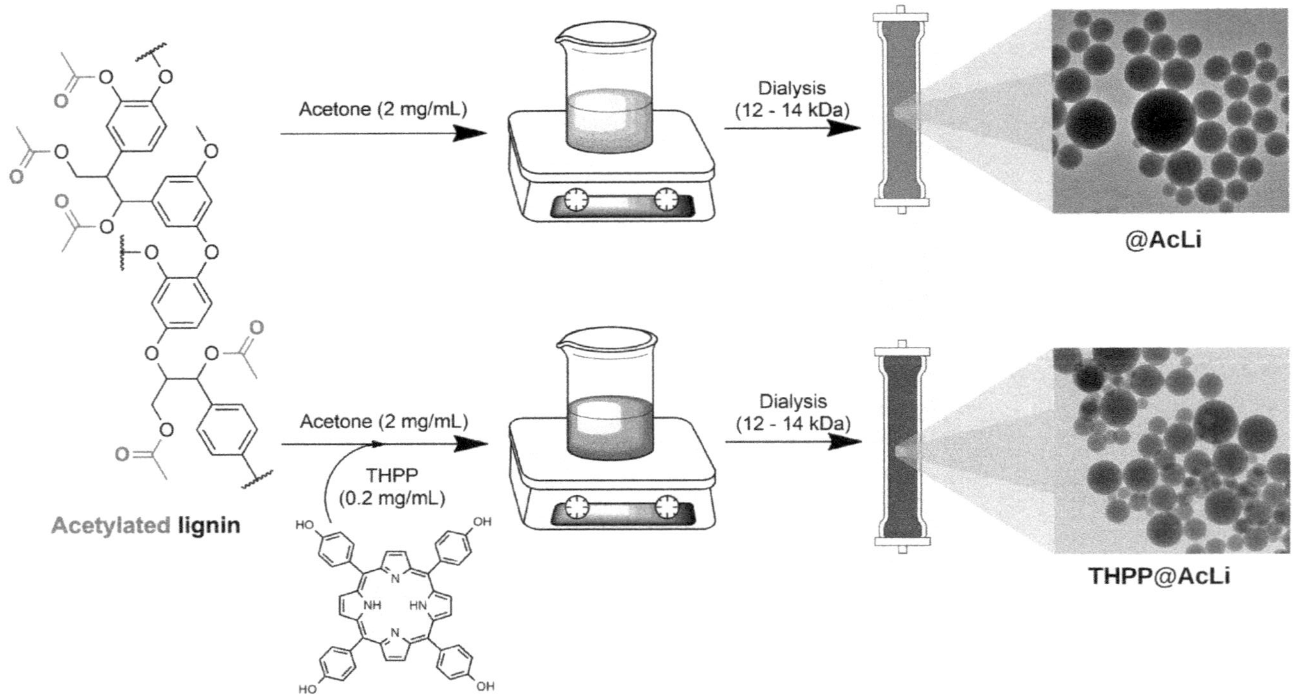

Figure 3.1 Preparation of acetylated lignin nanoparticles (@AcLi) and 5,10,15,20-tetrakis(4-hydroxyphenyl)-21*H*,23*H*-porphyrin (THPP)-loaded @AcLi (THPP@AcLi). Reproduced from ref. 24, https://doi.org/10.3389/fmicb.2020.606185, under the terms of the CC BY 4.0 license https://creativecommons.org/licenses/by/4.0/.

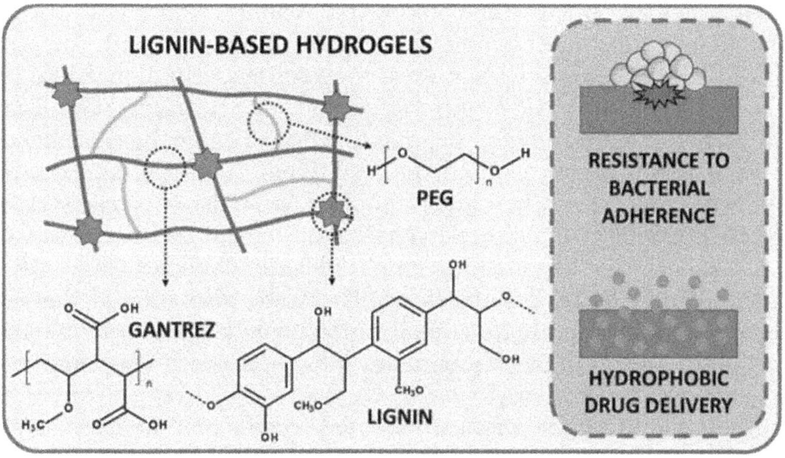

Figure 3.2 Schematic representation of a lignin hydrogel as a medical coating material (from Larraneta *et al.* 2018). Reproduced from ref. 26, https://doi.org/10.1021/acssuschemeng.8b01371, under the terms of the CC BY 4.0 license https://creativecommons.org/licenses/by/4.0/.

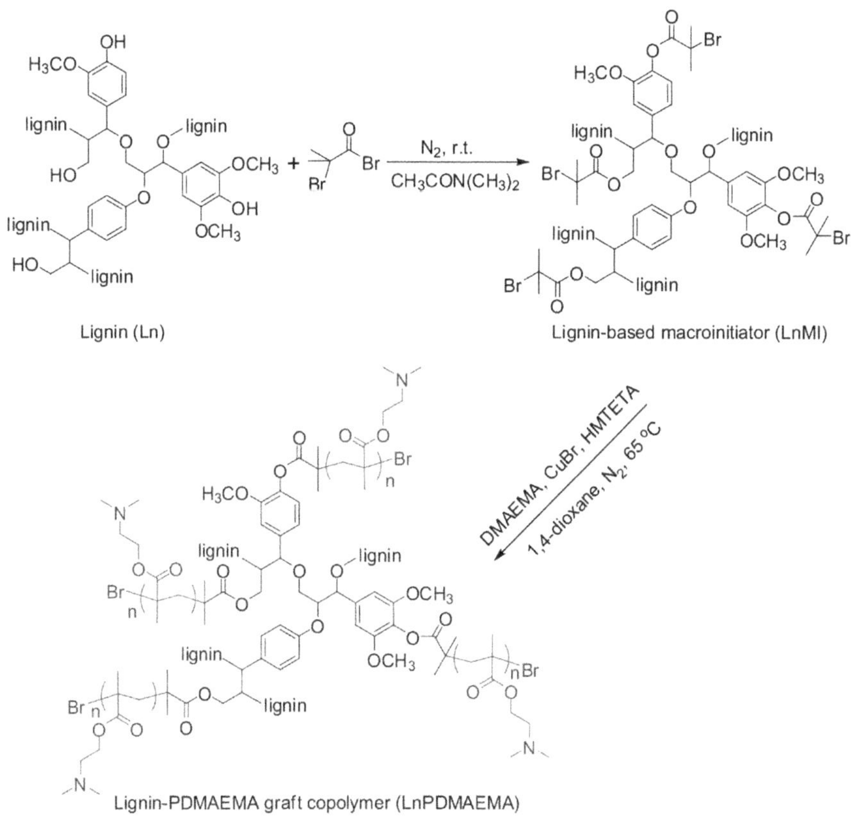

Figure 3.3 Synthetic procedures for a lignin-based macroinitiator (LnMI) and a lignin-PDMAEMA graft copolymer (LnPDMAEMA) *via* ATRP. Reproduced from ref. 28 with permission from Elsevier, Copyright 2015.

mimicry agent that was initially discovered in an *in vivo* screening of phage display libraries in tumor-bearing mice) were conjugated with the –COOH groups of lignin nanoparticles. To test the loading performance into lignin nanoparticles and the *in vitro* antiproliferation efficacy against several cancer cell lines, both before and after peptide functionalization, benzoazulene was used as a model chemical. Both the traditional 2D model and the three-dimensional (3D) cell spheroid morphology were used to test PC3-MM2, MDA-MB-231, and A549 cultures. When a 3D cell-culture model was used, the adsorption of the peptide-functionalized lignin nanoparticles was higher than that of the lignin nanoparticles. The study also showed that lignin nanoparticles with peptide functionalization could be employed to enhance the cellular adsorption of nanoparticles into various cell types and release diverse cargos intracellularly.

Another study revealed that carboxylated lignin could be used as an excipient in formulations comprising paracetamol.[30] Experimental results demonstrated that the release of hydrogen ions from the carboxylic group reduced the interaction between paracetamol and lignin, leading to a quicker disintegration and a softer tablet. As a consequence, a higher drug release was recorded.

3.2.4 Grafted Lignins in Biomedical Applications: New Insights

In order to create gels and materials that respond to temperature, lignin has also been grafted with other polymers. To develop biomaterials with specific features, readily available lignin has been called upon to serve as a responsive component and as a macro-cross-linker.

A new amphiphilic polymer was obtained[31] by grafting β-cyclodextrin (β-CD) onto enzymatically hydrolyzable lignin (EHL) (see Figure 3.4). It was discovered that β-CD extended the network structure of lignin molecules, making possible their interaction with an anticancer medicine called hydroxycamptothecin (HCPT). This interaction led to significant tumor inhibition. Also, the good sustained-release capability of this new material was demonstrated.

To increase the effectiveness of GDC-0623, a potent inhibitor of mitogen-activated protein kinase, in the treatment of triple-negative breast cancer (TNBC) with KRAS (Kirsten rat sarcoma viral oncogene homolog) mutations,[32] a novel alkaline-lignin-conjugated-poly(lactic-*co*-glycolic acid) (L-PLGA) nanoparticle drug delivery system was created. Increased efficacy in the delivery of the drug to cancer cells was demonstrated in multiple *in vitro* assays.

KL-poly(glycidyl methacrylate)-*co*-poly(ethylene glycol)methacrylate (PGMA–PEGMA) copolymers were investigated for possible application as gene delivery systems.[33] Some of the hydroxyl groups in lignin were turned into initiation sites, enabling side-chain growth. The copolymers were achieved by grafting glycidyl methacrylate (GMA) and poly(glycidyl methacrylate) (PEGMA)

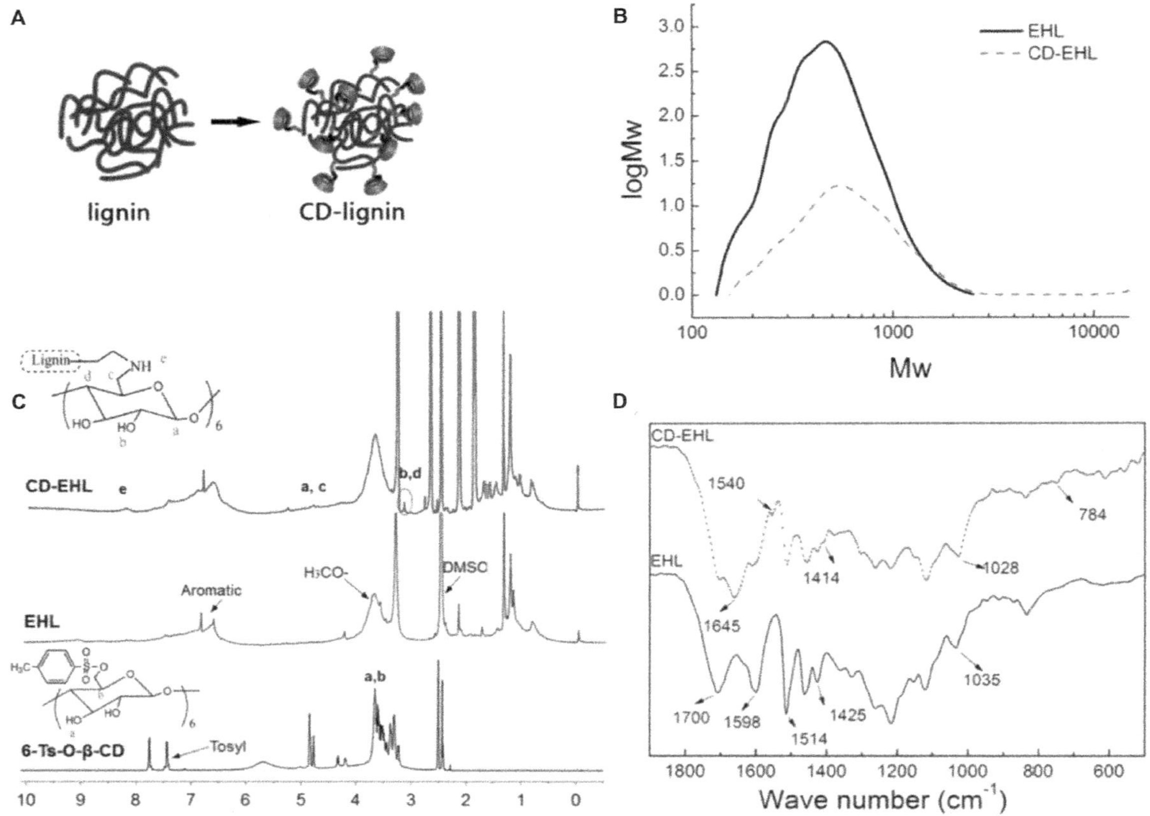

Figure 3.4 Cyclodextrin (CD)-enzymatically hydrolyzable lignin (EHL) (CD-EHL) system: (A) schematic diagram of β-CD conjugated with EHL; (B) molecular weight distribution of EHL and CD-EHL; (C) 1H-NMR spectra of 6-TsO-β-CD, EHL, and CD-EHL; and (D) infrared spectrograms of EHL and CD-EHL. Reproduced from ref. 31, https://doi.org/10.3390/nano9070997, under the terms of the CC BY 4.0 license https://creativecommons.org/licenses/by/4.0/.

onto the lignin macroinitiator, using different molar ratios of GMA, PEGMA, and initiator sites (see Figure 3.5).

The ring-opening reaction of the epoxide groups of PGMA produced water-soluble, cationic lignin–PGEA–PEGMA copolymers.

The effectiveness of these novel lignin–PGEA–PEGMA copolymers' *in vitro* gene transfection was assessed using the reporter gene luciferase in HEK 293T, a human cell line that expresses a mutant version of the SV40 large T antigen, and nontumorigenic Hep G2 cells with high proliferation rates and an epithelial-like morphology. The outcomes demonstrated low cytotoxicity and high gene transfection effectiveness in the cell lines for the copolymers with long enough PGEA side chains. Strong antioxidant activity was also demonstrated.

A novel medication to treat atherosclerosis was created by loading betulinic acid into poly(vinyl alcohol)/ethylacrylate grafted lignin polymer[34] (PVA/Lig-*g*-MA). The results of the tests showed that the levels of lipids and interleukins dropped, while the levels of vasodilators increased, limiting the deterioration of cardiac cells. A high-fat diet was also found to have a hepatoprotective effect.

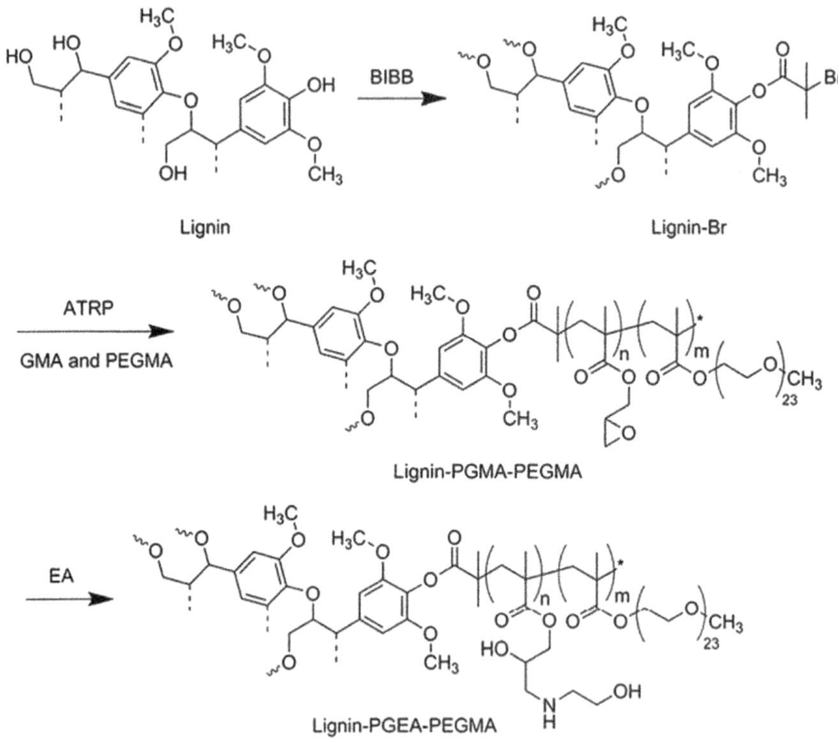

Figure 3.5 Schematic diagram illustrating the preparation processes of lignin–PGEA–PEGMA carriers *via* ATRP. Reproduced from ref. 33 with permission from the Royal Society of Chemistry.

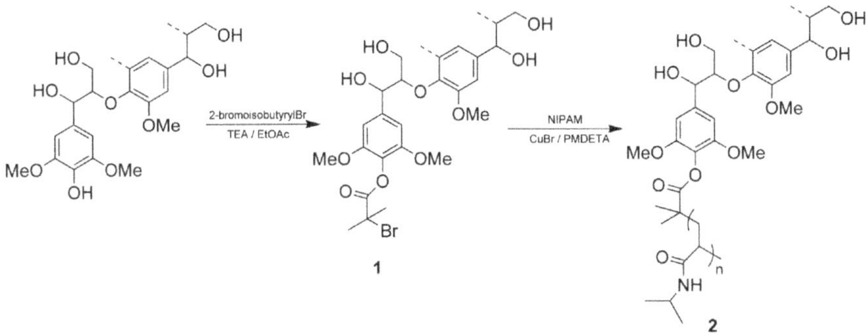

Figure 3.6 Synthetic scheme for the preparation of lignin-*g*-polyNIPAM copolymers. Reproduced from ref. 35 with permission from American Chemical Society, Copyright 2010.

Following the procedure illustrated in Figure 3.6, Kim and Kadla[35] used ATRP to copolymerize *N*-isopropylacrylamide (NIPAM) with hardwood KL (HWKL), utilizing a selectively modified lignin-based macroinitiator.

To produce selectively phenol-modified lignin macroinitiators, HWKL was combined with 2-bromoisobutyryl bromide in EtOAc with triethylamine (TEA). Due to polyNIPAM's temperature sensitivity, this lignin copolymer might be used in biomedical applications.

By plasticizing caprolactone and polyacrylic acid, a lignin/caprolactone biodegradable film was created. Carbonyldiimidazole (CDI) was used to activate the hydroxyl groups of the PAA/PCL polymer, and LL-37 peptides were then immobilized on the material.[36] Both Gram-positive (*S. aureus*) and -negative (*E. coli*) bacteria were resistant to the substance. Additionally, 3T3 murine fibroblasts were used to identify their capacity to enable tissue cell integration.

3.2.5 Lignin Micelles and Their Potential Use

Lignin-based micelles have drawn increased attention due to their potential for application in the pharmaceutical area, being able to modulate the biodegradability and stability of enzymes.

Using alkali lignin (AL), Cheng *et al.*[37] prepared acryloyl AL-*g*-polymethacrylate that was transformed into pH-sensitive nanomicelles by solvent exchange. The cytotoxicity studies performed on a human colon cancer cell line, a human bone marrow stromal cell line, and a mouse embryonic fibroblast cell line demonstrated the potential of the lignin-based nanomicelles as a pH-responsive carrier for oral drugs.

As shown in Figure 3.7, lignin-based complex micelles were created by quaternizing AL and then allowing it to self-assemble with sodium dodecyl benzene sulfonate (SDBS) in an ethyl alcohol/water mixture. Spherical micelles were formed due to the hydrophobic interactions between the quaternized lignin and SDBS complexes, supported by the nearby hydrophilic chains.[38]

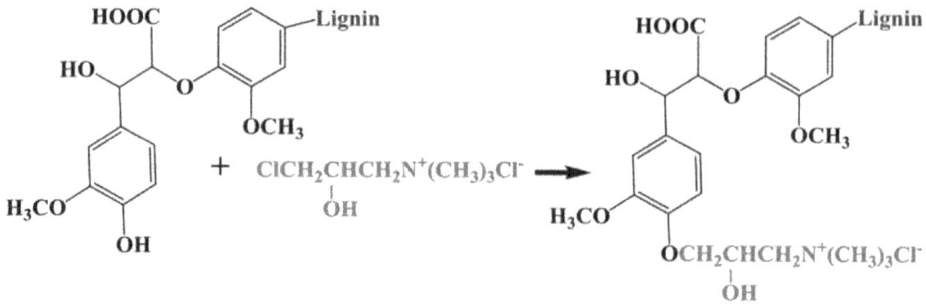

Figure 3.7 Synthetic route to obtain quaternized AL. Reproduced from ref. 38 with permission from Elsevier, Copyright 2017.

Quaternized lignin presented some weakly basic groups (quaternary ammonium groups) as well as ionizable acidic groups (carboxyl and phenolic hydroxyl groups).

Ibuprofen was encapsulated *via* hydrophobic interactions into lignin micelles. It was found that the ionization of carboxyl groups improved the release of the drug in the intestinal tract. Based on the results, the authors concluded that this formulation is suitable for the controlled delivery of hydrophobic oral drugs.

3.2.6 Applications of Lignin/Lignin Derivative-based Gels

The process of polymerizing and simultaneously cross-linking one or more monomers results in hydrogels. They are cross-linked hydrophilic polymeric materials that take the shape of a 3D network and can be either natural or synthetic in origin. While there are many diverse uses for hydrogels, the ones that are most important include the capture of heavy metal ions, medical and therapeutic uses, and the controlled release of pharmaceuticals.

A new nanogel was obtained (see Figure 3.8) by graft copolymerization of *N*,*N*-(dimethylamino)ethyl methacrylate (DMAEMA) and NIPAM from brominated lignin.[39]

The hydrophobic properties of lignin have a positive effect on the release process and the diffusion of curcumin in cellular environments. The insignificant cell toxicity and the behavior of this nanogel in the pH range of 6–8 make it suitable for drug delivery applications.

Composite hydrogels can be created using lignin and its derivatives with 3D network architectures. KL was functionalized using the Mannich reaction with glycine in the presence of formaldehyde. The water-soluble lignin derivative was further used to obtain a cross-linked Na hyaluronan/lignin (NaHy/lignin) hydrogel. *N*-(3-Dimethylaminopropyl)-*N*-ethylcarbodiimide hydrochloride (EDC) was used[40] as a cross-linking agent. Hydroxyl and carboxyl groups can be found in lignin as well as in NaHy. It appears that the phenolic OH groups of lignin and the activated –COOH groups of NaHy may interact to form a hetero-cross-linkage (see Figure 3.9).

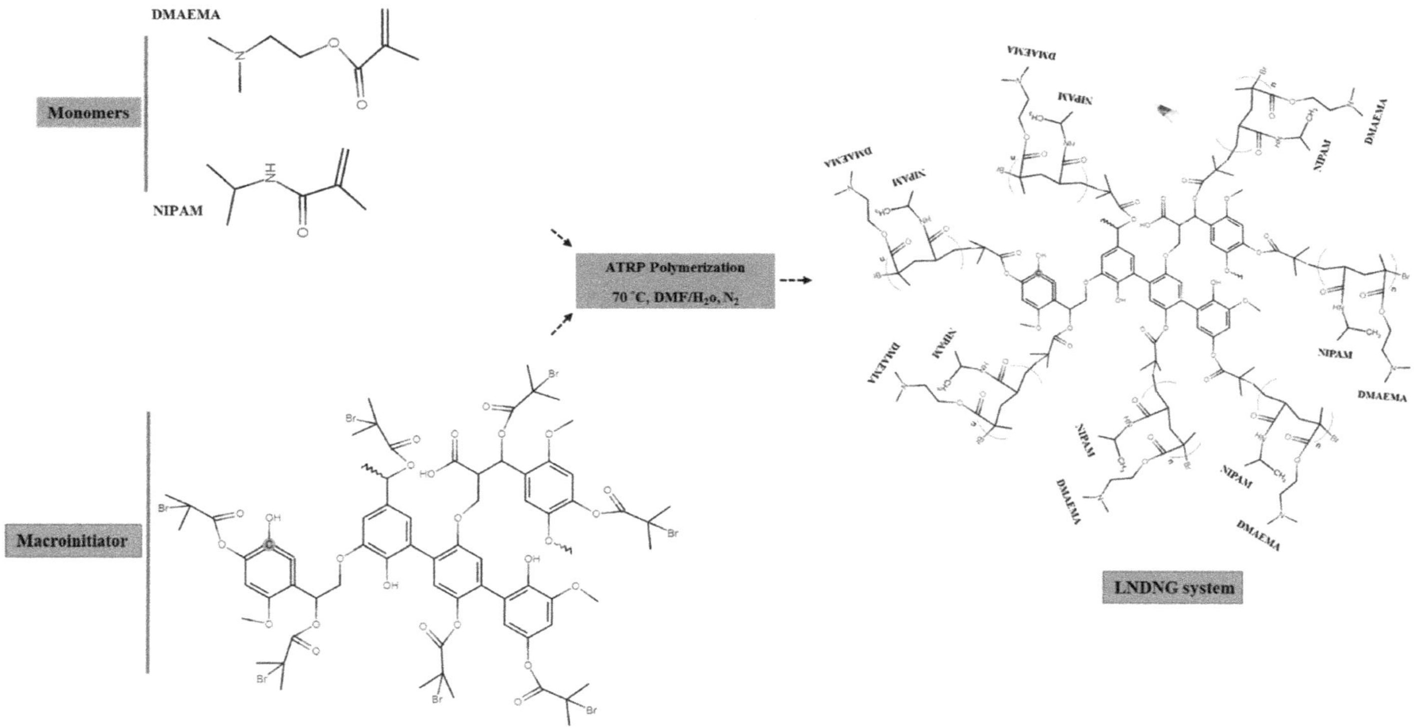

Figure 3.8 Obtaining a lignin-*g*-P(NIPAM-*co*-DMAEMA) nanogel (LNDNG) *via* "grafting from" atom transfer radical polymerization. Reproduced from ref. 39, https://doi.org/10.1038/s41598-021-81393-3, under the terms of the CC BY 4.0 license https://creativecommons.org/licenses/by/4.0/.

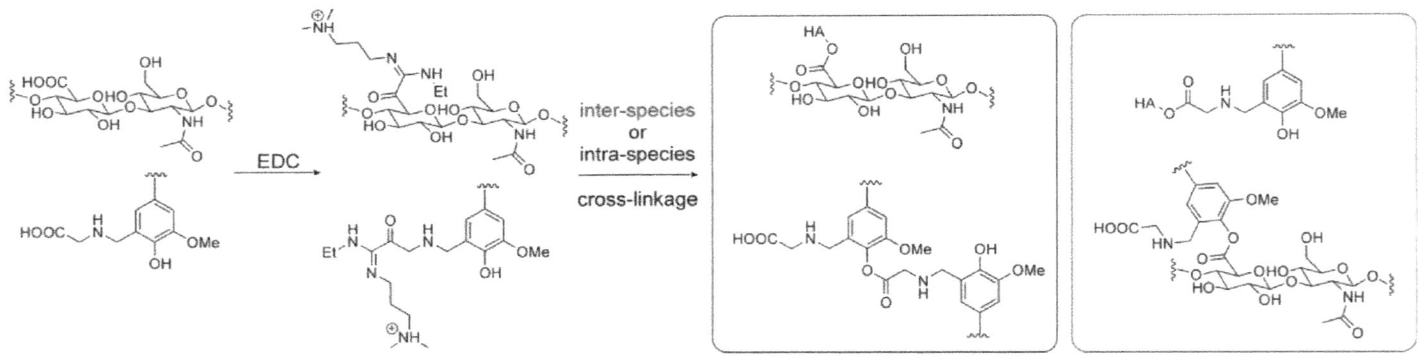

Figure 3.9 Cross-linkage formation from activated lignin and/or hyaluronan. Reproduced from ref. 40 with permission from Elsevier, Copyright 2018.

Additionally, it may be assumed that both lignin and NaHy can undergo self-cross-linking if the two reactive functional groups are derived from the corresponding materials.

The stability, swelling ability, and viscoelastic properties of the NaHy/lignin hydrogels were acceptable. The proposed hydrogels were shown to be non-cytotoxic substances, appropriate for tissue engineering or other biomedical applications, after cytotoxicity studies on a mouse embryonic fibroblast cell line.

A new polyacrylamide/lignin hydrogel was synthesized by *in situ* free radical polymerization[41] (see Figure 3.10).

Using H_2O_2/ascorbic acid as a redox initiator and *N*,*N*′-methylenebisacrylamide (MBAAm) as a cross-linking agent, the reaction between an acrylamide monomer and a lignin nano-dispersion was carried out. Human esophageal squamous carcinoma cells were co-cultured with the PAM/LNP hydrogel for 36 hours. This study showed that the PAM/LNP hydrogel may be used in biological applications, such as tissue engineering and regeneration, due to the development of 3D aggregated cells and the multiplication of the cells.

Zhang *et al.*[42] prepared (see Figure 3.11) a new lignin–chitosan–PVA hydrogel with protein adsorption capacity. The sulfonate groups in the lignin structure established ionic bonds with the amino groups in the chitosan chain, thus improving the mechanical strength of the hydrogel.

The presence of lignin led to an improvement in the protein's capacity for adsorption. Due to the antioxidant activity of lignin and its capacity to scavenge free radicals, the produced material also assisted in accelerating wound healing in a mouse skin wound model. The performed investigations demonstrated the inhibitory effect of the developed hydrogel on *S. aureus* due to penetration into its cell membrane, which decreased with increasing lignin content.

By radical polymerization of acrylic acid in the presence of acid-activated lignosulfonate and poly(vinyl pyrrolidone), some authors[43] produced doubly cross-linked hydrogels. Amoxicillin was loaded as a model drug into the synthesized hydrogels. Small amounts of the medication were lost in the stomach, but overall, the system managed to favourably transport the drug to the intestine. It was observed that, as a result of the grafting reaction, the amount of available binding sites increased, and a more porous 3D network of the hydrogel was formed.[44]

3.2.7 Applications of Lignin-based Fibers

The use of lignin-based high-performance fibers should be emphasized as being of great interest due to the mechanical performance of the fibers resulting from the polymer orientation along the axial direction. Electrospun poly(glycerol sebacate)–poly(vinyl alcohol) lignin fibers were obtained by the electrospinning technique and assessed for neural applications.[45] The results revealed that the addition of 5% of lignin is promising for the development of materials intended for nerve tissue engineering, with the

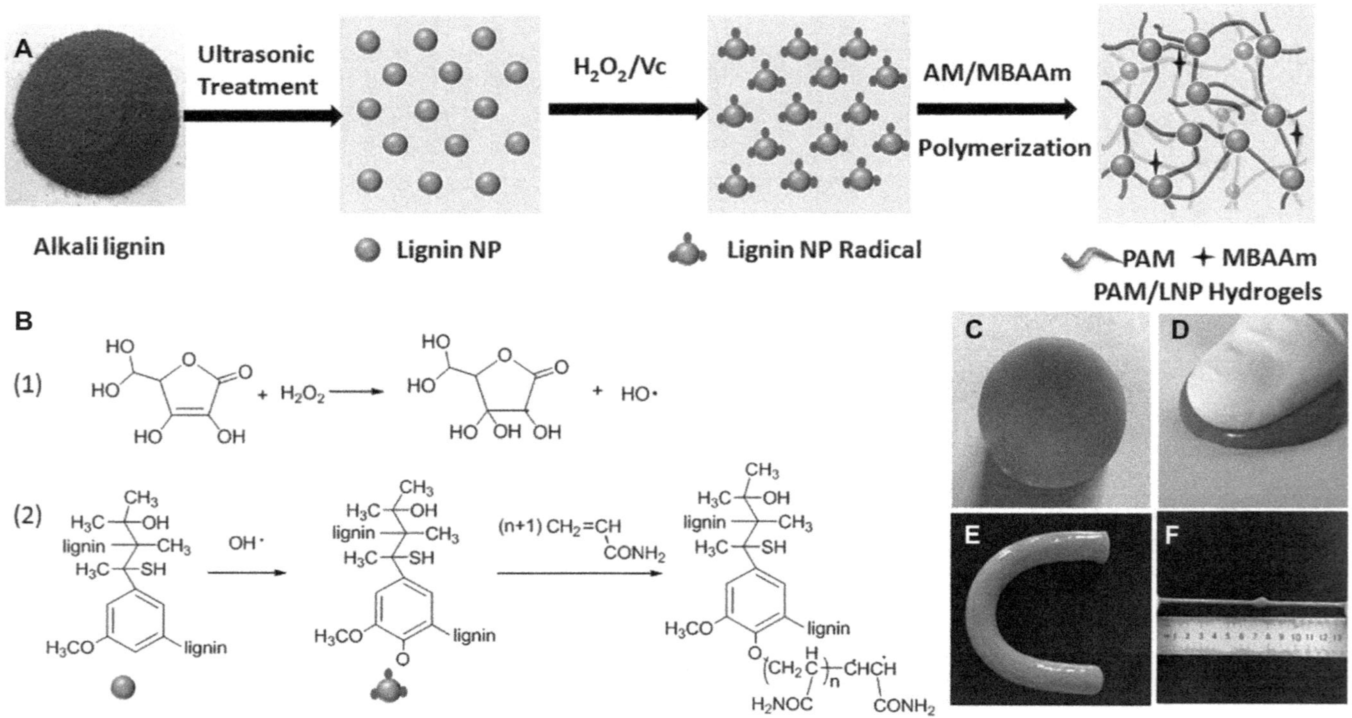

Figure 3.10 (A) Schematic illustration of the preparation of a PAM/LNP hydrogel. (B) The reaction of forming lignin NP free radicals (1) and the PAM/LNP hydrogel (2). (C–F) Photographs of the PAM/LNP hydrogel: (D) compressing; (E) bending; and (F) knotting. Reproduced from ref. 41 with permission from Elsevier, Copyright 2019.

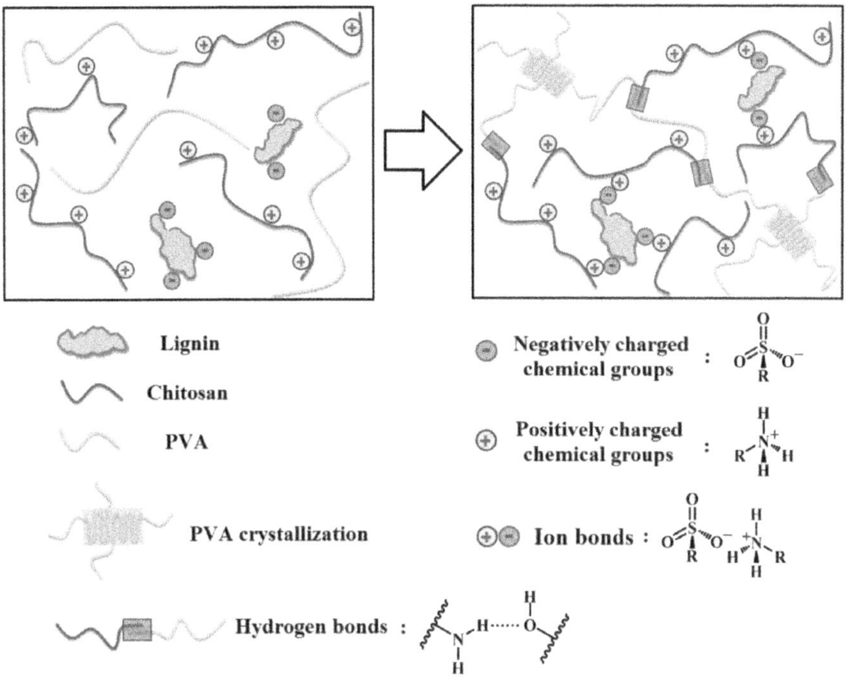

Figure 3.11 The cross-linking mechanism of a lignin–chitosan–PVA hydrogel. Reproduced from ref. 42 with permission from Elsevier, Copyright 2019.

role of lignin in the proliferation and differentiation of rat pheochromocytoma cells being well demonstrated. Hybrid fibers comprising a chitin-lignin-based sol–gel mixture and elastomeric poly(glycerol sebacate) proved their potential in advanced biomedical applications, such as wound care products.[46] Liang *et al.*[47] produced a novel material with antioxidant activity, low cytotoxicity, and outstanding anti-inflammatory properties by grafting poly(ε-caprolactone) (PCL) onto the hydroxyl groups of lignin using solvent-free ring-opening polymerization (Figure 3.12).

It was found that electrospun PCL-lignin nanofibers are suitable for the treatment of osteoarthritis due to their preventive effects against oxidative stress. In chondrocytes under oxidative stress, the newly discovered PCL-lignin implantable membrane boosted the expression of antioxidant enzymes (see Figure 3.13).

Solvent-free ring-opening polymerization was used by other authors[48] to create lignin-PCL copolymers, which were then combined with PCL (see Figure 3.14).

The components were employed to create scaffolds for the growth of Schwann cells and neurons. The investigations performed revealed that the lignin copolymers promoted cell proliferation of both Schwann cells and bone marrow mesenchymal stem cells. These compounds have been recommended for use in nerve tissue repair and regeneration as they boost the resistance of nerve cells to oxidative stress and encourage neurite outgrowth.

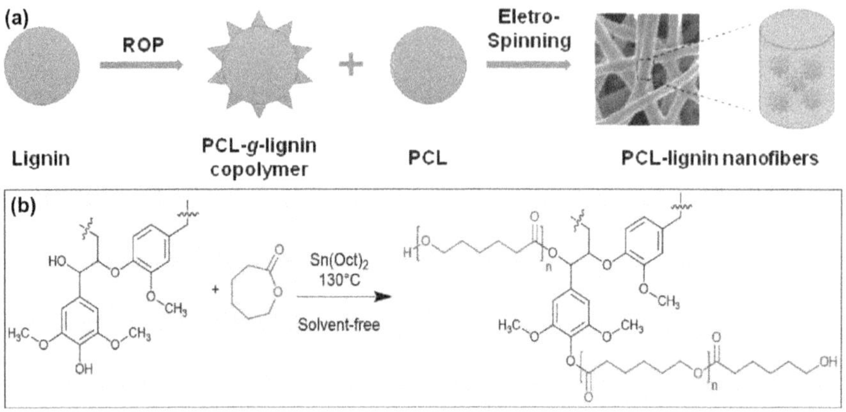

Figure 3.12 (a) Schematic illustration of PCL-*g*-lignin and PCL-lignin nanofibers. (b) Synthetic route for PCL-*g*-lignin copolymers *via* solvent-free ring-opening polymerization (ROP). Reproduced from ref. 47 with permission from Elsevier, Copyright 2020.

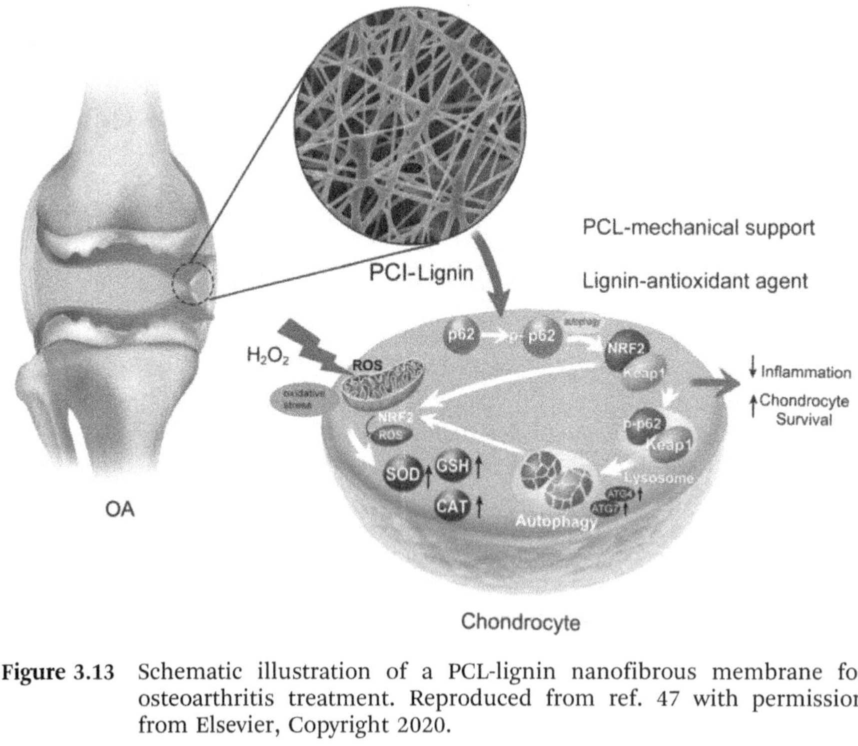

Figure 3.13 Schematic illustration of a PCL-lignin nanofibrous membrane for osteoarthritis treatment. Reproduced from ref. 47 with permission from Elsevier, Copyright 2020.

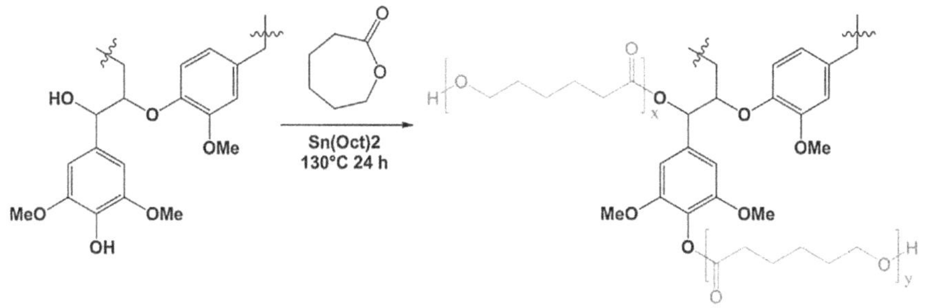

Figure 3.14 Synthetic route for PCL-lignin copolymers *via* solvent-free polymerization. Reproduced from ref. 48 with permission from Elsevier, Copyright 2018.

3.3 Conclusions

The current work provides evidence substantiating the idea of lignin being a platform macromolecule with huge potential for creating a wide variety of materials with medical uses.

Chemistry mechanisms are particularly important in the case of lignin because its composition and structure depend on the plant species it originates from and the delignification process used to extract it. A better understanding of these mechanisms would lead to its use as an alternative to obtain biomaterials, exploiting lignin's biocompatibility, antioxidant, antimicrobial, and anti-UV properties. Significant efforts have been put into the development of lignin-based medical devices, wound dressings, pharmaceuticals, drug delivery systems, and other materials. Besides the accessibility of lignin as a starting material, sustainable and successful innovative concepts for its use in value-added products with medical applications are expected from academia and industry. These should be translated into successful products after intensive techno-economic analyses based on market research, also considering their environmental and social implications.

References

1. W. Boerjan, J. Ralph and M. Baucher, *Annu. Rev. Plant Biol.*, 2003, **54**, 519.
2. E. Ten, C. Ling, Y. Wang, A. Srivastava, L. A. Dempere and W. Vermerris, *Biomacromolecules*, 2013, **15**, 327.
3. H. Wang, Y. Pu, A. Ragauskas and B. Yang, *Bioresour. Technol.*, 2019, **271**, 449.
4. B. Hansen, B. Kamm and M. Schulze, Qualitative and quantitative analysis of lignins from different sources and isolation methods for an application as a biobased chemical resource and polymeric material, in *Analytical Techniques and Methods for Biomass Products*, ed. S. Vaz Jr. and

P. Seidl, Springer, Berlin, Germany, 2017, pp. 15–44, ISBN 978-3-319-41414-0.
5. W. G. Glasser, *Front. Chem.*, 2019, **7**, 565.
6. J. J. Liao, N. H. Abd Latif, D. Trache, N. Brosse and M. H. Hussin, *Int. J. Biol. Macromol.*, 2020, **162**, 985.
7. P. Figueiredo, K. Lintinen, J. T. Hirvonen, M. A. Kostiainen and H. A. Santos, *Prog. Mater. Sci.*, 2018, **93**, 233.
8. P. Azadi, O. R. Inderwildi, R. Farnood and D. A. King, *Renewable Sustainable Energy Rev.*, 2013, **21**, 506.
9. P. Prinsen, J. Rencoret, A. Gutiérrez, T. Liitiä, T. Tamminen, J. L. Colodette, M. A. Berbis, J. Jimenes-Barbero, A. T. Martinez and J. C. Del Rio, *Ind. Eng. Chem. Res.*, 2013, **52**, 15702.
10. A. L. Holmberg, N. A. Nguyen, M. G. Karavolias, K. H. Reno, R. P. Wool and T. H. Epps, *Macromolecules*, 2016, **49**, 1286.
11. I. Spiridon, *Environ. Chem. Lett.*, 2020, **18**, 771.
12. M. Witzler, A. Alzagameem, M. Bergs, B. El Khaldi-Hansen, S. E. Klein, D. Hielscher, B. Kamm, J. Kreyenschmidt, E. Tobiasch and M. Schulze, *Molecules*, 2018, **23**, 1885.
13. I. Spiridon, *Cellul. Chem. Technol.*, 2018, **52**, 543.
14. M. P. Vinardell and M. Mitjans, *Int. J. Mol. Sci.*, 2017, **18**, 1219.
15. R. Liu, L. Dai, C. Xu, K. Wang, C. Zheng and C. Si, *ChemSusChem*, 2020, **13**, 4266.
16. X. Xie, Q. Tao, Y. Zou, F. Zhang, M. Guo, Y. Wang, H. Wang, Q. Zhou and S. Yu, *J. Agric. Food Chem.*, 2011, **59**, 9280.
17. A. P. Richter, J. S. Brown, B. Bharti, A. Wang, S. Gangwal, K. Houck, E. A. Cohen Hubal, V. N. Paunov, D. Stoyanov and O. D. Velev, *Nat. Nanotechnol.*, 2015, **10**, 817.
18. M. B. Marulasiddeshwara, S. S. Dakshayani, M. N. Sharath Kumar, R. Chethana, P. Raghavendra Kumar and S. Devaraja, *Mater. Sci. Eng., C*, 2017, **81**, 182.
19. P. Figueiredo, K. Lintinen, A. Kiriazis, V. Hynninen, Z. Liu, T. Bauleth-Ramos, A. Rahikkala, A. Correia, T. Kohout, B. Sarmento, J. Yli-Kauhaluoma, J. Hirvonen, O. Ikkala, M. A. Kostiainen and H. A. Santos, *Biomaterials*, 2017, **121**, 97.
20. L. Dai, R. Liu, L. Q. Hu, Z. F. Zou and C. L Si, *ACS Sustainable Chem. Eng.*, 2017, **5**, 8241.
21. M. S. Alqahtani, M. Kazi, M. Z. Ahmad, R. Syed, M. A. Alsenaidy and S. A. Albraiki, *Int. J. Biol. Macromol.*, 2020, **163**, 1314.
22. E. Capecchi, D. Piccinino, E. Tomaino, B. M. Bizzarri, F. Polli, R. Antiochia, F. Mazzei and R. Saladino, *RSC Adv.*, 2020, **10**, 29031.
23. C. Yuan, Z. Lou, W. Wang, L. Yang and Y. Li, *Int. J. Mol. Sci.*, 2019, **20**, 741.
24. N. Maldonado-Carmona, G. Marchand, N. Villandier, T. S. Ouk, M. M. Pereira, M. J. F. Calvete, C. A. Calliste, A. Żak, M. Piksa, K. J. Pawlik, K. Matczyszyn and S. Leroy-Lhez, *Front. Microbiol.*, 2020, **11**, 606185.

25. N. Maldonado-Carmona, T. S. Ouk, N. Villandier, C. A. Calliste, M. J. F. Calvete, M. M. Pereira and S. Leroy-Lhez, *Antibiotics*, 2021, **10**, 513.
26. E. Larraneta, M. Imizcoz, T. J. Xi, N. J. Irwin, A. Ripolin, A. Perminova, J. Domínguez-Robles, A. Rodríguez and R. F. Donnelly, *ACS Sustainable Chem. Eng.*, 2018, **6–7**, 9037.
27. T. R. R. Singh, P. A. McCarron, A. D. Woolfson and R. F. Donnelly, *J. Appl. Polym. Sci.*, 2009, **112**, 2792.
28. X. Liu, H. Yin, Z. Zhang, B. Diao and J. Li, *Colloids Surf., B*, 2015, **125**, 230.
29. P. Figueiredo, M. H. Sipponen, K. Lintinen, A. Correia, A. Kiriazis, J. Yli-Kauhaluoma, M. Osterberg, A. George, J. Hirvonen, M. A. Kostiainen and H. A. Santos, *Small*, 2019, **15**, 1901427.
30. M. Pishnamazi, H. Hafizi, S. Shirazian, M. Culebras, G. M. Walker and M. N. Collins, *Polymers*, 2019, **11**, 1059.
31. Y. Zhou, Y. Han, G. Li, S. Yang and F. Chu, *Nanomaterials*, 2019, **9**, 997.
32. E. Byrnel, C. E. Astete, M. Vaithiyanathan, A. T. Melvin, M. Moradipour, S. E. Rankin, B. L. Knutson, C. M. Sabliov and E. C. Martin, *Nanomedicine*, 2020, **15**, 981.
33. S. Jiang, D. Kai, Q. Q. Dou and X. J. Loh, *J. Mater. Chem. B*, 2015, **3**, 6897.
34. J. Zhang, H. S. Almoallim, S. A. Alharbi and B. Yang, *Arabian J. Chem.*, 2021, **14**, 102934.
35. Y. S. Kim and J. F. Kadla, *Biomacromolecules*, 2010, **11**, 981.
36. A. Ogan, B. Yüce-Dursun, D. Abdullah, A. Beyler-Çigil, M. V. Kahraman, P. Çaglayan, M. Birbir, Ö. Mutlu and N. Gülsoy, *Polym. Adv. Technol.*, 2020, **31**, 2222.
37. L. Cheng, B. Deng, W. Luo, S. Nie, X. Liu, Y. Yin, S. Liu, Z. Wu, P. Zhan, L. Zhang and J. Chen, *J. Agric. Food Chem.*, 2020, **68**, 5249.
38. Y. Li, X. Qiu, Y. Qian, W. Xiong and D. Yang, *Chem. Eng. J.*, 2017, **327**, 1176.
39. A. Dinari, M. Abdollahi and M. Sadeghizadeh, *Sci. Rep.*, 2021, **11**(1), 1962.
40. L. Musilová, A. Mráček, A. Kovalcik, P. Smolka, A. Minařík, P. Humpolíček, R. Vích and P. Ponížil, *Carbohydr. Polym.*, 2018, **181**, 394.
41. Y. Chen, K. Zheng, L. Niu, Y. Zhang, Y. Liu, C. Wang and F. Chu, *Int. J. Biol. Macromol.*, 2019, **128**, 414.
42. Y. Zhang, M. Jiang, Y. Zhang, Q. Cao, X. Wang, Y. Han, G. Sun, Y. Li and J. Zhou, *Mater. Sci. Eng., C*, 2019, **104**, 110002.
43. X. Wang, Z. Zhou, X. Guo and Q. He, *RSC Adv.*, 2016, **6**, 35550.
44. V. N. Bykov, I. S. Drachev, S. Y. Kraev, M. A. Maydin, E. A. Gubarev, S. E. Pigarev, V. N. Anisimov, I. A. Balduev, E. I. Fedoros and A. V. Panchenko, *Int. J. Radiat. Biol.*, 2017, **94**, 114.
45. A. Saudi, S. Amini, N. Amirpour, M. Kazemi, A. Zargar Kharazi, H. Salehi and M. Rafienia, *Mater. Sci. Eng., C*, 2019, **104**, 110005.

46. T. Abudula, L. Gzara, G. Simonetti, A. Alshahrie, N. Salah, P. Morganti, A. Chianese, A. Fallahi, A. Tamayol, S. A. Bencherif and A. Memic, *Materials*, 2018, **11**, 451.
47. R. Liang, J. Zhao, B. Li, P. Cai, X. J. Loh, C. Xu, P. Chen, D. Ki and L. Zheng, *Biomaterials*, 2020, **230**, 11960.
48. J. Wang, L. Tian, B. Luo, S. Ramakrishna, D. Kai, X. J. Loh, I. H. Yang, G. R. Deen and X. Mo, *Colloids Surf., B*, 2018, **169**, 356.

CHAPTER 4

Technical Lignin Fractionation: A Powerful Tool for Lignin Structure Homogenization and Its Application

HASAN SADEGHIFAR*[a] AND ARTHUR RAGAUSKAS[a,b,c,d]

[a] R&D Laboratory, Hollingsworth & Vose Company, Groton, Massachusetts 01450, USA; [b] Center for Renewable Carbon, Department of Forestry, Wildlife and Fisheries, University of Tennessee (UT), Knoxville, TN, USA; [c] Joint Institute for Biological Sciences, Biosciences Division, Oak Ridge National Laboratory (ORNL), Oak Ridge, TN, USA; [d] Department of Chemical and Biomolecular Engineering, University of Tennessee, Knoxville, TN, USA, Email: aragausk@utk.edu
*Email: hasan.sadeghifar@hovo.com

4.1 Introduction

Lignin is the primary bio-phenolic chemical resource in nature and forms around 20–30% of all woody materials.[1] The major proportion of technical lignin in the marketplace is a by-product of chemical pulping and bio-refining processes.[2] It is well known that lignin is composed of three main precursors, including *p*-coumaryl alcohol (H type), coniferyl alcohol (G type), and sinapyl alcohol (S type).[3] The main structural lignin differences among woody materials are attributed to the ratio of the mentioned precursors and the chemical bonds connecting them[4] (Scheme 4.1). However, our primary

Biomaterials Science Series No. 16
Lignin-based Materials: Health Care and Medical Applications
Edited by Kuruvilla Joseph, Runcy Wilson, Gejo George and Saritha Appukuttan

Published by the Royal Society of Chemistry, www.rsc.org

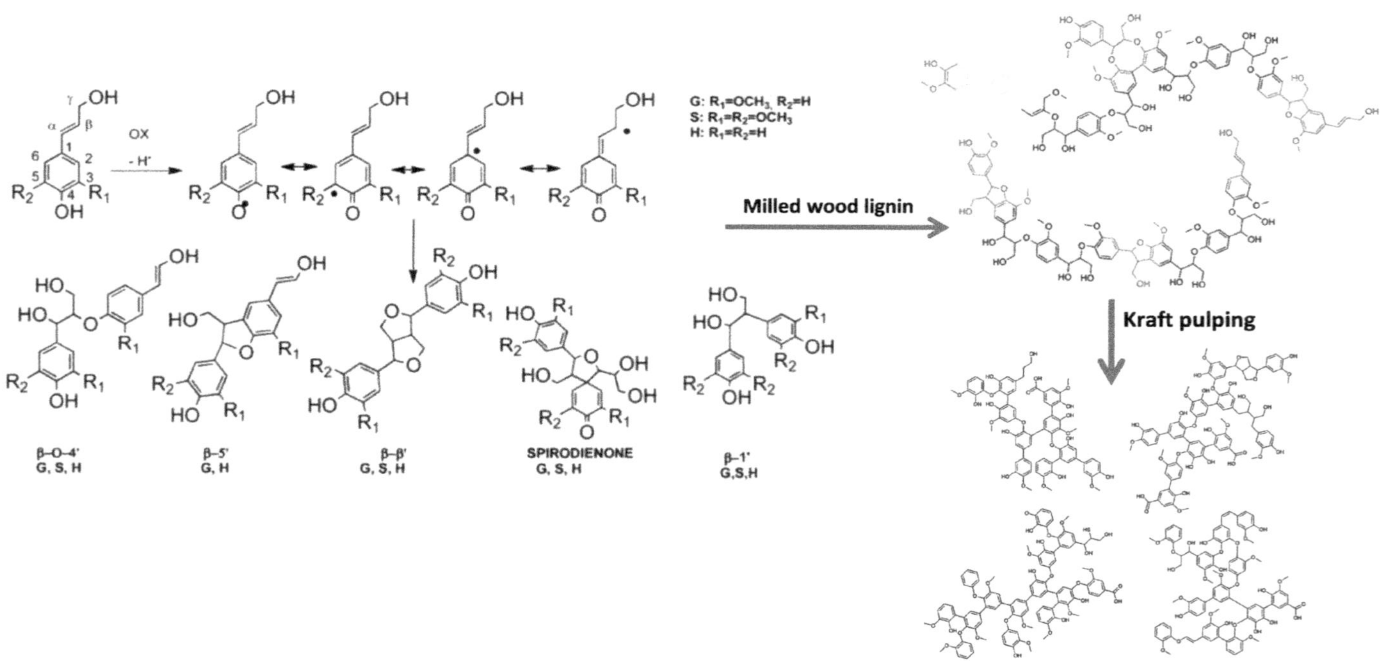

Scheme 4.1 Lignin biosynthesis: coupling mode for monolignols and oligomeric lignin chains, proposed lignin structure in the wood (milled wood lignin). Adapted from ref. 7 with permission from American Chemical Society, Copyright 2011; adapted from ref. 8 with permission from the Royal Society of Chemistry.

knowledge of the lignin chemical structure is based on technical or extracted lignin from wood, which is different from *in situ* lignin in the wood structure. Although advances in whole-cell NMR analysis have begun to address this issue,[5,6] recent studies indicate that lignin in the wood has a more linear structure in comparison to the cross-linked, highly branched structure in the extracted lignin.[7,8]

The β-O-4 ether bond is the primary linkage in non-extracted lignin (*in situ* lignin) (more than 50%) and has a key role in lignin depolymerization during pulping and bio-refining processes.[8] β-5, β-β, 5-5, and 4-O-5 are the other important linkages (Scheme 4.1).

More than 70 Mt of technical lignin from the chemical pulping process is available annually, which will increase in the future due to new cellulosic ethanol production processes.[1,9] About 2% of lignin is used commercially for low-value products. The remaining lignin is burned in a pulping recovery system to recover energy and recycle pulping chemicals. The main goal of chemical pulping is to protect cellulosic fibers while degrading lignin as much as possible. The main pulping processes are alkaline, sulfite, and organosolv pulping. Kraft pulping is the prominent and dominant pulping process and uses NaOH and Na_2S as pulping chemicals at an elevated temperature and potentially is the main source of lignin in the market. Black liquor after the cooking process contains dissolved lignin, carbohydrates, and minerals and is a highly alkaline (pH 13–14) solution with a brown to black color.[2]

4.1.1 Lignin Recovery and Purification from Black Liquor

In the pulping process, the term "recovery" refers to a complex set of process operations involving the concentration of black liquor, burning the concentrated lignin and other organic materials to recycle the cooking chemicals, and generating energy/steam. However, in this chapter, the term "recovery" considers technologies to separate lignin from "black liquor" for further higher value-added applications. Pulping black liquor is not a pure solution of lignin. On top of the soluble lignin, it contains dissolved carbohydrates, minerals, sulfur, and chemical extractives.[9,10] The concentration of black liquor is around 12–18% depending on the wood species and delignification conditions.[11–13] To recover lignin from black liquor, different methods have been developed. In all recovery processes, reducing the solubility of lignin through pH reduction using mainly carbon dioxide and sulfuric acid is the first step, followed by lignin precipitation and separation. Minerals and carbohydrates remain soluble during black liquor pH reduction and are retained in the filtrate after recovering the precipitated lignin.

The kraft pulping recovery system is an expensive component of the overall pulp factory. It is often a bottleneck for expanding the pulping production capacity without increasing the recovery furnace capacity. "Debottlenecking" of the process by using part of black liquor for other applications instead of burning is one of the strongest motivations for lignin recovery and further application. It allows higher pulp production at a given pulp mill without new capital investment in the recovery boiler system.

The typical kraft pulping process contains sodium and sulfur as cooking chemicals. It is important to avoid disrupting the balance between these two elements in a system when planning to separate part of the lignin from black liquor.[14]

CO_2 is the only acid source that can do lignin phase separation in black liquor without disturbing the cooking chemical balance. After pre-separation, other mineral acids, such as H_2SO_4, further reduce the pH and complete lignin recovery. Usually, the fully protonated lignin has low ash content after rinsing with water; HCl can accomplish the same effect; however, its industrial application is limited due to the corrosive effect of chloride ions. The current pulping process uses black liquor to recover cooking chemicals and produce steam in the recovery system. Therefore, the amount of lignin separation for other applications is limited. Using about 5% of the total lignin reported to be a good target value without interfering with the recovery process.[15] LignoBoost[16] and LignoForce[17] are the current main processes for lignin separation from black liquor. The Sequential Liquid–Lignin Recovery and Purification (SLRP) system[18] is another system currently being pursued on a pilot scale.

4.1.1.1 LignoBoost System

The Inventia research group and Chalmers University in Sweden[16] developed the LignoBoost process. Valmet Corporation implemented two full-scale processes as of 2016.[19] The LignoBoost process at the Plymouth, NC, USA, pulp mill is the main installed line. The basic process steps of the LignoBoost system[20] are illustrated in Figure 4.1. A pH reduction of around 10 using carbon dioxide gas is the first step in the process. Precipitated lignin, known as cake, is re-slurred with a sulfuric acid solution with pH in the range of 2–4. A further pH reduction to around 1.5 provides a final product with less ash. Precipitated lignin after the second step of acidification goes to a second filter press followed by a wash with dilute acid and water. The final product is relatively pure lignin with 50–60% solid content.

4.1.1.2 LignoForce System

FPInnovations in Canada developed the LignoForce system.[17] A schematic diagram of the LignoForce process is illustrated in Figure 4.2. Oxidation with oxygen at 80 °C of black liquor is the first step to reduce hydrogen sulfide, and other non-odorous compounds are the critical feature of the process. It is reported that oxidation also improves the filterability of lignin during the whole process.[21] After oxidation, the liquor pH is reduced to around 9–10 using pressurized CO_2 injection at 65–70 °C to convert lignin to solid particles followed by filtration and pressing. The resulting lignin is in sodium form and is preferred for applications like phenolic resins and activated

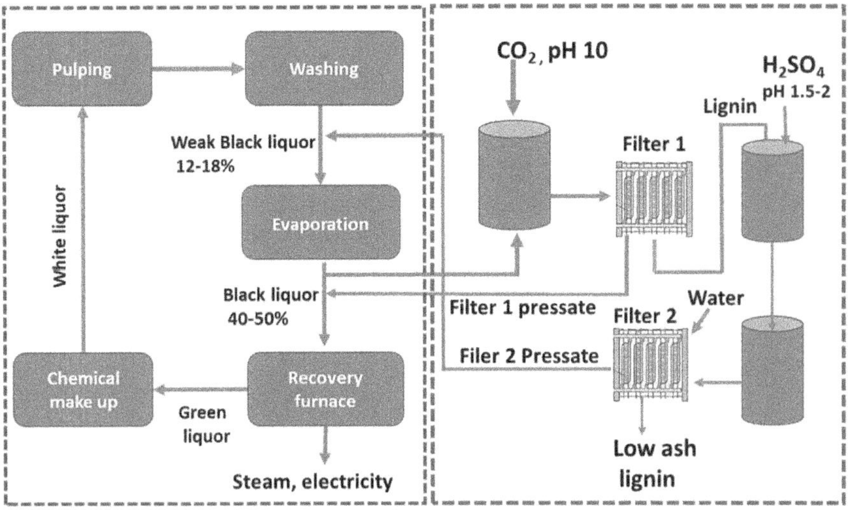

Figure 4.1 LignoBoost process.[20] Data from ref. 20.

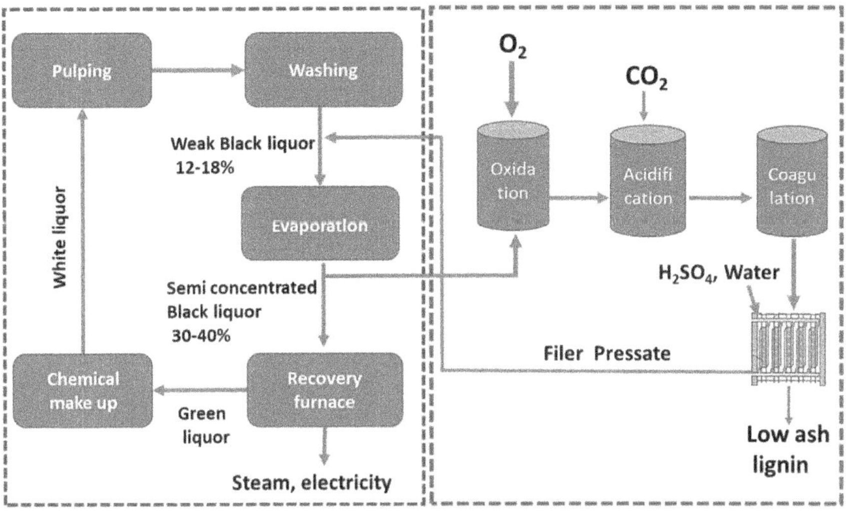

Figure 4.2 LignoForce process.[22] Data from ref. 22.

carbon. More steps of acid and water wash and filter press are required to make low ash and sodium-free products in this process. It is reported that recovered lignin has larger particle size in this process, which is desirable for obtaining higher filtration rates.[22]

4.1.1.3 SLRP Process

Clemson University, Lake, and Blackburn developed and patented the SLRP process in 2016.[19,23,24] The SLRP process is carried out at a temperature higher than the melting point of hydrated lignin. Under high temperature and pressure conditions (100 to 150 °C at 500 to 800 kPa), precipitated lignin behaves as a free-flowing liquid rather than as a "gummy" material. This gravity separation behavior negates the need for a filtration stage before washing. In the first step of the process, black liquor is acidified with CO_2 to reduce the pH to around 9 to 10.5 at a pressure between 50 and 200 psig (345 to 1480 kPa).[24] The highly solvated lignin is pumped into an acidification reactor with a pH of ~2.5 to precipitate it as a solid. Using sulfuric acid results in the release of CO_2 and H_2S, which can be reused in the first step to decrease the consumption of CO_2. The solid-lignin slurry is then filtered in a filter press to yield a lignin sample with around 1% ash and a solid content of ~65%.

4.1.2 Lignin Fractionation

Recovered lignin from black liquor is an amorphous and highly branched polymer with a complex chemical structure. Kraft lignin has an average molecular weight (M_w) of 1 kDa to 100 kDa, with the polydispersity index (PDI, M_w/M_n) ranging from 1.8 to 3.6[3,25,26] depending on wood sources and pulping process conditions.[25,27–29] Recovered lignin is a mixture of different fractions with different molecular weights and functional groups.[25]

The chemical complexity of this technology is one of the main challenges associated with its application as a value-added product. Lignin fractionation is one of the best solutions to obtain a lignin polymer with more uniform properties.[30] Many unique lignin properties such as resistance to decay and biological attacks, UV absorbance, high stiffness, and antioxidant activity are related to its molecular weight and chemical structure.[31,32] Different fractionation strategies developed have their own advantages and disadvantages. They can roughly be grouped into three categories: precipitation with an acidic solution, solvent extraction, and membrane filtration.

Lignin precipitation and fractionation can be divided into four categories, including acidic fractionation, multiple solvent extractions, solvent/water extraction, and membrane fractionation,[33,34] as illustrated in Table 4.1.

4.1.2.1 Lignin Fractionation Using pH Control

Kraft lignin with phenolic hydroxyl and carboxyl groups acts as a weakly acidic polyelectrolyte in an aqueous solution. In an alkaline solution, lignin ionizes (dissociates) and dissolves in the solution in colloidal form.[36–42] The colloidal lignin has a compact core and a loose surface layer.[36,43] The kraft lignin macromolecule is surrounded by a double layer of cations (mainly Na^+ and K^+).[44] With the addition of acid into a lignin solution, the H^+ neutralizes the charge on the colloid surface, and lignin starts to precipitate.[37,38] Due to the negative surface charge, lignin macromolecules repel each other because of the electrostatic repulsive forces, thereby stabilizing lignin in solution. The repulsive forces between the lignin molecules are

Table 4.1 Different lignin fractionation methods.[34,35]

Fractionation method	Procedure	Example	Advantages	Disadvantages
pH control	Changing the pH of alkali-soluble lignin from >12 to 1	Gradually adding H_2SO_4 acid	Easy procedure Low energy consumption	Colloid formation High impurity fractions Higher lignin–carbohydrate complexes
Solvent fractionation	Sequential fractionation in different organic solvents with different polarities	Alcohols, chloroform, ether, DMSO, DCM, THF, hexane, ketones	Easy to operate High efficiency Relatively uniform fractions Solvent recovery	Use large amounts of organic solvents High-cost solvents
Solvent/water fractionation	Sequential fractionation with different solvent/water ratios	Acetone/water Alcohol/water THF/water AcOH/water	Easy to operate High efficiency Relatively uniform fractions Solvent recovery	Use large amounts of organic solvents High-cost solvents
Membrane fractionation	Using membranes	Microfiltration Ultrafiltration Nanofiltration	Continuous operation Low energy consumption Control of the molar mass and homogeneity	Performance potential over time High capital cost

reduced with protonation and help lignin precipitation due to lignin aggregation.[25,28]

An old and straightforward method for lignin fractionation by pH control is by using sequential precipitation with H_2SO_4[25,28,39,40,45,46] or HCl in the pH range of 12 to 1.[47] In general, the pH of black liquor or soluble lignin is reduced in multiple steps with direct acid addition and separation of precipitated lignin in each step by centrifugation or filtration. The separated fraction at each pH usually undergoes further acid washing and a final DI water neutralization. In general, the lower pH fractions have smaller molecular weights with higher hydrophilicity,[39,40,48] lower polydispersity and higher acidic groups,[41,46] and higher zeta-potentials.[36,42,43,49] The higher molecular weight fractions of lignin with smaller pK_a tend to precipitate first due to their lower solubility and weaker acid groups.[39,40,47] The sulfur content of kraft lignin is reported to be present in precipitated fractions at lower pH.[50,51]

Precipitation and fractionation of lignin by pH control through direct mineral acid addition are suitable for lab-scale studies. However, on a commercial production scale and as a part of the current pulping recovery process, this will disrupt the balance between cooking chemicals in black liquor.[14] CO_2 is the only acid source that can do lignin phase separation in black liquor without disturbing the cooking chemical balance. One of the improved methods, which fractionate lignin directly from black liquor,[52] is referred to as "liquid lignin" fractionation. The method acidifies black liquor with CO_2 at elevated temperature and relatively high pressure (115 °C and 600 kPa). In the first step of the process, the pH of the black liquor is lowered to 12.8 to separate part of the lignin using liquid–liquid separation. The pH of the separated lignin in this step is reduced to around 0.5 using H_2SO_4, and the precipitated lignin is separated by filtration using liquid–solid separation as fraction 1. In the first step, the remaining black liquor acidified more with CO_2 to pH 12.1, and fraction 2 was recovered using the same method as fraction 1. This sequential fractionation can be continued by reducing the pH by 0.5 unit in each step. The main part of lignin separates in the pH range of 11.1 to 10. A schematic of the process is illustrated in Figure 4.3. Lignin fractions had the same aliphatic OH (1.2–1.3 mmol mol^{-1}) content for all fractions. The lignin fraction at pH 11.6 had around 0.45 mmol mol^{-1} phenolic OH, which gradually increased to 0.7 mmol mol^{-1} in the fraction separated at 9.5. The molecular weight of the fraction reduced from high pH recovery (7500 Da at pH 12.8) to low pH recovery (5500 Da at pH 9.5).

4.1.2.2 Lignin Fractionation in Organic Solvents

The Hildebrand solubility parameter (HSP)[53] describes the effect of non-polar solvent properties with the solubility of a substrate (eqn (4.1)):

$$\delta = \left(\frac{E}{V_m}\right)^{1/2}, \ (\text{cal cm}^{-3})^{1/2} \tag{4.1}$$

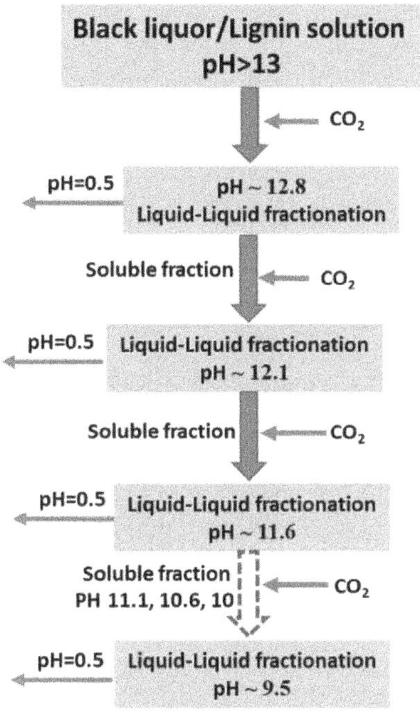

Figure 4.3 "Lignin" fractionation from black liquor.[52] Data from ref. 52.

where E is the solvent energy of vaporization and V_m is its molar volume. Using this equation, a solvent dissolves a substrate with the same δ or a value close to δ.

After introducing this parameter, the HSPs for lignin were calculated and reported.[54] The solubility parameter of lignin was reported to be in the range of 12 to 28.0 MPa$^{1/2}$ depending on the lignin source and pulping process.[29,39,55–59]

In the last decade, HSPs were developed for polar systems to extend Hildebrand's approach.[60] The cohesive energy in this equation correlated with three major parameters including dispersion forces (δ_D), permanent dipole–permanent dipole interactions (δ_P) (*i.e.*, polar forces), and hydrogen bonding (δ_H) (eqn (4.2)).

$$\delta^2 = {\delta_D}^2 + {\delta_P}^2 + {\delta_H}^2 \tag{4.2}$$

Hydrogen bonding plays the leading role for alcohols,[54] while solvent polarity (δ_P) is important for other solvents. High polarity solvents like DMAc, DMF, and DMSO with δ_P of 11.5, 13.7, and 16.4 MPa$^{1/2}$, respectively, can dissolve lignin completely. However, lignin can dissolve in a few low polarity solvents like EtOAc and pyridine. Indeed, all three parameters δ_D, δ_P, and δ_H play a synergistic role in solubility and should be considered together rather than individually.

Hansen's space is another equation to calculate lignin solubility in a solvent.[60] The equation uses single solvent points of coordinates (δ_D, δ_P, δ_H) and a space solute sphere as an interaction radius (R_0). The "solubility parameter difference" (R_a) is used to compare two substances within Hansen's space.

$$R_a{}^2 = 4(\delta_{D1} - \delta_{D2})^2 + (\delta_{P1} - \delta_{P2})^2 + (\delta_{H1} - \delta_{H2})^2$$

$$\text{RED} = R_a/R_0$$

Lignin solubility theoretically requires $R_a < R_0$; hence it can be helpful for solubility prediction to consider the relative energy difference (RED). In this case, solubility should be achieved for RED < 1. The solubility parameters of important lignin solvents are provided in Table 4.2.[60]

Recent studies using lignin model compounds and using principal component analysis (PCA), partial least squares discriminant analysis (PLS-DA), and density functional theory (DFT) calculations[61–63] indicated more details about their solubility parameters. They indicated that solvents with high polarity, such as ethanol, strongly interact with the lignin model more than other solvents like diethyl ether and hexane. In contrast, non-polar solvents like hexane exhibit the least interaction energy. Hydrogen bonding and non-covalent interaction showed the same results and trend. Ethanol makes the shortest hydrogen bond with the lignin model. Hydrogen bonding of lignin with ethanol and diethyl ether was reported to happen through aliphatic γ-OH. In contrast, acetone forms hydrogen bonds with the methoxyl group and α-O atom in the lignin.

4.1.2.2.1 Fractionation of Lignin Using Organic Solvents. Lignin fractionation using organic solvents is popular because it is a low cost, fast, and straightforward technique with easy solvent recovery. As already mentioned, the lignin solubility depends on the solvation power of solvent (cohesive). Initial reports for lignin fractionation using organic solvents date back to more than 80 years ago.[54,64] Raw lignin from the pulping or bio-refinery process contains different small molecules of organic impurities. In general, before solvent fractionation, lignin is washed with water

Table 4.2 Hansen solubility parameters for solvents.[60]

Solvent	δ_D	δ_P	δ_H
Toluene	18	1.4	2.0
Acetone	15.5	10.4	7
Ethyl acetate (EtOAc)	15.8	5.3	7.2
1,4-Dioxane	19	1.8	7.4
Chloroform	17.8	3.1	5.7
Methanol	15.1	12.3	22.3
Ethanol	15.8	8.8	19.4
DMSO	18.4	16.4	10.2
Water	15.5	16	42.3

(usually at a low pH) to remove minerals, sugars, and other water-soluble materials, followed by extraction with MTBE (methyl tertiary butyl ether) or other ether solvents to remove extractive materials. After this cleaning process, a sequential fractionation of lignin is applied using low to high polarity solvents. An example of the general process is illustrated in Figure 4.4.[34,51,59,65–75]

Good lignin solvents have HSPs from 11 to 14 $(\text{cal cm}^{-3})^{1/2}$. Considering the energy requirements for the solvent recovery after fractionation, it is important to usc solvents with relatively low boiling points (under 100 °C) with relatively low heats of evaporation (Figure 4.5). Acetone, ethanol, EtOAc, methanol, and ethanol are examples of proper solvents.

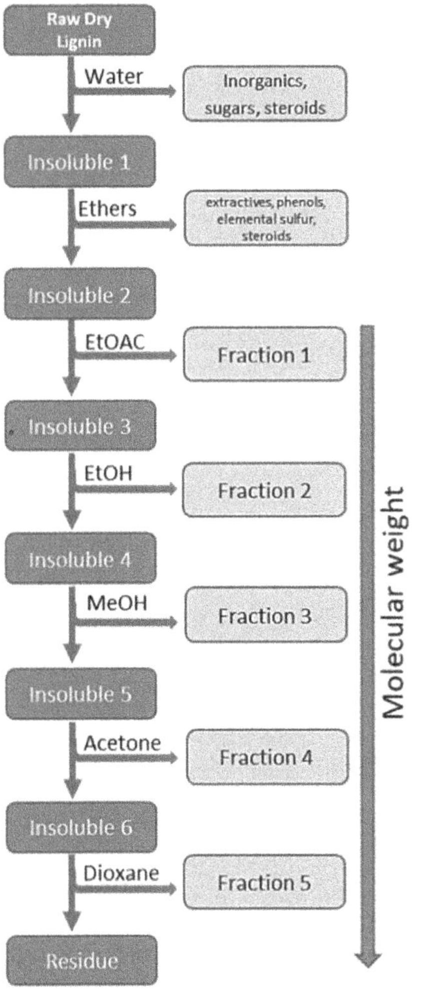

Figure 4.4 Sequential organic solvent lignin fractionation.

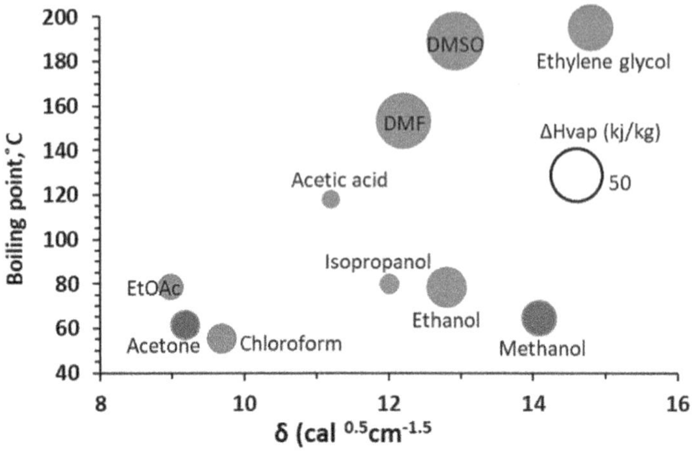

Figure 4.5 Proper organic solvents for lignin fractionation. Adapted from ref. 76 with permission from the Royal Society of Chemistry.

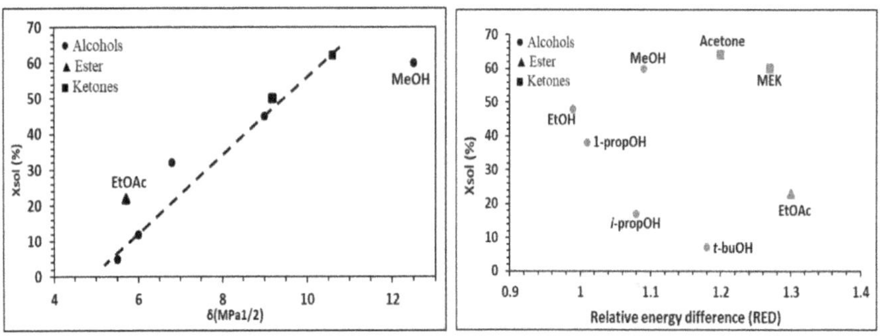

Figure 4.6 Yield of the soluble fraction as a function of solvent polarity, as expressed by HSP δ_P (left) and RED (right).[58] Adapted from ref. 58 with permission from Walter de Gruyter, Copyright 2016.

The potential of different alcohols, ketones, and ester for kraft lignin fractionation was investigated.[58] The solubility parameters of solvents with fractionated lignin properties are reported in Figure 4.6, which shows yields of the soluble fractions as a function of solvent polarity, with RED solvents with higher solubility parameters dissolving more lignin.

Sequential extraction and fractionation of softwood kraft lignin (UPM BioPiva 350) was reported[65] recently using solvents from low to high polarity, EtOAc, EtOH, MeOH, acetone, and dioxane. The molar mass of fractionated lignin (M_w, M_n) increased for the fractions obtained from higher polarity solvents. The M_w for the EtOAc fraction was reported to be around 2800 g mol^{-1}, while that for the dioxane extract was more than 27k g mol^{-1}. The polydispersities of all fractions were around 1.3–1.7. The total carbohydrate content was more for the extracted lignin from higher polarity solvents.

Table 4.3 Chemical and thermal properties of kraft lignin and its fractions obtained using acetone.[79–80]

Sample	Aliphatic OH ($mmol\,g^{-1}$)	Phenolic OH ($mmol\,g^{-1}$)	COOH ($mmol\,g^{-1}$)	M_w ($g\,mol^{-1}$)	PDI	T_g (°C)	Yield (%)
Indulin AT kraft lignin	2.24	4.24	0.42	6300	5.7	153	
Soluble fraction (ASKL)	1.63	5.76	0.57	3500	3.7	112	65
Insoluble fraction (AIKL)	2.71	3.08	0.35	14 000	7	173	35

Lignin fractions with higher molecular weights had lower aliphatic OH and carboxylic OH contents.

Sequentially, fractionation of bamboo organosolv lignin,[77] mixed wheat straw, and grass soda lignin[66] was reported using ether, EtOAc, methanol, acetone, and dioxane/water (9/1, v/v). Lignin fractions collected with higher polarity solvents had higher molecular weights and lower antioxidant capabilities.

Organosolv lignin from wheat straw was fractionated sequentially using dichloromethane, methanol, and methanol/dichloromethane (7/3, v/v).[67] The total phenolic OH groups in all fractions were reported to be the same. The aliphatic group was lowest in the dichloromethane fraction (0.94 $mmol\,g^{-1}$) and highest in the methanol fraction (2.5 $mmol\,g^{-1}$).

Sequential solvent fractionation was applied to yellow poplar milled wood lignin and organosolv lignin (OL) using EtOAc, 2-butanone, methanol, acetone, and dioxane:water.[68] The lignin fractions isolated from higher polarity solvents were shown to exhibit higher molecular weights and methoxyl group content, while the phenolic hydroxyl groups decreased.

A four-step sequential fractionation of softwood (Norway spruce) and hardwood (*Eucalyptus grandis*) lignin using EtOAc, EtOH, MeOH, and acetone, including the final insoluble residue, was reported.[69] For both lignins, the lowest M_w value was recorded for the EtOAc fraction, and the highest M_w values (8k–20k $g\,mol^{-1}$) were for the insoluble residue fraction.

Fractionation of softwood kraft lignin using acetone and hexane solution was also investigated.[70] The lignin was first mixed with pure acetone, the insoluble fraction was discarded, and the soluble part was used for further fractionation. Hexane was gradually added into the acetone-soluble fraction (acetone : hexane of 1 : 1 to 1 : 4 v/v), and the insoluble part was used for the following fractionation stage. They reported lignin fractionation with a narrow PDI close to one. In addition, they found that almost all metals (Na, Fe, Mn, and Cu) remained in the acetone-insoluble fraction, and the acetone-soluble fraction was reported to be almost metal-free.

Kraft lignin sequential fractionation was investigated using water, dichloromethane, *n*-propanol, methanol, and a methanol/dichloromethane mixture.[78] Like in other reports, the final residue and the recovered fractions with higher polarity (*i.e.*, MeOH/CH_2Cl_2) had higher molecular weights.

Using just acetone to fractionate softwood kraft lignin a simple and effective process was reported to isolate acetone-soluble (ASKL) and acetone-insoluble (AIKL) fractions.[26,79–82] The effects of this fractionation on the lignin properties were investigated. The effect of kraft lignin fractionation on its antioxidant activities and their effect on polyethylene protection against thermal degradation are reported in Table 4.3. The acetone-soluble fraction with lower molecular weight showed better antioxidant properties and polyethylene thermal stabilization. Acetone-soluble lignin showed better reactivity with poly(arylene ether sulfone) to form a copolymer. The final product had a higher molecular weight in comparison with the acetone-insoluble fraction. The acetone-soluble fraction also demonstrated better results when used for polymer preparation through chain extension chemistry[26] and preparation of carbon fibers from lignin.[83]

A recent study on single-step fractionation of industrial hardwood kraft lignin reported using solvents with different chemical nature and polarity based on the Hansen solubility parameters.[61] Ethanol ($\delta_P = 8.8$; $\delta_H = 19.4$) was used as a polar protic solvent, and acetone ($\delta_P = 10.4$; $\delta_H = 7.0$) was selected as a polar aprotic solvent. Moreover, non-polar solvents diethyl ether ($\delta_P = 2.9$; $\delta_H = 5.1$) and hexane ($\delta_P = 0$; $\delta_H = 0$) were used for lignin fractionation. The dissolved fractions in ethanol and acetone were more than 70% with molecular weights (M_w) of 2000 and 2250 g mol^{-1}, respectively, while for the insoluble fractions, the molecular weight was 8400 and 10 300, respectively. Ponnuchamy *et al.* also reported PCA, PLS-DA, and DFT calculations on the fractionated lignin. They found that ethanol with high polarity strongly interacted with the lignin more than the other examined solvents (*i.e.*, acetone, diethyl ether, and hexane), and non-polar hexane had the least interaction energy. The α-O and γ-O linkages in the lignin showed a large variation in the bond lengths with ethanol and acetone solvents. Ethanol showed the same trend mainly due to the shortest hydrogen bond with a lignin model. The main mechanism of hydrogen bonding between ethanol and diethyl ether with lignin is through aliphatic γ-OH, while acetone makes hydrogen bonds with the methoxyl group and α-O atom. Ethanol had a lower bandgap with the lignin model than the other solvents.

4.1.2.2.2 Lignin Fractionation Using Organic Solvent/Water Mixtures. Mixing water in the range of 10–40% with organic solvents like methanol, acetone, toluene, THF, and acetic acid can dissolve lignin completely.[33,41,84–86]

The addition of water into organic solvent changes the mixture polarity, hydrogen-bonding capacity, and other HSP parameters. With the addition of extra water in the soluble lignin solution, lignin starts to precipitate gradually and can fractionate into fractions with different molecular weights.

Fractionation of extracted pine and switchgrass OL using aqueous ethanol (65 : 35, v/v) with sulfuric acid (0.9%, w/w as the catalyst) was reported using an acetone–water mixture.[33] Both lignin samples showed complete dissolution in a 60–90% acetone–water mixture (Figure 4.7). The acetone concentration was reduced gradually from 60%, and the precipitated fraction was filtered. Lignin fractionated at lower acetone concentrations showed lower molecular weight and smaller polydispersity. The aliphatic hydroxyl group content was almost the same for all fractions, in agreement with other reported articles.[46,52] However, the phenolic OH concentration increased in fractions collected at lower acetone concentrations in both lignins.

Alcell lignin was fully soluble in 75–85% ethanol–water.[29] The lignin solubility parameter for Alcell lignin was calculated to be around 14 $(\mathrm{cal\,cm^{-3}})^{1/2}$.

The effect of the acetone–water ratio on different lignins was investigated.[57] The 70–90% acetone–water mixtures had the potential to solve lignin completely. The differences between the solubility parameters of solvent and lignin ($\delta_{\mathrm{lignin}} - \delta_{\mathrm{solvent}}$) in this range of acetone–water mixtures were around −2 to +2. The results indicated how important the solubility parameter is to predict lignin solubility in a specific solvent. The solubility parameters of lignin were calculated to be around 13.5–14.4 $(\mathrm{cal\,cm^{-3}})^{1/2}$, which is in the solubility parameter range (12–15.7 $(\mathrm{cal\,cm^{-3}})^{1/2}$) for 70–90% acetone–water mixtures. They indicated a positive correlation of the lignin antioxidant activity with its phenolic hydroxyl groups ($Y = 0.797X + 0.99$, $R^2 = 0.978$) and a negative correlation with molecular weight ($Y = -0.462X + 4.921$, $R^2 = 0.999$) in agreement with another report.[79]

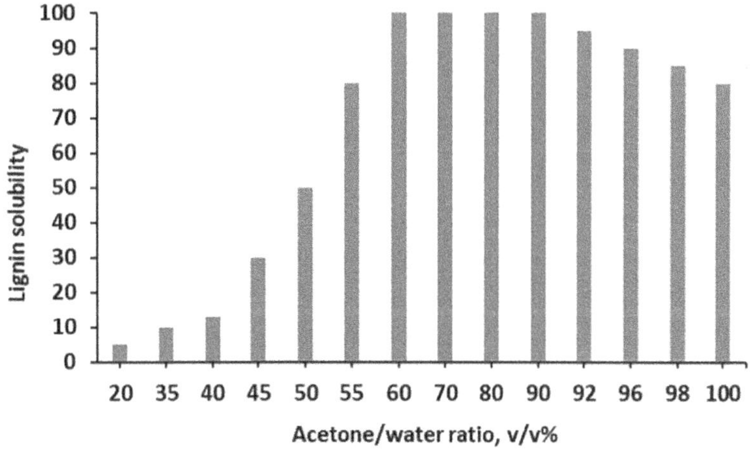

Figure 4.7 Effect of the acetone–water ratio on softwood organosolv lignin.[33] Reproduced from ref. 33 with permission from American Chemical Society, Copyright 2017.

The glass transition temperatures (T_g) for the starting switchgrass and pine OL were determined to be approximately 105 °C and 130 °C, respectively. The high molecular weight fractions (precipitated in 55% acetone) of switchgrass and pine lignin had higher T_g of about 127 °C and 150 °C, respectively.

Fractionation of softwood kraft lignin using acetone–water, ethanol–water, and PGME–water mixtures was also reported.[87] The lignin studies showed full solubilization in 80% ethanol, 60% acetone, and 60% PGME solutions. Different fractions were collected by gradually adding water into the lignin solutions. In all cases, with decreasing solvent concentration, the molecular weight decreased.

Kraft, organosolv, and soda lignin fractionation was reported using acetone–water.[85] All lignin samples had maximum solubility for 70–90% acetone concentrations. Fractionation was carried out by adding water and decreasing the acetone concentration to 50% and 40%. The insoluble fractions had higher molecular weights than the soluble fractions in all samples. Higher molecular weight lignin had lower phenolic OH and COOH group contents.

Fractionation of organosolv lignin from Populus using THF/water solution was reported.[86] The lignin was dissolved fully in a 40% THF/water solution. Water was then added to the solution to reduce the THF concentration by 5 wt% at each step. The dissolved lignin at a lower concentration of THF had a lower molecular weight.

Using acetic acid/water solution was reported as an effective method for ash-free lignin fractionation.[88–91] Lignin in hot acetic acid–water solution under appropriate conditions had two equilibrium liquid phases, including the solvent-rich phase and ultraclean lignin-rich phase. Different mass ratios of the lignin fractions could be obtained by changing the AcOH/H_2O feed ratio.[88]

Lignin fractionation using acetic acid/water and CO_2 in a pressurized vessel was introduced as the gas-expanded liquid (GXL) method.[92] With the changing CO_2 pressure in the solvent, the solubility of lignin changed. Thus, this multi-component separation system could be performed with changing pressure without changing the solvent. In a 90/10 wt/wt AcOH/water composition, around 58% of the starting lignin was dissolved in the starting AcOH water solution. Lignin dissolved 25% and 50% when the CO_2 pressure was 100 and 200 psi, respectively.

4.1.3 Fractionation Using Membrane Filtration

Membrane technology, including microfiltration (MF), ultrafiltration (UF), and nanofiltration (NF), has been employed for lignin fractionation. The general process of the method is illustrated in Figure 4.8.

Membrane processes have been applied in the chemical pulping industry since the early 1970s for wastewater treatment.[93] Industrial application of membrane technology to recover lignin from thermomechanical pulping of black liquor was reported more recently.[94,95] Currently, membrane technology is one of the most popular methods for lignin fractionation.[1,95]

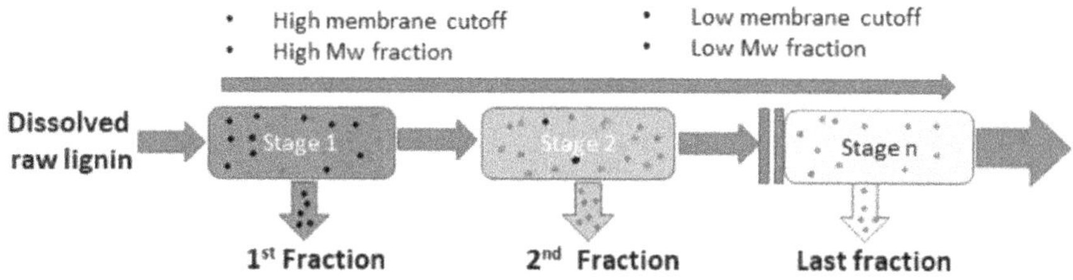

Figure 4.8 Schematic representation of a multistage membrane fractionation system.

In general, membrane filtration is carried out under pressure and in multiple steps using different membrane cutoffs depending on the fractionation plan. Usually, membranes are arranged from a high to low range of molecular weight cutoffs (MWCOs). Diafiltration is the state-of-the-art process to separate lignin into high and low molecular weight fractions.[95,96] Ceramic[97] and polymeric[98,99] membranes are most popular for lignin recovery and fractionation due to their high stability at extreme pHs and temperatures.

Fractionated lignin using membrane technology results in less contamination[95,100] and better lignin thermal properties compared to traditional fractionation due to the reduced presence of carbohydrates and inorganics in their composition.[101,102] One of the most significant advantages of membrane fractionation is achieving lignin fractions with high homogeneity, which is required for industrial applications of lignin.[103]

Direct lignin fractionation from kraft black liquor *Eucalyptus globulus* using a sequence of three stages 50, 15, and 5 kDa ceramic membranes was reported.[104] The fractions separated using the lower cutoff membranes had lower dissolved solids and higher inorganic content.

The content of carboxylic groups was reported to be the same in all fractions, while the amount of sulfur was significantly higher in the lower M_w. Sugars (3–5%) were only present in the higher M_w fractions. In agreement with standard lignin fractionation methods,[33,68,79] lignin fractions with higher M_w values had higher contents of aliphatic hydroxyl groups and lower contents of phenolic groups.

A pilot-scale crossflow diafiltration with a 1 kDa cutoff was employed to fractionate softwood kraft lignin.[50] The temperature and ionic strength of solvent are important for membrane fractionation of lignin. Lignin showed the same trend for molecular weight and functional groups for different fractions as fractionated lignin using pH control and solvent fractionation.[104–108]

4.1.4 Conclusion and Future Perspectives

Available technical lignin as a by-product of pulping and the bio-refinery process is a highly branched and relatively cross-linked polyphenolic macromolecule with a wide range of molecular weights and functional groups. The major part of technical lignin is currently used as a resource for energy generation through burning in the pulping recovery system. A small part of the extracted lignin, mainly lignosulfonates, is used as dispersants, concrete admixtures, binders in mining operations, and adhesives.[1,9] Having a phenolic structure with active functional groups makes lignin a good biopolymer candidate for green polymer synthesis and application. However, the heterogeneous and polydisperse structure of technical lignin is the bottleneck of lignin application for polymers, especially high-temperature applications. Lignin fractionation is one of the best strategies to create a uniform structure for polymer application. Lignin fractionation using pH control is one of the oldest methods, especially in lab-scale applications. Direct use of mineral acid changes the chemical balance in the black liquor recovery system. The current industrial-scale lignin recovery is based on injecting CO_2 to reduce the pH of black liquor to around 10 and collecting precipitated lignin using a filter press. The collected lignin was acidified further to pH around 2 and finally washed with DI water. Another proposed method is the fabrication of lignin sequentially using organic solvents or a mixture of solvent with water with different polarity and solubility parameters. Solvents with solubility parameters around 9–14 with low boiling points under 100 °C are the best solvents for lignin fractionation. Organic solvents with different polarities can dissolve only a fraction of the lignin; therefore they can be used as tools for lignin fractionation. Mixing water with organic solvents showed good potential for lignin fractionation. Water can change the organic solvent's polarity, hydrogen bonding, and other parameters, which can affect lignin solubility. Solvation power parameters, including cohesive energy and the hydrogen-bonding capacity of the solvent molecule, are the main parameters to select a solvent for lignin fractionation. On the other hand, multistage membrane filtration with a wide range of MWCOs showed good potential for lignin fractionation. Ceramic membranes with high thermal stability and corrosion resistance are good candidates for lignin fractionation.

After decades of investigation, it was revealed that the heterogeneity of technical lignin is the main challenge and bottleneck as a raw material for value-added naturally based chemicals. Each fraction of lignin in the raw technical lignin has its own physical and chemical properties. Future application of lignin as a pulping process by-product is highly dependent on the extraction of a fraction with more uniformity with defined properties. Currently, lignin fractionation is the main method that can provide a specific lignin fraction for a given application. However, all of the mentioned lignin fractionation methods using different strategies are on the lab scale yet and need to be scaled up in the future. Fractionation costs, environmental concerns about the process, and target-specific application will be the future aspects to focus.

References

1. A. J. Ragauskas, G. T. Beckham, M. J. Biddy, R. Chandra, F. Chen and M. F. Davis, *et al.*, Lignin valorization: Improving lignin processing in the biorefinery, *Science*, 2014, **344**, 6185.
2. S. Y. Lin and C. W. Dence, *Method in lignin chemistry*, Springer, 1992.
3. E. Sjöström, *Wood Chemistry: Fundamentals and Applications*, Academic Press, San Diego, CA, 2nd edn, 1993.
4. E. Dorrestijn, L. Laarhoven, I. Arends and P. Mulder, The occurrence and reactivity of phenoxyl linkages in lignin and low rank coal, *J. Anal. Appl. Pyrolysis*, 2000, **54**, 153–192.
5. C. G. Yoo, Y. Pu, M. Li and A. J. Ragauskas, Elucidating Structural Characteristics of Biomass using Solution-State 2D NMR with a Mixture of Deuterated Dimethylsulfoxide and Hexamethylphosphoramide, *ChemSusChem*, 2016, **9**, 1090–1095.
6. S. D. Mansfield, H. Kim, F. Lu and J. Ralph, Whole plant cell wall characterization using solution-state 2D NMR, *Nat. Protoc.*, 2012, 7(9), 1579–1590.
7. C. Crestini, F. Melone, M. Sette and R. Saladino, Milled wood lignin: A linear oligomer, *Biomacromolecules*, 2011, **12**(11), 3928–3935.
8. C. Crestini, H. Lange, M. Sette and D. S. Argyropoulos, On the structure of softwood kraft lignin, *Green Chem.*, 2017, **19**(17), 4104–4121.
9. G. Gellerstedt, Softwood kraft lignin: raw material for the future, *Ind. Crops Prod.*, 2015, 77, 845–854.
10. M. Nagy, M. Kosa, H. Theliander and A. J. Ragauskas, Characterization of CO_2 precipitated kraft lignin to promote its utilization, *Green Chem.*, 2010, **12**(1), 31–34.
11. J. Ralph and L. Landucci, NMR of Lignin, in *Lignin and Lignans Advances in Chemistry*, CRC Press, 2010, pp. 137–243.
12. L. X. Jiang, S. Fang, J. W. Mao and J. H. Xie, Progress in Fractionation Processes of Lignocellulosic Biomass for Biorefinery: A Review, *Mater. Sci. Forum*, 2016, **852**, 1175–1181.
13. J. C. Dean, P. Navotnaya, A. P. Parobek, R. M. Clayton and T. S. Zwier, Ultraviolet spectroscopy of fundamental lignin subunits: Guaiacol, 4-methylguaiacol, syringol, and 4-methylsyringol, *J. Chem. Phys.*, 2013, **139**, 144313.
14. H. Wallmo, T. Richards and H. Theliander, An investigation of process parameters during lignin precipitation from kraft black liquors: A step towards an optimised precipitation operation, *Nord. Pulp Pap. Res. J.*, 2009, **24**, 158–164.
15. C. Culbertson, T. Treasure, R. Venditti, H. Jameel and R. Gonzalez, Life Cycle Assessment of lignin extraction in a softwood kraft pulp mill, *Nord. Pulp Pap. Res. J.*, 2016, **31**, 30–40.
16. P. Tomani, The lignoboost process, *Cellul. Chem. Technol.*, 2010, **44**, 53–58.

17. L. Kouisni, P. Holt-Hindle, K. Maki and Michael Paleologou, The LignoForce System: A new process for the production of high-quality lignin from black liquor, *Pulp Pap. Can.*, 2012, **115**, 18–22.
18. J. Velez and M. C. Thies, Solvated liquid-lignin fractions from a Kraft black liquor, *Bioresour. Technol.*, 2013, **148**, 586–590.
19. J. Kihlman, The sequential liquid-lignin recovery and purification process: Analysis of integration aspects for a kraft pulp mill, *Nord. Pulp Pap. Res. J.*, 2016, **31**, 573–582.
20. G. Gellerstedt, P. Tomani, P. Axegård and B. Birgit, in *Integrated Forest Biorefineries – Challenges and Opportunities*, ed. L. Christopher, Royal Society of Chemistry, Cambridge, 2013, pp. 180–210.
21. M. A. Hubbe, R. Alén, M. Paleologou, M. Kannangara and J. Kihlman, Lignin recovery from spent alkaline pulping liquors using acidification, membrane separation, and related processing steps: A review, *BioResources*, 2019, **14**(1), 2300–2351.
22. L. Kouisni, P. Holt-Hindle, K. Maki and M. Paleologou, The LignoForce System™: A new process for the production of high-quality lignin from black liquor, *Pulp Pap. Can.*, 2014, **115**(1), 18–22.
23. M. A. Lake and J. C. Blackburn, SLRP™ – An innovative Lignin-Recovery technology, *Cellul. Chem. Technol.*, 2014, **48**(9–10), 799–804.
24. M. A. Lake and J. C. Blackburn, *Process for recovering lignin*, US9260464, 2016.
25. S. Y. Lin and C. W. Dence, *Methods in Lignin Chemistry*, Springer, 1992.
26. S. Sanghamitra, H. Sadeghifar and D. S. Argyropoulos, Kraft lignin chain extension chemistry via propargylation, oxidative coupling, and Claisen rearrangement, *Biomacromolecules*, 2013, **14**(10), 3399–3408.
27. T. Allison, A. Hannah, K. Ratayakorn, K. N. Amit and A. J. Ragauskas, Characterization and analysis of the molecular weight of lignin for biorefining studies, *Biofuels, Bioprod. Biorefin.*, 2012, **6**(3), 246–256.
28. M. Alekhina, O. Ershova, A. Ebert, S. Heikkinen and H. Sixta, Softwood kraft lignin for value-added applications: fractionation and structural characterization, *Ind. Crops Prod.*, 2015, **66**, 220–228.
29. Y. Ni and Q. Hu, Alcell® lignin solubility in ethanol–water mixtures, *J. Appl. Polym. Sci.*, 1995, **57**(12), 1441–1446.
30. L. Hui and A. G. McDonald, Fractionation and characterization of industrial lignins, *Ind. Crops Prod.*, 2014, **62**, 67–76.
31. S. Laurichesse and L. Avérous, Chemical modification of lignins: Towards biobased polymers, *Prog. Polym. Sci. Prog Polym Sci.*, 2014, **39**, 1266–1290.
32. W. O. S. Doherty, P. Mousavioun and C. M. Fellows, Value-adding to cellulosic ethanol: Lignin polymers, *Ind. Crops Prod.*, 2011, **13**, 259–276.
33. H. Sadeghifar, T. Wells, R. K. Le, F. Sadeghifar, J. S. Yuan and A. J. Ragauskas, Fractionation of Organosolv Lignin Using Acetone:-Water and Properties of the Obtained Fractions, *ACS Sustainable Chem. Eng.*, 2017, **5**(1), 580–587.

34. J. S. Rodrigues, V. Lima, L. C. P. Araújo and V. R. Botaro, Lignin Fractionation Methods: Can Lignin Fractions Be Separated in a True Industrial Process?, *Ind. Eng. Chem. Res.*, 2021, **60**(30), 10863–10881.
35. H. Sadeghifar and A. Ragauskas, Perspective on Technical Lignin Fractionation, *ACS Sustainable Chem. Eng.*, 2020, **8**(22), 8086–8101.
36. A. Rezanowich and D. I. Goring, Polyelectrolyte expansion of a lignin sulfonate microgel, *J. Colloid Sci.*, 1960, **15**, 452–471.
37. M. Norgren and B. Lindström, Dissociation of phenolic groups in kraft lignin at elevated temperatures, *Holzforschung*, 2000, **54**(5), 519–527.
38. T. Lindström and L. Westman, The colloidal behaviour of kraft lignin. Part III. Swelling behaviour and mechanical properties of kraft lignin gels, *Colloid Polym. Sci.*, 1980, **258**, 390–397.
39. E. I. Evstigneev, Specific features of lignin dissolution in aqueous and aqueous-organic media, *Russ. J. Appl. Chem.*, 2010, **83**(3), 509–513.
40. E. I. Evstigneev, Factors affecting lignin solubility, *Russ. J. Appl. Chem.*, 2011, **84**(6), 1040–1045.
41. G. Wang and H. Chen, Fractionation and characterization of lignin from steam-exploded corn stalk by sequential dissolution in ethanol–water solvent, *Sep. Purif. Technol.*, 2013, **120**, 402–409.
42. S. R. Sen, Y. L. Sen and D. L. Woerner, Association of kraft lignin in aqueous solution, in Lignin Properties and Materia, *ACS Symp. Ser.*, 1989, **397**, 145–254.
43. T. Lindström, The colloidal behaviour of kraft lignin. Part I. Association and gelation of kraft lignin in aqueous solutions, *Colloid Polym. Sci.*, 1979, 277–285.
44. S. M. Notley and M. Norgren, Measurement of interaction forces between lignin and cellulose as a function of aqueous electrolyte solution conditions, *Langmuir*, 2006, **22**(26), 11199–11204.
45. A. Arshanitsa, J. Ponomarenko, T. Dizhbite, A. Andersone, R. J. A. Gosselink and J. Van Der Putten, *et al.*, Fractionation of technical lignins as a tool for improvement of their antioxidant properties, *J. Anal. Appl. Pyrolysis*, 2013, **103**, 78–85.
46. L. B. Brenelli, F. Mandelli, A. Z. Mercadante, G. J. de Moraes Rocha, S. A. Rocco and A. F. Craievich, *et al.*, Acidification treatment of lignin from sugarcane bagasse results in fractions of reduced polydispersity and high free-radical scavenging capacity, *Ind. Crops Prod.*, 2016, **83**, 94–103.
47. T. V. Lourençon, F. A. Hansel, T. A. Da Silva, L. P. Ramos, G. I. B. De Muniz and W. L. E. Magalhães, Hardwood and softwood kraft lignins fractionation by simple sequential acid precipitation, *Sep. Purif. Technol.*, 2015, **154**, 82–88.
48. J. I. Santos, R. Martín-Sampedro, Ú. Fillat, J. M. Oliva, M. J. Negro and M. Ballesteros, *et al.*, Evaluating lignin-rich residues from biochemical ethanol production of wheat straw and olive tree pruning by FTIR and 2D-NMR, *Int. J. Polym. Sci.*, 2015, **2015**, 314891.

49. J. Ropponen, L. Räsänen, S. Rovio, T. Ohra-Aho, T. Liitiä and H. Mikkonen, *et al.*, Solvent extraction as a means of preparing homogeneous lignin fractions, *Holzforschung*, 2011, **65**(4), 543–549.
50. M. Helander, H. Theliander, M. Lawoko, G. Henriksson, L. Zhang and M. E. Lindström, Fractionation of technical lignin: Molecular mass and pH effects, *BioResources*, 2013, **8**(2), 2270–2282.
51. A. P. Dodd, J. F. Kadla and S. K. Straus, Characterization of fractions obtained from two industrial softwood kraft lignins, *ACS Sustainable Chem. Eng.*, 2015, **3**(1), 103–110.
52. J. S. Ryan, J. Velez, S. Kelkar, C. M. Saffron, M. C. Thies and D. B. Hodge, Correlating lignin structural features to phase partitioning behavior in a novel aqueous fractionation of softwood Kraft black liquor, *Green Chem.*, 2013, **15**, 2904–2912.
53. J. H. Hildebrand and R. L. Scott, *The solubility of nonelectrolytes*, Reinhold Publishing Corporation, New York, 1950.
54. C. Schuerch, The solvent properties of liquids and their relation to the solubility, swelling, isolation and fractionation of lignin, *J. Am. Chem. Soc.*, 1952, **74**, 5061–5067.
55. J. Quesada-Medina, F. J. López-Cremades and P. Olivares-Carrillo, Organosolv extraction of lignin from hydrolyzed almond shells and application of the δ-value theory, *Bioresour. Technol.*, 2010, **101**, 8252–8260.
56. Q. Wang, K. Chen, J. Li, G. Yang, S. Liu and J. Xu, The solubility of lignin from bagasse in a 1,4-butanediol/water system, *BioResources*, 2011, **6**, 3034–3043.
57. C. G. Boeriu, F. I. Fiţigău, R. J. A. Gosselink, A. E. Frissen, J. Stoutjesdijk and F. Peter, Fractionation of five technical lignins by selective extraction in green solvents and characterisation of isolated fractions, *Ind. Crops Prod.*, 2014, **62**, 481–490.
58. A. Duval, F. Vilaplana, C. Crestini and M. Lawoko, Solvent screening for the fractionation of industrial kraft lignin, *Holzforschung*, 2016, **70**(1), 11–20.
59. V. Passoni, C. Scarica, M. Levi, S. Turri and G. Griffini, Fractionation of Industrial Softwood Kraft Lignin: Solvent Selection as a Tool for Tailored Material Properties, *ACS Sustainable Chem. Eng.*, 2016, **4**(4), 2232–2242.
60. C. M. Hansen, *Hansen Solubility Parameters: A User's Handbook*, CRC Press, Boca Raton, 2007.
61. V. Ponnuchamy, O. Gordobil, R. H. Diaz, A. Sandak and J. Sandak, Fractionation of lignin using organic solvents: A combined experimental and theoretical study, *Int. J. Biol. Macromol.*, 2021, **168**, 792–805.
62. H. Ji and P. Lv, Mechanistic insights into the lignin dissolution behaviors of a recyclable acid hydrotrope, deep eutectic solvent (DES), and ionic liquid (IL), *Green Chem.*, 2020, **22**(4), 1378–1387.
63. Y. Zhang, H. He, K. Dong, M. Fan and S. A. Zhang, DFT study on lignin dissolution in imidazolium-based ionic liquids, *RSC Adv.*, 2017, **7**(21), 12670–12681.

64. R. F. Patterson, K. A. West, E. L. Lovell, W. L. Hawkins and H. Hibbert, Studies on Lignin and Related Compounds. LI. The Solvent Fractionation of Maple Ethanol Lignin, *J. Am. Chem. Soc.*, 1941, **63**(8), 2065–2070.
65. R. Liu, A. Smeds, L. Wang, A. Pranovich, J. Hemming and S. Willför, *et al.*, Fractionation of Lignin with Decreased Heterogeneity: Based on a Detailed Characteristics Study of Sequentially Extracted Softwood Kraft Lignin, *ACS Sustainable Chem. Eng.*, 2021, **9**(41), 13862–13873.
66. J. Y. Kim, S. Y. Park, J. H. Lee, I. G. Choi and J. W. Choi, Sequential solvent fractionation of lignin for selective production of monoaromatics by Ru catalyzed ethanolysis, *RSC Adv.*, 2017, 7(84), 53117–53125.
67. A. Arshanitsa, J. Ponomarenko, T. Dizhbite, A. Andersone, R. J. A. Gosselink and J. Van Der Putten, *et al.*, Fractionation of technical lignins as a tool for improvement of their antioxidant properties, *J. Anal. Appl. Pyrolysis*, 2013, **103**, 78–85.
68. S. Y. Park, J.-Y. Kim, H. J. Youn and J. W. Choi, Fractionation of lignin macromolecules by sequential organic solvents systems and their characterization for further valuable applications, *Int. J. Biol. Macromol.*, 2018, **106**, 793–802.
69. A. Tagami, C. Gioia, M. Lauberts, T. Budnyak, R. Moriana and M. E. Lindström, *et al.*, Solvent fractionation of softwood and hardwood kraft lignins for more efficient uses: Compositional, structural, thermal, antioxidant and adsorption properties, *Ind. Crops Prod.*, 2019, **129**, 123–134.
70. C. Cui, R. Sun and D. S. Argyropoulos, Fractional precipitation of softwood kraft lignin: Isolation of narrow fractions common to a variety of lignins, *ACS Sustainable Chem. Eng.*, 2014, **2**(4), 959–968.
71. J. A. Hemmingson, Steam-explosion lignins: Fractionation, composition, structure and extractives, *J. Wood Chem. Technol.*, 1987, 7(4), 527–553.
72. R. W. Thring, M. N. Vanderlaan and S. L. Griffin, Fractionation of ALCELL® lignin by sequential solvent extraction, *J. Wood Chem. Technol.*, 1996, **16**(2), 139–154.
73. M. N. Vanderlaan and R. W. Thring, Polyurethanes from Alcell® lignin fractions obtained by sequential solvent extraction, *Biomass Bioenergy*, 1998, **14**(5–6), 525–531.
74. D. van de Pas, A. Hickson, L. Donaldson, G. Lloyd-Jones, T. Tamminen and A. Fernyhough, *et al.*, Characterization of fractionated lignins polymerized by fungal laccases, *BioResources*, 2011, **6**(2), 1105–1121.
75. T. Q. Yuan, J. He, F. Xu and R. C. Sun, Fractionation and physicochemical analysis of degraded lignins from the black liquor of Eucalyptus pellita KP-AQ pulping, *Polym. Degrad. Stab.*, 2009, **94**(7), 1142–1150.
76. X. Jiang, C. Abbati De Assis, M. Kollman, R. Sun, H. Jameel and H. M. Chang, *et al.*, Lignin fractionation from laboratory to commercialization: Chemistry, scalability and techno-economic analysis, *Green Chem.*, 2020, **22**(21), 7448–7459.

77. M. F. Li, S. N. Sun, F. Xu and R. C. Sun, Sequential solvent fractionation of heterogeneous bamboo organosolv lignin for value-added application, *Sep. Purif. Technol.*, 2012, **101**, 18–25.
78. R. J. A. Gosselink, J. E. G. Van Dam, E. De Jong, E. L. Scott, J. P. M. Sanders and J. Li, *et al.*, Fractionation, analysis, and PCA modeling of properties of four technical lignins for prediction of their application potential in binders, *Holzforschung*, 2010, **64**(2), 193–200.
79. H. Sadeghifar and S. Dimitris, Argyropoulos. Correlations of the antioxidant properties of softwood kraft lignin fractions with the thermal stability of its blends with polyethylene, *ACS Sustainable Chem. Eng.*, 2015, **3**(2), 349–356.
80. H. Sadeghifar and D. S. Argyropoulos, Macroscopic Behavior of Kraft Lignin Fractions: Melt Stability Considerations for Lignin-Polyethylene Blends, *ACS Sustainable Chem. Eng.*, 2016, **4**(10), 5160–5166.
81. D. S. Argyropoulos, H. Sadeghifar, C. Cui and S. Sen, Synthesis and characterization of poly(arylene ether sulfone) kraft lignin heat stable copolymers, *ACS Sustainable Chem. Eng.*, 2014, **2**(2), 264–271.
82. C. Cui, H. Sadeghifar, S. Sen and D. Argyropoulos, Toward Thermoplastic Lignin Polymers; Part II: Thermal & Polymer Characteristics of Kraft Lignin & Derivatives, *BioResources*, 2013, **8**(1), 864–886.
83. H. Sadeghifar, S. Sen, S. V. Patil and D. S. Argyropoulos, Toward Carbon Fibers from Single Component Kraft Lignin Systems: Optimization of Chain Extension Chemistry, *ACS Sustainable Chem. Eng.*, 2016, **4**(10), 5230–5237.
84. W. M. Goldmann, J. Ahola, M. Mikola and J. Tanskanen, Solubility and fractionation of Indulin AT kraft lignin in ethanol–water media, *Sep. Purif. Technol.*, 2019, **209**, 826–832.
85. J. Domínguez-Robles, T. Tamminen, T. Liitiä, M. S. Peresin, A. Rodríguez and A. S. Jääskeläinen, Aqueous acetone fractionation of kraft, organosolv and soda lignins, *Int. J. Biol. Macromol.*, 2018, **106**, 979–987.
86. X. Meng, A. Parikh, B. Seemala, R. Kumar, Y. Pu and C. E. Wyman, *et al.*, Characterization of fractional cuts of co-solvent enhanced lignocellulosic fractionation lignin isolated by sequential precipitation, *Bioresour. Technol.*, 2019, **272**, 202–208.
87. A. S. Jääskeläinen, T. Liitiä, A. Mikkelson and T. Tamminen, Aqueous organic solvent fractionation as means to improve lignin homogeneity and purity, *Ind. Crops Prod.*, 2017, **103**, 51–58.
88. A. Klett, P. Chappell and M. Thies, Recovering ultraclean lignins of controlled molecular weight from Kraft black-liquor lignins, *Chem. Commun.*, 2015, **51**(64), 12855–12858.
89. S. Kubo, Y. Uraki and Y. Sano, Preparation of carbon fibers from softwood lignin by aqueous acetic acid pulping, *Holzforschung*, 1998, **49**(4), 343–350.
90. M. F. Li, S. N. Sun, F. Xu and R. C. Sun, Mild acetosolv process to fractionate bamboo for the biorefinery: structural and antioxidant

properties of the dissolved lignin, *J. Agric. Food Chem.*, 2012, **60**(7), 1703–1712.
91. D. Lachenal, G. Mortha, R. Sevillano and M. Zaroubine, Isolation of residual lignin from softwood kraft pulp. Advantages of the acetic acid acidolysis method, *C. R. Biol.*, 2004, **327**, 911–916.
92. A. S. Klett, A. M. Payne, T. Phongpreecha, D. B. Hodge and M. C. Thies, Benign Fractionation of Lignin with CO_2-Expanded Solvents of Acetic Acid + Water, *Ind. Eng. Chem. Res.*, 2017, **56**(34), 9778–9782.
93. W. Guanhua, One-Pot Lignin Depolymerization and Activation by Solid Acid Catalytic Phenolation for Lightweight Phenolic Foam Preparation, *Ind. Crops Prod.*, 2018, **24**, 216–225.
94. P. Emilia, T. Tommaso and M. Francesco, The rise of lignin biorefinery, *Curr. Opin. Green Sustainable Chem.*, 2020, **24**, 1–6.
95. D. Humpert, M. Ebrahimi and P. Czermak, Membrane technology for the recovery of lignin: a review, *Membranes*, 2016, **6**(3), 1–13.
96. M. Olivares, J. Guzmán, A. Natho and A. Saavedra, Kraft lignin utilization in adhesives, *Wood Sci. Technol.*, 1988, **22**, 157–165.
97. V. Uloth and J. T. Wearing, Kraft lignin recovery. Acid precipitation versus ultrafiltration, Part 1: Laboratory test results, *Pulp Pap. Can.*, 1989, **90**, 67–71.
98. O. Wallberg, A. S. Jönsson and R. Wimmerstedt, Fractionation and concentration of kraft black liquor lignin with ultrafiltration, *Desalination*, 2003, **154**(2), 187–199.
99. S. Gillet, M. Aguedo, L. Petitjean, A. R. C. Morais, A. M. Da Costa Lopes and R. M. Łukasik, *et al.*, Lignin transformations for high value applications: Towards targeted modifications using green chemistry, *Green Chem.*, 2017, **19**(18), 4200–4233.
100. P. Tairan, W. Guanhua, S. Hao, S. Wenjie and S. Chuanling, Lignin fractionation: Effective strategy to reduce molecule weight dependent heterogeneity for upgraded lignin valorization, *Ind. Crops Prod.*, 2021, **165**, 113442.
101. A. Toledano, L. Serrano, A. Garcia, I. Mondragon and J. Labidi, Comparative study of lignin fractionation by ultrafiltration and selective precipitation, *Chem. Eng. J.*, 2010, **157**(1), 93–99.
102. S. F. Mendes, J. S. Rodrigues, V. H. de Lima, V. R. Botaro, V. L. Cardoso and M. H. M. Reis, Forward Black Liquor Acid Precipitation: Lignin Fractionation by Ultrafiltration, *Appl. Biochem. Biotechnol.*, 2021, **193**(10), 3079–3097.
103. S. Aminzadeh, M. Lauberts, G. Dobele, J. Ponomarenko, T. Mattsson and M. E. Lindström, *et al.*, Membrane filtration of kraft lignin: Structural characteristics and antioxidant activity of the low-molecular-weight fraction, *Ind. Crops Prod.*, 2018, **112**, 200–209.
104. C. A. E. Costa, P. C. R. Pinto and A. E. Rodrigues, Separation and Purification Technology Lignin fractionation from E. Globulus kraft liquor by ultrafiltration in a three stage membrane sequence, *Sep. Purif. Technol.*, 2018, **192**, 140–151.

105. W. Zhu, G. Westman and H. Theliander, Lignin separation from kraft black liquor by combined ultrafiltration and precipitation: a study of solubility of lignin with different molecular properties, *Nord. Pulp Pap. Res. J.*, 2016, **31**(2), 270–278.
106. C. Huang, J. He, R. Narron, Y. Wang and Q. Yong, Characterization of Kraft Lignin Fractions Obtained by Sequential Ultrafiltration and Their Potential Application as a Biobased Component in Blends with Polyethylene, *ACS Sustainable Chem. Eng.*, 2017, **5**(12), 11770–11779.
107. V. Rohde, S. Böringer, B. Tübke, C. Adam, N. Dahmen and D. Schmiedl, Fractionation of three different lignins by thermal separation techniques – A comparative study, *GCB Bioenergy*, 2019, **11**(1), 206–217.
108. A. Chiara, F. Simon, K. Yannick, L. Matthias, S. Alberto, T. Julien, T. Stefano, G. Gianmarco and D. Paola, Fractionation of Soda Pulp Lignin in Aqueous Solvent through Membrane-Assisted Ultrafiltration, *ACS Sustainable Chem. Eng.*, 2018, **6**(7), 9056–9064.

CHAPTER 5

Lignin-based Materials for Biomedical Applications: Basic Requirements and Properties

KSENIJA RADOTIĆ,*[a] DRAGICA SPASOJEVIĆ[a] AND DANICA ZMEJKOSKI[b]

[a] Institute for Multidisciplinary Research, University of Belgrade, Kneza Višeslava 1, 11000 Belgrade, Serbia; [b] Vinča Institute of Nuclear Sciences – National Institute of the Republic of Serbia, University of Belgrade, Mihaila Petrovića Alasa 12–14, 11000 Belgrade, Serbia
*Email: xenia@imsi.rs

5.1 Lignin Structure and Modifications

Lignin is one of the richest organic compounds on Earth. It is one of the main components of plant cell walls providing mechanical support to plant tissues and protection from various kinds of stress. In plants, lignin is produced by oxidative coupling of three phenolic alcohols/acids (coniferyl, *p*-coumaryl, and sinapyl) catalysed by the enzyme peroxidase.[1] In plant cells, lignin is intertwined and covalently bound with molecules of cellulose and hemicellulose, forming supramolecular structures as constituents of the plant cell walls.

Thus it is impossible to extract unmodified lignin from plant tissues since severe conditions connected with lignin isolations[2] lead to lignin structural modification. For these reasons, lignin structure (Figure 5.1) has not been completely understood yet, which impacts its further applications since the

Biomaterials Science Series No. 16
Lignin-based Materials: Health Care and Medical Applications
Edited by Kuruvilla Joseph, Runcy Wilson, Gejo George and Saritha Appukuttan

Published by the Royal Society of Chemistry, www.rsc.org

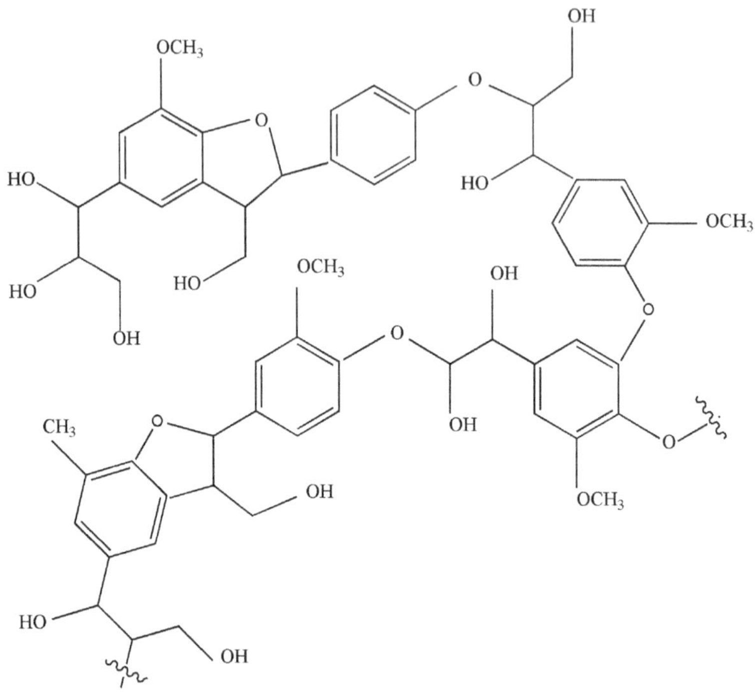

Figure 5.1 Structure of the lignin fragment.

knowledge of the chemical structure and macromolecular organization of material is crucial to understanding its properties. Thus in basic studies of lignin molecular structure, *in vitro* synthesized lignin model compounds are more suitable than extracted polymers. One of the most common lignin model compounds is enzymatically polymerized poly-coniferyl alcohol,[3] well known as a dehydrogenase polymer (DHP). The DHP has been used in many experiments since it is the best lignin substitute.[4]

Most of our knowledge of lignin structures comes from spectroscopic and microscopy studies. Based on scanning tunneling microscopy (STM),[5] environmental scanning electron microscopy (ESEM),[6,7] and atomic force microscopy (AFM),[8] the DHP tends to organize itself into semi-regular spherulitic assemblies, which further form supramolecular structures resembling various geometrical structures in plant cell walls having possible physiological significance. A lignin globule is a layered, shell-like structure, as seen in AFM-based force-spectroscopy studies[8] and Langmuir–Blodgett studies.[9] The smallest globule consists of about 20 units.[5] The layered lignin globules consist of regions with different rheological properties, stiffness, and mobility, as documented by near-field scanning optical microscopy (NSOM), AFM, and electron paramagnetic resonance spectroscopy (EPR).[10,11] These properties are important for potential biomedical applications of lignin.

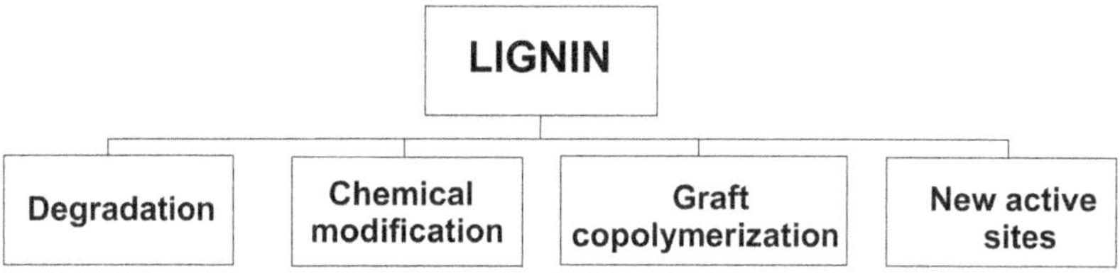

Figure 5.2 Schematic representation of different modifications of lignin to increase its applicability.

Lignin has a complex chemical structure consisting of various functional groups such as phenolic, methoxy, and aliphatic hydroxyl groups. The aliphatic hydroxyl group is the most common of all groups in lignin.[12] Lignin is an amphiphilic molecule, whose hydrophobic parts are oriented inwards in a polar solvent and thus form aggregates. This leads to insolubility in aqueous media and hinders most of its applications. Lignin can be modified to serve as applicable materials by some modifications: (1) lignin degradation into oligomers, (2) chemical modification of reactive groups, (3) creation of new active sites, and (4) fabrication of lignin graft copolymers (Figure 5.2).[13,14] A promising modification of lignin is its hybridization with inorganic nanomaterials to create hybrid nanocomposites. Hybridization modifies the chemical properties and reactivity of lignin.[15,16] The active functional groups (phenol, hydroxyl, and carboxyl groups) are involved in the binding of lignin with metal ions and nanomaterials, and in chelating and reducing reactions.[13,17]

5.2 Properties of Lignin as a Biomaterial

Biomaterials can be defined as nonviable materials that interact with human or animal cells and tissues without causing any harm. The prefix bio in the word "biomaterial" refers to biocompatibility as its most important property.[18] Other properties required for successful medical application include hardness, tensile strength, modulus of elasticity, resistance to abrasion and wear, long-term dimensional stability, swelling in aqueous media, and permeability to gases, water, and small biomolecules.[19]

Although the use of products from natural origins as biomaterials has been known for centuries, in the last few decades lignin has attracted novel attention. Biomaterial research is one of the most important areas of research in modern medicine, especially in the terms of tissue engineering, wound healing, and drug delivery.[20]

5.2.1 Biocompatibility

Biomaterials are used to cure or to replace a component of the human body, or to support physiological functions. Thus, biomaterials must be biocompatible with human cells, tissues, or organs,[21] be stable in physiological media (pH, temperature), and not form metabolic residues in the biodegradation process. The need for mechanical stability depends on their specific application, while metallic constituents must be resistant to corrosion.[22]

Biomaterials obtained from natural sources usually have satisfactory biocompatibility. However, there have not been numerous studies on lignin biocompatibility.[23] Cytotoxicity is an essential issue related to biocompatibility, which should be studied on the cell level and in *in vivo* studies. An example is presented in Figure 5.3 showing that a lignin-based hydrogel does not have harmful effects on an injured animal, but promotes wound healing.[24] The cytotoxicity of lignin depends on its chemical extraction process, which changes its physical and chemical properties and affects its solubility, aromatic unit content, and impurity level.[25] Ugartondo *et al.* studied the cytotoxicity of four lignins (lignosulfonates, bagasse lignin, steam explosion lignin, and kraft lignin), and found that lignosulfonates were the least cytotoxic to the HaCaT and 3T3 cell lines, while the most cytotoxic was kraft lignin on HaCaT cells and bagasse lignin on 3T3 cells.[26] On the other hand, the lignin model compound, synthesized under controlled conditions, was not toxic to human epithelial cells.[24] Another aspect of biosafety is the biodegradability of a biomaterial since it can be decomposed into toxic degradation products. Lignin biodegradability stays unexplored.[27] Lignin can be partially decomposed by enzymes;[28] chemically modified lignin may be less biodegradable due to the absence of phenolic

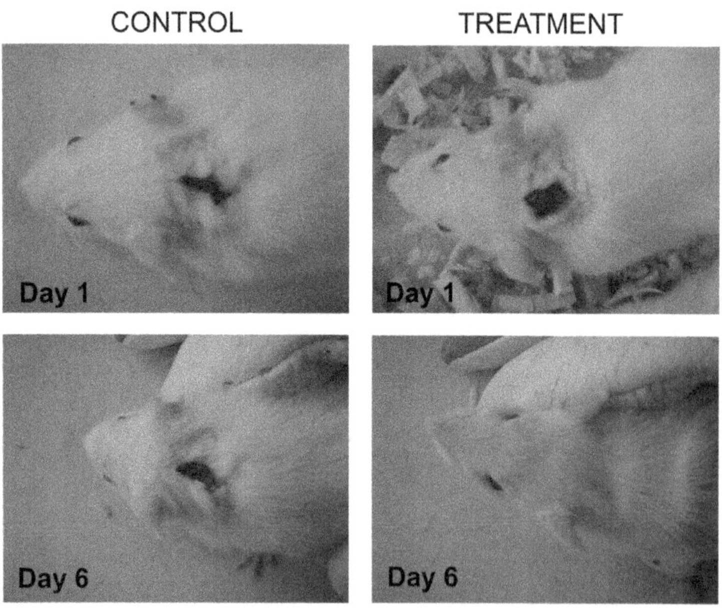

Figure 5.3 *In vivo* wound healing treatment of an injured animal with an artificial lignin-based hydrogel.

hydroxyl groups and thus less accessible to enzymatic degradation processes. Therefore, it is crucial to conduct biocompatibility and biodegradability studies for each type of lignin.

5.2.2 Antioxidant and Antimicrobial Properties

Antioxidant and antimicrobial properties are a good precondition for biomedical applications of a material. The antioxidant and antimicrobial properties of lignin depend on the extraction method and post-treatment reactions,[29,30] as well as on the lignin source.[31] The radical scavenging properties of lignin have been documented.[32–34] In most extraction procedures for lignin, severe conditions are applied which modify the chemical profile of lignin. As a result, its functional properties are also modified, which could provide potential for new applications.[35] The antioxidant activity of lignin may be strengthened by various modifications.[36,37]

Due to the increasing bacterial resistance caused by the overuse of antibiotics,[38] new natural antimicrobial agents are needed. Lignin, as a natural compound, may be a good candidate. Spasojević *et al.*[24] found that hydrogels consisting of artificial lignin and natural polysaccharides (alginate, biocellulose[39]) were effective against several Gram(+) and Gram(−) bacterial strains (Figure 5.4), as well as clinical isolates from patients' wounds and biofilms. These results indicate the potential of the lignin-based hydrogels in wound healing. Antimicrobial activity depends on the type of lignin and the microbial species.[24,35,40] It can be strengthened by certain lignin modifications.[41] Lignin antimicrobial properties also depend on the extraction or preparation conditions and plant origin.[29,42,43]

5.2.3 Ultraviolet (UV)-protective Activity

Due to the increase in the UV-B (280–320 nm) and UV-A (320–400 nm) regions of solar radiation on Earth and their damaging effects on DNA and proteins in living beings, there is increasing interest in the development of natural UV protection materials.[44,45] Thus lignin has been explored as a UV shield for biomedical applications,[46] which is reviewed by Sadeghifar

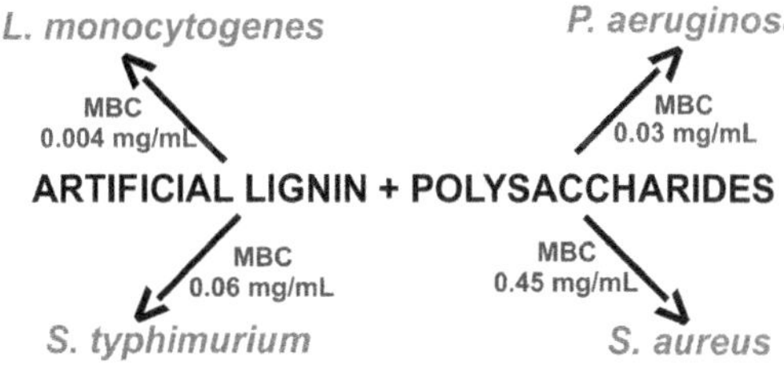

Figure 5.4 Antibacterial activity expressed as minimum bactericidal concentrations (MBCs) of artificial lignin-based hydrogels on selected bacterial strains frequently present in wounds.

Table 5.1 Mechanical properties of PCL–lignin.

Weight percentage of PCL–lignin	Tensile strength (MPa)	Young modulus (MPa)	Elongation at break (%)
PCL–lignin5	4.45 ± 0.26	9.5 ± 1.0	150 ± 19
PCL–lignin50	0.66 ± 0.13	3.30 ± 0.70	67 ± 4

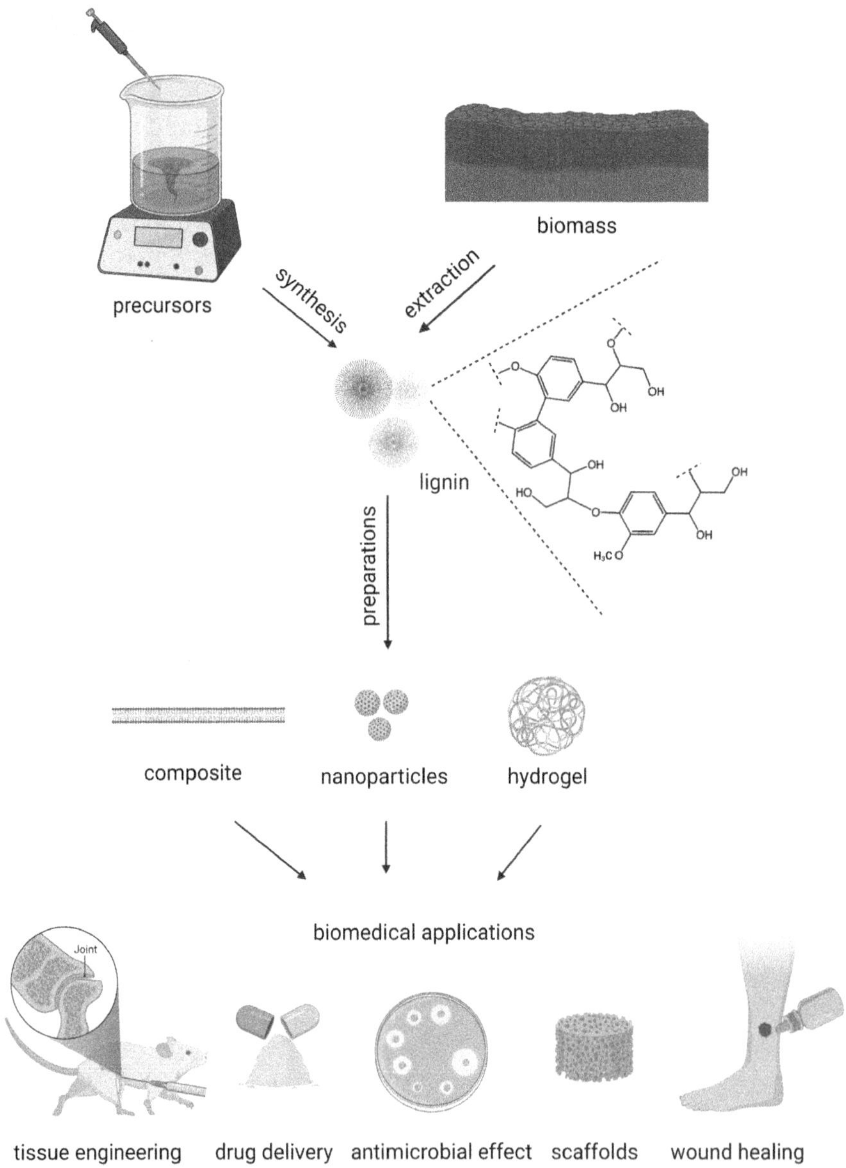

Figure 5.5 Overview of lignin-based biomaterials and their biomedical applications.

and Ragauskas. Non-modified and modified lignin improved the UV protection of sunscreen products. In mixtures with other synthetic UV screening products, lignin enhanced the UV shielding capacity.[47]

5.2.4 Mechanical Compatibility

This property is mostly related to bioimplants. Bioimplants should have the appropriate mechanical strength to resist all forces and loads leading to fractures. Satisfactory mechanical properties and slow degradation were found for PCL (ε-caprolactone)-lignin, making it suitable for nerve application,[48] implantation, and regeneration of tissue (Table 5.1).[49] Lignin–poly(lactic acid) and lignin–chitosan gave good results for mechanical properties.[50,51]

5.2.5 Corrosion Resistance

Biomaterials are normally exposed to the critical level of humidity and to the environment with corrosion potential. This is especially relevant to the materials used as implants or scaffolds which contain metals. Released ions may cause allergic and toxic reactions.[52] Although lignin was found to be an efficient anti-corrosion coating for metal surfaces,[53] there are no data on its corrosion resistance as a biomaterial; thus this property should be studied more thoroughly.

In Figure 5.5, an overview of the pathway from lignin (obtained by either extraction or synthesis) to various biomaterials and their further use in biomedical applications is given.

5.3 Lignin in Wound Healing

For wound healing, lignin is mostly applied in the form of hydrogels or films. A hydrogel is a three-dimensional (3D) polymeric network that absorbs and retains a large amount of fluid.[54] Providing elimination of wound exudates, controlling the wound moisture, and having an antibacterial effect, hydrogels are high potential wound dressings.[55–57]

Lignin, as a natural polymer with antioxidant and antimicrobial properties, was explored in combination with other compounds to form hydrogels with wound healing potential. On comparing lignin with frequently used materials for hydrogel preparations (*i.e.* polyethylene glycol (PEG)), it was observed that it had increased mechanical strength and adhesiveness,[58] which is desirable for wound treatments. Lignin from different sources, combined with other polymers or Ag nanoparticles, has been used to prepare hydrogels for wound dressings.[59–61] Chitosan–alkali lignin hydrogels showed scratch wound healing capacity and were not toxic to mesenchymal stem cells and to zebrafish at concentrations up to 100 $\mu g\,mL^{-1}$.[62] The hydrogels also exhibited infection prevention and antioxidant capacity. Copolymerized lignin–triazole hydrogels showed *in vivo* capacity to prevent wound infection, aid healing, and exhibit anti-inflammatory activity.[63]

The highly complex structure of lignin and broad MW distribution of the fractions within the polymer are limitations to the standardized utilization of lignin. These limitations may be effectively improved by the use of lignin fractions with a narrow MW range for the production of lignin-based materials. This can be achieved by the synthesis of artificial lignin and separation of particular fractions, which enables control and standardization of lignin structure, and the purity of the obtained polymer and its fractions. Thus, the low MW lignin fractions inhibited the growth of MRC5 cells at lower concentrations than the whole polymer including all fractions.[64] The hydrogels based on the synthesized lignin, DHP, and sodium alginate showed antimicrobial activity and wound healing potential while being non-toxic to human epithelial cells.[24] It was found that in the hydrogels consisting of bacterial cellulose and the synthesized lignin, the low MW polymer fractions were most active in antimicrobial activity.[39] The specificity of the antimicrobial effects of the purified low MW DHP fractions on particular bacterial strains depended on the fractions' MW and specific molecular groups.[65]

5.4 Lignin in Tissue Engineering

Tissue engineering, as a part of regenerative medicine, involves the design and development of biomaterials intended for the regeneration of damaged tissues or organs.[66] There have been several reports on the tissue engineering applications of lignin-based materials, because of their antioxidant, antibacterial, and UV shielding activities. For tissue engineering applications, in recent years, the most studied material forms of lignin have been hydrogels and nanofibers, and this area is rapidly growing.[67,68]

Hydrogels have been promising for tissue engineering due to their ability to absorb a large quantity of water in their polymer network, and thus they can mimic the tissue environment.[69] Synthetic polymers have been complexed with lignin as a natural polymer to obtain hydrogels for potential use in tissue engineering.[58] Aerogels are highly porous, light materials obtained from different gels, in which gas replaces the solvent phase. They are suitable for tissue engineering scaffolds due to their flexibility, high porosity, large surface area, and high mechanical strength.[13] Alginate–lignin aerogels were developed, where rapid expansion of CO_2 created pores of few hundred microns in the aerogels. The obtained alginate–lignin aerogels are suitable for tissue engineering applications due to their structural properties, cytocompatibility, and good cell adhesion.[70] Hydrogels were prepared from lignin-grafted copolymers with poly(ethylene glycol) methacrylate (PEGMA). After mixing with a-cyclodextrin (a-CD), self-assembling of lignin copolymer into supramolecular hydrogels was observed. The lignin core was important in the gelation process, controlling the oscillation strain, important in the transition from a solid into a liquid state. The hydrogels showed satisfactory biocompatibility and thus have the potential for tissue engineering.[71] Lignin nanoparticles (LNPs) and polyacrylamide were used to create nanocomposite hydrogels. The LNPs improved the hydrogels' mechanical properties due to the hydrogen bonds between polyacrylamide chains and LNPs. The simple synthesis of polyacrylamide/LNP nanocomposite hydrogels together with their non-toxicity and good mechanical properties make them potent agents for tissue engineering and regeneration and creation of artificial muscles.[72] An aerogel was prepared using lignin-containing arabinoxylan, cross-linked using citric acid, and reinforced with cellulose nanofibers, intended for tissue engineering.[73] An adhesive and tough hydrogel, Ag–LNPs–pectin–polyacrylic acid (NPs–P–PAA), was shown to be antiseptic and exhibited outstanding mechanical and adhesive properties. Also, the hydrogel showed durable and repeatable adhesiveness to various material surfaces and tissues. The catechol groups of lignin and the carboxyl groups of PAA synergistically contributed to the strong adhesion of the hydrogel. The *in vivo* and *in vitro* tests showed that the hydrogel showed favorable cell affinity and accelerated cell proliferation, thus being suitable for wound healing and skin tissue regeneration.[74]

In tissue engineering, scaffolds are widely used to imitate the native extracellular matrix with regard to chemical composition and physical properties. Lignin has been explored for obtaining polymeric nanofiber matrices to mimic the extracellular matrix for the development of tissue engineering scaffolds.[75] A lignin-based copolymer (lignin–PCL-*co*-lactide), lignin–PCLLA) was synthesized and blended with PCL and poly(L-lactic acid) affording nanofibers whose mechanical properties were substantially enhanced. The nanofibers showed antioxidant activity and *in vitro* cytocompatibility, indicating that they may be used in tissue engineering as scaffolds for protecting cells from oxidative stress.[76] Electrospun fibers with high surface area/volume ratios have potential to be used for tissue engineering scaffolds.[77] Aligned electrospun poly(vinyl alcohol) (PVA)–poly(glycerol sebacate) (PGS) fibers with different percentages of lignin were fabricated using the electrospinning technique for nerve tissue engineering. The effect of various contents of lignin on the physicochemical properties of the produced fibers was reported. The results revealed that the fibers were smooth and uniform in diameter. The modulus of elasticity and fiber diameter changed in proportion to the percentage of lignin. Lignin had positive effects on neural cell proliferation and differentiation of rat pheochromocytoma cells.[78] Several studies reported that mixing lignin with PCL affords an effective material for nerve tissue regeneration,[79] and as a model for the design of soft tissue engineering scaffolds.[80] PCL/lignin–PCL nanofibers preserved cells from oxidative stress and had a free radical scavenging effect. Thus lignin was considered to be suitable for tissue regeneration applications where cellular oxidative stress occurs.[81]

Bone regeneration scaffolds, one of the key applications of tissue engineering, must provide adequate mechanical stability after implantation, in combination with controlled degradability, and they need to be replaced by natural bone to avoid harmful products of degradation.[82] A role of lignin in facilitating the construction of artificial hydroxyapatite bone was reported.

A lignin–PCL fibrous matrix was developed; it induced the formation of hydroxyapatite in an artificial body fluid. The phenolic and aliphatic hydroxyl groups in lignin are proposed as reactive sites for chelation of metal ions and growth of hydroxyapatite *via* co-precipitation of calcium and phosphate ions, which is similar to the natural process of bone formation. Hybrid lignin biominerals are promising for the future development of biomaterials in bone healing and hard tissue engineering.[83] The *in vitro* study indicated that the nanofibers attach to human dermal fibroblasts and interact with them. However, the mechanism of supporting cell growth by lignin copolymers was not explored.[84] The mixing of lignin with other materials such as hydroxyapatite, calcium phosphate, and PCL has been explored for bone regeneration potential. Hydroxyapatite in combination with nanofibrous lignin serves as a suitable scaffold for *in situ* cell proliferation and osteogenesis. The lignin/PCL and hydroxyapatite blends had potential for bone cures.[85,86] The mechanical behavior of hydrogels made of agarose and lignin was studied. Lignin from oil palm fruit peels was used to produce a hydrogel with agarose, which was cross-linked with epichlorohydrin. The mechanical properties of the obtained hydrogel were studied by texture profile analysis.[87] Chitin and lignin were explored for possible uses as artificial extracellular matrices. The composites made of nanofibers of lignin and chitin were studied as potential scaffolds.[88] Erakovic *et al.* used lignin as an active coating for implants made of titanium.[89] Various ratios of hydroxyapatite and organosolv lignin were deposited onto the implants using electrophoresis and further were sintered. The coatings were biocompatible and after adding silver during deposition showed antibacterial activity against *Staphylococcus aureus*. Lignin in combination with alginate was an appropriate milieu for cell adhesion, growth, and differentiation. Adding lignin to alginate minimizes the constraints of alginate due to its hydrophilicity, like decreased protein adsorption and cell adhesion. Also, the lignin stability decreases the scaffold degradation rate and helps to coordinate the degradation rate with the rate of regeneration of bone tissue.[90,91] Rekola *et al.* made an artificial composite bone material consisting of reinforced fibers without metal by heat treatment of wood, based on the similarity between the structures of wood and bone.[92] In the rabbit model, it was reported that the heat treatment positively affected the osteoconductive properties of the artificial wood implant, showing the good biocompatibility of lignin.

The new technology of 3D printing brings a revolution in tissue engineering.[93] 3D printing enables specialists to provide personalized treatment for injured individuals. Lignin as a natural polymer shows good potential to be used in 3D printing technologies. Polyurethanes (PU) have been studied as a component in tissue engineering. The PU molecular structure and morphology can be altered to mimic natural bone and stimulate tissue ingrowth.[94] 3D printing was used to produce biodegradable water-based PU scaffolds for bone regeneration. These materials induced osteogenesis after incorporation of superparamagnetic iron oxide nanoparticles.

The composites of lignin and PU may have future potential for drug delivery and tissue engineering, since their various compositions differing in mechanical properties and degradability are available.[95,96]

5.5 Lignin in Drug Delivery

Drug delivery comprises targeted and controlled release of active substances which enhance the therapeutic effects of drugs. There is increasing scientific interest in biomaterials which can be used in controlled drug delivery to specific areas. The development of drug delivery agents is presently focused on the use of micro- and nanoparticles and encapsulated forms. Several methods to prepare LNPs loaded with drugs, including milling, precipitation, solvent exchange, high-pressure homogenization, the pH shifting method, the ice-segregation-induced self-assembly process, the aerosol process, electrospinning, supercritical fluid technology, the solvent anti-solvent precipitation, preparation of nanoparticles by acoustic cavitation, dialysis, and ultrasonic radiation, have been reported.[97–99]

Lignin has a high ability to encapsulate hydrophobic/lipophilic drugs; this is promising in controlled drug release having potential in oncological therapy. However, the utility of lignin in drug delivery is suppressed by its insolubility in aqueous media. This problem can be solved by lignin chemical alterations through its active functional groups such as phenolic/alcoholic hydroxyl groups, aromatic units, and carbon–carbon double bonds.[100,101] Hybridization of inorganic compounds with lignin is another possible method for the creation of valuable materials.[97] The obtained lignin-based nanomaterials can be applied for drug/gene delivery. The property of lignin to bind nanomaterials or metal ions is based on its active groups (phenol, hydroxyl, and carboxyl groups), which may have chelating and reducing characteristics in the reaction with the metal center. The type and chemical properties of lignin determine its chelating ability.[17] Colloidal cationic LNPs with encapsulated enzymatic biocatalysts were synthesized for esterification in aqueous media.[102] LNPs with integrated iron or iron oxide were synthesized by means of self-assembly during the dialysis process. They were used for drug delivery of the drugs Sorafenib® and Benzazulene® with low water solubility. The particles were not cytotoxic. For hydrophobic drugs, the release kinetics depends on the pH and water content of the release medium.[103]

Magnetic nanocarriers are considered to be important vehicles in drug delivery due to numerous benefits such as targeted delivery, control of the target site, and maintenance of the therapeutic concentration of drugs. There has been a considerable development of magnetic delivery vehicles,[104] for example, LNPs as a matrix for Resveratrol®. The hydrophobic drug Resveratrol® was encapsulated in LNPs together with nano-Fe_3O_4, as a possible cancer therapeutic agent. Both *in vitro* and *in vivo* experiments showed that particles with and without magnetic properties had no harmful effects on cell lines or tested animals.[105]

A composite produced from lignin and cellulose cross-linked with epichlorohydrin was synthesized as a porous material. A study of drug release from lignin-based gels, using different polyphenols as model drugs, reported that drug release is dependent on the lignin ratio in the gel. A higher swelling rate and a faster release rate of drugs were obtained with the higher lignin content in the gel. However, the maximal drug release after 10 h was 30%.[106] The esterification of lignin's hydroxyl groups with PEG and a polymer of carboxylic acid produced hydrogels with antibacterial activity and capability for hydrophobic drug delivery. In comparison to the GANPEG (Gantrez S-97 + poly(ethylene glycol) hydrogel used as a control, the lignin hydrogel was more resistant to two tested pathogens.[107] The potential of cross-linked organosolv LNP application for oral drug delivery was reported.[108] The LNPs with an average size of 104 nm encapsulated successfully a lipophilic drug model based on curcumin with 92% encapsulation efficiency. The *in vitro* release study observed the high stability of the LNPs under artificial gastric conditions and a slow release under intestinal conditions. The curcumin-loaded LNPs were shown to be non-toxic, and curcumin cellular uptake and intestinal permeation were higher than those for free curcumin. A study was done on the assessment of the potential of self-assembling blank kraft LNPs (BLNPs) whose size was controlled and optimized. The hemocompatibility, cytotoxicity, and genotoxicity of BLNPs as a drug carrier were studied on the *Drosophila melanogaster* model organism; the main advantage of BLNPs over metallic and inorganic nanoparticles was their biocompatibility. The BLNPs loaded with irinotecan as a model drug had a satisfactory drug loading capacity and high encapsulation efficiency, and reduced the IC50 value of irinotecan 3 times.[109] Siddiqui *et al.* studied lignin's potential as a drug nanocarrier and its antioxidant activity in cancer therapy and prevention of its relapse. They synthesized LNPs with irinotecan and the P-gp modulator, functionalized with CD44 receptors for the treatment of colon cancer.[110] Lignin extracted from sugarcane was studied for potential use as a carrier of methotrexate in the treatment of rheumatoid arthritis.[111] In another attempt, sodium lignosulfonate-grafted poly(acrylic acid-*co*-poly(vinyl pyrrolidone)) hydrogels were synthesized by radical polymerization combined with ultrasonic treatment.[112] Amoxicillin was used as a model drug, due to its pH sensitivity and the possibility to control its release *in vitro*. The amoxicillin-loaded hydrogels had higher percentages of release in simulated intestinal fluids than those in simulated gastric fluids. It was shown that functionalization of lignin into carboxy lignin was promising for drug delivery. A pH-responsive drug delivery system, based on introduction of new functional groups in carboxy lignin with PEG and poly(histidine), was produced for controlled delivery of anticancer drugs. The produced nanocarriers did not show adverse effects in different cancer cell lines as compared with the control – normal endothelial cells.[113]

Nanocapsules, with a polymeric shell and an inner core containing active substances, have been studied as possible drug delivery vehicles.

The reduction of side effects in targeted drug delivery was reported by alterations in the shell material composition and functionalization of the shell with the ligand.[114] Another way was obtaining lignin microcapsules and nanocapsules through the oil in water emulsion and by ultrasound-promoted cross-linking of lignin in the water/oil interface.[115] It was described that the method for obtaining pH-sensitive lignin nanocapsules was based on interfacial miniemulsion polymerization. The controlled release in a pH-changing environment and satisfactory drug loading capacity for coumarin-6 were obtained using the nanocapsules.[116]

Nanomicelles, as well as nanoscaled colloidal amphiphilic monomers, have been studied as potential agents for drug delivery. Lignin complexed with quaternized 3-chloro-2-hydroxypropyltrimethylammonium chloride (CHMAC) was used to synthesize pH-sensitive lignin micelles loaded with ibuprofen intended for oral delivery, suggesting a novel pathway for obtaining lignin nanomicelles for oral drug delivery.[117]

5.6 Pharmaceuticals

Lignin can be used as a base to obtain pharmaceuticals potentially used to cure diseases. The usual way is to depolymerize lignin and obtain derivatives to create biologically active compounds, which can be used to treat diseases caused by viruses such as the herpes simplex virus and influenza virus. Since the toxicity of drugs is an important factor, regarding the cytotoxicity of lignin it was shown that lignins are generally safe to consume and do not disrupt cell viability.[118] The high antioxidant potential of these drugs is also an important parameter. Lignin may also be helpful in controlling obesity. Lignin was shown to be a strong bile acid adsorbent, which may be attributed to its methoxyl and carbonyl groups. Thus after lignin binding to the bile, micelles cannot be produced, resulting in decreased lipid absorption.[118] Lignosulfonic acid taken with glucose causes a delay in the uptake of glucose; thus it may be potentially used to treat diabetes.[119] Several studies describing the products synthesized from technical lignins, mostly through their depolymerization, have shown the potential of such obtained compounds on several biological diseases.[120,121] An artificial lignin, synthesized by polymerization of ferulic acid, was shown to inhibit tumor growth in mouse skin.[122] Synthetic lignins have been found to show anti-HIV activity.[123]

5.7 Technical Challenges in Lignin-based Materials for Biomedical Applications

The biomedical application of natural polymers is desirable in comparison with synthetic polymers due to their favorable characteristics (*e.g.* biocompatibility, biodegradability, and natural abundance).[124] The studies on the biomedical applications of lignin are increasing; however, there are still important challenges that need to be addressed with regard to lignin structure, degradability, and toxicology. Larger use of lignin-based materials in clinical studies, including systematic toxicological studies, and understanding of the structure and chemistry of lignin would meet this challenge. Most toxicological studies of lignin have been performed through simple cytotoxicity testing, which is not an appropriate simulation of the human body environment. More parameters for lignin performance in the human body milieu need to be explored, such as accumulation, degradation, and elimination of lignin from the body, the interaction of lignin with different cellular structures and metabolites, long-term stability, and potential harmful effects.

Another difficulty for the development of lignin-based biomaterials is the heterogeneity of lignin structure. The lignin chemical structure varies depending on the extraction method and the plant origin of lignin. There are numerous studies on the structure, chemistry, pre-treatment, and processing techniques of lignin, and also its antioxidant and antimicrobial activities, yet there are still problems like its broad MW distribution and extremely complex structures, which are obstacles for standardization and scale-up. Various chemical alterations of lignin have been explored to increase its potential for medical applications. The existing limitations might be overcome by the fabrication of narrow lignin fractions, either during its extraction process or by controlled synthesis of lignin model compounds. Such studies require an interdisciplinary approach from the scientific, medical, and public health points of view.

Different studies on lignin-based materials show the capacity of lignin for scaffolds and drug release systems. The antioxidant and antimicrobial properties of lignin are good for various applications, *i.e.* as a cancer therapeutic agent. Novel biomaterials are needed for stem cell-based methodologies in tissue engineering. Thus, more work is needed in lignin-based scaffold development for bone regeneration, probably towards hybrid nanomaterials based on hydroxyapatite similar to natural bone.

As for lignin-based materials for drug release applications, the present knowledge indicates that pH is one of the main stimuli; thus it could be used to regulate the selective release of drugs.[125] However, other stimuli in biological systems might be exploited such as the redox potential and certain enzymes, but the data of such reports are lacking on lignin-based materials. It seems that there is potential in structural modifications or functionalization of lignin, providing a base for the production of biomaterials for drug delivery.[126,127]

A direction in medical applications of lignin-based materials is the synthesis of LNPs, where an important issue is the solubility of nanoparticles in aqueous media which considerably affects their biomedical applications. Studies on the potential of lignin applications in the creation of nanomaterials for drug/gene delivery are still limited. Although some experiments have been performed,[128,129] more knowledge is needed about the colloidal stability and aggregation of LNPs.

5.8 Future Perspectives and Conclusion

The increasing studies of lignin as a natural and widely present compound revealed its potential in creation of materials for biomedical applications, such as wound healing, tissue engineering, and drug delivery. Future perspectives, as well as challenges to be more thoroughly studied, can be addressed in lignin-based biomaterial research. Even though there are some promising methods that use lignin as a biomaterial such as lignin-based hydrogels, lignin-based biomaterials need to be further advanced. The mechanisms of the lignin biocompatibility are still poorly understood. More studies are needed to obtain the basic facts of lignin's toxicity and effects on cells, proteins, and genetic material. The same holds for various forms of lignin modified for improvement of its water solubility. In nature, lignin is depolymerized by certain bacteria and fungi, but the degradation of lignin in the human body remains an arguable topic that is to be more studied in the future. The use of LNPs in biomaterials has been insufficiently studied for medical applications, so this direction needs to be developed more thoroughly. Also, more work on comparison with other biomaterials is needed.

Acknowledgements

We acknowledge funding from the Ministry of Education, Science and Technological Development of the Republic of Serbia, Contract No. 451-03-68/2022-14/200053 and 451-03-2/2022-14/20-0302202.

References

1. W. Boerjan, J. Ralph and M. Baucher, *Annu. Rev. Plant Biol.*, 2003, **54**, 519.
2. K. Radotić and M. Mićić, in *Sample Preparation Techniques for Soil, Plant, and Animal Samples*, ed. M. Mićić, Springer Science + Business Media, New York, 2016, vol. 26, pp. 365–376.
3. M. Wayman and T. I. Obiaga, *Can. J. Chem.*, 1974, **52**, 2100.
4. K. Radotic, M. Micic and M. Jeremic, *Ann. N. Y. Acad. Sci.*, 2005, **1048**, 215.
5. K. Radotic, J. Simic-Krstic, M. Jeremic and M. Trifunovic, *Biophys. J.*, 1994, **66**, 1763.
6. M. Micic, M. Jeremic, K. Radotic and R. M. Leblanc, *J. Colloid Interface Sci.*, 2000, **231**, 190.
7. M. Micic, M. Jeremic, K. Radotic and R. M. Leblanc, *Scanning*, 2000, **22**, 288.
8. M. Micic, I. Benitez, M. Ruano, M. Mavers, M. Jeremic, K. Radotic, V. Moy and R. M. Leblanc, *Chem. Phys. Lett.*, 2001, **347**, 41.
9. D. Pasquini, D. T. Balogh, P. A. Antunes, C. J. L. Constantino, A. A. S. Curvelo, R. F. Aroca and O. N. Oliveira, *Langmuir*, 2002, **18**, 6593.

10. M. Mićić, K. Radotić, M. Jeremić, D. Đikanović and S. Kämmer, *Colloids Surf., B*, 2004, **34**, 33.
11. M. Mićić, J. Orbulescu, K. Radotić, M. Jeremić, G. Sui, Y. Zheng and R. M. Leblanc, *Biophys. Chem.*, 2002, **99**, 55.
12. C. Chio, M. Sain and W. Qin, *Renewable Sustainable Energy Rev.*, 2019, **107**, 232.
13. R. Liu, L. Dai, C. Xu, K. Wang, C. Zheng and C. Si, *ChemSusChem*, 2020, **13**(17), 4266.
14. P. Figueiredo, K. Lintinen, J. T. Hirvonen, M. A. Kostiainen and H. A. Santos, *Prog. Mater. Sci.*, 2018, **93**, 233.
15. J. Domínguez-Robles, Á. Cárcamo-Martínez, S. A. Stewart, R. F. Donnelly, E. Larrañeta and M. Borrega, *Sustainable Chem. Pharm.*, 2020, **18**, 100320.
16. M. H. Sipponen, H. Lange, C. Crestini, A. Henn and M. Österberg, *ChemSusChem*, 2019, **12**(10), 2039.
17. C. Thulluri, S. R. Pinnamaneni, P. R. Shetty and U. Addepally, *Ind. Biotechnol.*, 2016, **12**(3), 153.
18. S. K. Kiran and S. Ramakrishna, *An Introduction to Biomaterials Science and Engineering*, World Scientific Publishing, Singapore, 2021, vol. 1.
19. Encyclopedia Britannica, https://www.britannica.com/technology/materials-science (accessed January 2022).
20. O. O. Ige, L. E. Umoru and S. Aribo, *ISRN Mater. Sci.*, 2012, 983062.
21. S. J. Gobbi, V. J. Gobbi and Y. Rocha, *Biomed. J. Sc.i Tech. Res.*, 2019, **14**(3), 1.
22. M. Witzler, A. Alzagameem, M. Bergs, B. El Khaldi-Hansen, S. E. Klein, D. Hielscher, B. Kamm, J. Kreyenschmidt, E. Tobiasch and M. Schulze, *Molecules*, 2018, **23**, 1885.
23. S. Iravani and R. S. Varma, *Green Chem.*, 2019, **21**(18), 4839.
24. D. Spasojević, D. Zmejkoski, J. Glamočlija, M. Nikolić, M. Soković, V. Milošević, I. Jarić, M. Stojanović, E. Marinković, T. Barisani-Asenbauer, R. Prodanović, M. Jovanović and K. Radotić, *Int. J. Antimicrob. Agents*, 2016, **6**(48), 732.
25. E. Melro, A. Filipe, D. Sousa, B. Medronho and A. Romano, *New J. Chem.*, 2021, **45**, 6986.
26. V. Ugartondo, M. Mitjans and M. P. Vinardell, *Bioresour. Technol.*, 2008, **99**(14), 6683.
27. A. Moreno and M. H. Sipponen, *Mater. Horiz.*, 2020, 7, 2237.
28. R. Hilgers, G. Van Erven, V. Boerkamp, I. Sulaeva, A. Potthast, M. A. Kabel and J. P. Vincken, *Green Chem.*, 2020, **22**, 1735.
29. A. Alzagameem, S. E. Klein, M. Bergs, X. T. Do, I. Korte, S. Dohlen, C. Huwe, J. Kreyenschmidt, B. Kamm, M. Larkins and M. Schulze, *Polymers*, 2019, **11**(4), 670.
30. G. Wang, Y. Xia, B. Liang, W. Sui and C. Si,*J. Chem. Technol. Biotechnol.*, 2018, **93**(10), 2977.
31. S. Sugiarto, Y. Leow, C. Li Tan, G. Wang and D. Kai, *Bioact. Mater.*, 2022, **8**, 71.

32. F. J. Lu, L. H. Chu and R. J. Gau, *Nutr. Cancer*, 1998, **30**, 31.
33. T. Dizhbite, G. Telysheva, V. Jurkjane and U. Viesturs, *Bioresour. Technol.*, 2004, **95**(3), 309.
34. H. Faustino, N. Gil, C. Baptista and A. P. Duarte, *Molecules*, 2010, **15**(12), 9308.
35. X. Dong, M. Dong, Y. Lu, A. Turley, T. Jin and C. Wu, *Ind. Crops Prod.*, 2011, **34**, 1629.
36. D. Kai, K. Zhang, L. Jiang, H. Z. Wong, Z. Li, Z. Zhang and X. J. Loh, *ACS Sustainable Chem. Eng.*, 2017, **5**(7), 6016.
37. A. Bado-Nilles, P. Pandard, T. Jayabalan, V. Aguie-Beghin, P. H. Ducrot, C. Lapierre, G. Marlair, R. J. A. Gosselink, S. Baumberger and B. Cottyn, *ChemSusChem*, 2019, **12**(21), 4799.
38. P. Grenni, V. Ancona and A. Barra Caracciolo, *Microchem. J.*, 2018, **136**, 25.
39. D. Zmejkoski, D. Spasojević, I. Orlovska, N. Kozyrovska, M. Soković, J. Glamočlija, S. Dmitrović, B. Matović, N. Tasić, V. Maksimović, M. Sosnin and K. Radotić, *Int. J. Biol. Macromol.*, 2018, **118**, 494.
40. E. Lee, Y. Song and S. Lee, *Text. Res. J.*, 2017, **89**(1), 3.
41. S. Erakovic, A. Jankovic, G. C. P. Tsui, C. Y. Tang, V. Miskovic-Stankovic and T. Stevanovic, *Int. J. Mol. Sci.*, 2014, **15**, 12294.
42. B. Ndaba, A. Roopnarain, M. O. Daramola and R. Adeleke, *Sustainable Chem. Pharm.*, 2020, **18**, 100342.
43. G. Wang, T. Pang, Y. Xia, X. Liu, S. Li, A. M. Parvez, F. Kong and C. Si, *Int. J. Biol. Macromol.*, 2019, **133**, 156.
44. F. Solano, *Molecules*, 2020, **25**(7), 1537.
45. A. L. Mitrović, J. S. Radosavljević, M. Prokopijević, D. Spasojević, J. Kovačević, O. Prodanović, B. Todorović, B. Matović, M. Stanković, V. Maksimović, D. Mutavdžić, M. Skočić, M. Pešić, L. Prokić and K. Radotić, *Plant Physiol. Biochem.*, 2021, **161**, 176.
46. F. Xiong, Y. Wu, G. Li, Y. Han and F. Chu, *Ind. Eng. Chem. Res.*, 2018, **57**(4), 1207.
47. H. Sadeghifar and A. Ragauskas, *Polymers*, 2020, **12**(5), 1134.
48. Y. Ito, M. Shimazawa, Y. Akao, M. Shimazawa, N. Seki, Y. Nozawa and H. Hara, *J. Pharmacol. Sci.*, 2006, **102**(2), 196.
49. R. Liang, J. Zhao, B. Li, P. Cai, X. J. Loh, C. Xu, P. Chen, D. Kai and L. Zheng, *Biomaterials*, 2019, **230**(4), 119601.
50. K. Wang, L. S. Loo and K. L. Goh, *Mater. Res. Express*, 2016, **3**, 035301.
51. I. Spiridon and C. E. Tanase, *Int. J. Biol. Macromol.*, 2018, **114**, 855.
52. M. Geetha, A. K. Singh, R. Asokamania and A. K. Gogia, *Prog. Mater. Sci.*, 2009, **54**(3), 397.
53. J. C. de Haro, L. Magagnin, S. Turri and G. Griffini, *ACS Sustainable Chem. Eng.*, 2019, 7(6), 6213.
54. B. Choudhary, S. R. Paul, S. K. Nayak, D. Qureshi and K. Pal, *Polymeric Gels – Characterization, Properties and Biomedical Applications*, Woodhead Publishing, 1st edn, 2018.

55. S. Jain, A. J. Domb and N. Kumar, *Focal Controlled Drug Delivery*, Springer, Boston, US, 2014.
56. Z. M. Zmejkoski, N. M. Zdravković, D. D. Trišić, M. D. Budimir, Z. M. Marković, N. O. Kozyrovska and B. M. Todorović Marković, *Int. J. Bio. Macromol.*, 2021, **191**, 315.
57. D. Z. Zmejkoski, Z. M. Marković, M. D. Budimir, N. M. Zdravković, D. D. Trišić, N. Bugarova, M. Danko, N. O. Kozyrovska, Z. Spitalský, A. Kleinova, S. B. Kuzman, V. B. Pavlović and B. M. Todorović Marković, *Mater. Sci. Eng., C*, 2021, **122**, 111925.
58. D. Rico-García, L. Ruiz-Rubio, L. Perez-Alvarez, S. L. Hernandez-Olmos, G. L. Guerrero-Ramírez and J. L. Vilas-Vilela, *Polymers*, 2020, **12**(1), 23.
59. Y. Zhang, M. Jiang, Y. Zhang, Q. Cao, X. Wang, Y. Han, G. Sun, Y. Li and J. Zhou, *Mater. Sci. Eng., C*, 2019, **104**, 110002.
60. J. Xu, J. J. Xu, Q. Lin, L. Jiang, D. Zhang, Z. Li, B. Ma, C. Zhang, L. Li, D. Kai, H.-D. Yu and X. J. Loh, *ACS Appl. Bio Mater.*, 2021, **4**(1), 3.
61. M. Li, X. Jiang, D. Wang, Z. Xu and M. Yang, *Colloids Surf., B*, 2019, **177**, 370.
62. K. Ravishankar, M. Venkatesan, R. P. Desingh, A. Mahalingam, B. Sadhasivam, R. Subramaniyam and R. Dhamodharan, *Mater. Sci. Eng., C*, 2019, **102**, 447.
63. D. Mahata, M. Jana, A. Jana, A. Mukherjee, N. Mondal, T. Saha, S. Sen, G. B. Nando, C. K. Mukhopadhyay, R. Chakraborty and S. M. Mandal, *Sci. Rep.*, 2017, 7(1), 1.
64. L. Andrijevic, K. Radotic, J. Bogdanovic, D. Mutavdzic and G. Bogdanovic, *J BUON*, 2008, **13**, 241.
65. Y. Xie, X. Chen, C. Yiang, H. Wu and Z. Ye, *BioResources*, 2019, **14**, 2842.
66. C. Jin, W. Song, T. Liu, J. Xin, W. C. Hiscox, J. Zhang, G. Liu and Z. Kong, *ACS Sustainable Chem. Eng.*, 2018, **6**, 1763.
67. D. Kai, M. J. Tan, P. L. Chee, Y. K. Chua, Y. L. Yap and X. J. Loh, *Green Chem.*, 2016, **18**(5), 1175.
68. D. Kai, K. Zhang, L. Jiang, H. Z. Wong, Z. Li, Z. Zhang and X. J. Loh, *ACS Sustainable Chem. Eng.*, 2017, **5**, 6016.
69. J. H. Lee and H. W. Kim, *J. Tissue Eng.*, 2018, **9**, 2041731418768285.
70. S. Quraishi, M. Martins, A. A. Barros, P. Gurikov, S. P. Raman, I. Smirnova, A. R. C. Duarte and R. L. Reis, *J. Supercrit. Fluids*, 2015, **105**, 1.
71. D. Kai, Z. W. Low, S. S. Liow, A. Abdul Karim, H. Ye, G. Jin, K. Li and X. J. Loh, *ACS Sustainable Chem. Eng.*, 2015, **3**, 2160.
72. Y. Chen, K. Zheng, L. Niu, Y. Zhang, Y. Liu, C. Wang and F. Chu, *Int. J. Biol. Macromol.*, 2019, **128**, 414.
73. L. Berglund, F. Forsberg, M. Jonoobi and K. Oksman, *RSC Adv.*, 2018, **8**(67), 38219.
74. D. Gan, W. Xing, L. Jiang, J. Fang, C. Zhao, F. Ren, L. Fang, K. Wang and X. Lu, *Nat. Commun.*, 2019, **10**, 1487.
75. Z. Ma, M. Kotaki, R. Inai and S. Ramakrishna, *Tissue Eng.*, 2005, **11**, 101.

76. D. Kai, K. Zhang, L. Jiang, H. Z. Wong, Z. Li, Z. Zhan and X. J. Loh, *ACS Sustainable Chem. Eng.*, 2017, **5**, 6016.
77. H. S. Yoo, T. G. Kim and T. G. Park, *Adv. Drug Delivery Rev.*, 2009, **61**, 1033.
78. A. Saudi, S. Amini, N. Amirpour, M. Kazemi, A. Z. Kharazi, H. Salehi and M. Rafienia, *Mater. Sci. Eng., C*, 2019, **104**, 110005.
79. S. Amini, A. Saudi, N. Amirpour, M. Jahromi, S. S. Najafabadi, M. Kazemi, M. Rafienia and H. Salehi, *Int. J. Biol. Macromol.*, 2020, **159**, 154.
80. M. A. Salami, F. Kaveian, M. Rafienia, S. Saber-Samandari, A. Khandan and M. Naeimi, *J. Med. Signals Sens.*, 2017, 7(4), 228.
81. J. Wang, L. Tian, B. Luo, S. Ramakrishna, D. Kai, X. J. Loh, I. Hong Yang, G. Roshan Deen and X. Mo, *Colloids Surf., B*, 2018, **169**, 356.
82. M. Witzler, A. Alzagameem, M. Bergs, B. El Khaldi-Hansen, S. E. Klein, D. Hielscher, B. Kamm, J. Kreyenschmidt, E. Tobiasch and M. Schulze, *Molecules*, 2018, **23**, 1885.
83. D. Wang, J. Jang, K. Kim, J. Kim and C. B. Park, *Biomacromolecules*, 2019, **20**(7), 2684.
84. D. Kai, S. Jiang, Z. W. Low and X. J. Loh, *J. Mater. Chem. B*, 2015, **3**(30), 6194.
85. S. Ramakrishna, *Biomaterials*, 2012, **33**(3), 846.
86. M. Kim, M. Yeo, M. Kim and G. Kim, *RSC Adv.*, 2018, **8**(37), 20637.
87. S. Sathawong, W. Sridach and K. Techato, *J. Environ. Chem. Eng.*, 2018, **6**(5), 5879.
88. P. Morganti, S. Danti and M. B. Coltelli, *Res. Clin. Dermatol.*, 2018, **1**, 5.
89. S. Erakovic, A. Jankovic, G. C. P. Tsui, C. Y. Tang, V. Miskovic-Stankovic and T. Stevanovic, *Int. J. Mol. Sci.*, 2014, **15**, 12294.
90. N. G. Genes, J. A. Rowley, D. J. Mooney and L. J. Bonassar, *Arch. Biochem. Biophys.*, 2004, **422**(2), 161.
91. J. A. Rowley, G. Madlambayan and D. J. Mooney, *Biomaterials*, 1999, **20**(1), 45.
92. J. Rekola, A. J. Aho, J. Gunn, J. Matinlinna, J. Hirvonen, P. Viitaniemi and P. K. Vallittu, *Acta Biomater.*, 2009, **5**, 1596.
93. Q. Yan, H. Dong, J. Su, J. Han, B. Song, Q. Wei and Y. Shi, *Engineering*, 2018, **4**(5), 729.
94. S. A. Guelcher, *Tissue Eng., Part B*, 2008, **14**, 3.
95. S. T. Witzleben, K. Walbrueck, S. E. Klein and M. Schulze, *J. Chem. Chem. Eng.*, 2015, **9**, 494.
96. J. M. Lang, U. M. Shrestha and M. Dadmun, *Front. Energy Res.*, 2018, **6**, 4.
97. C. Frangville, M. Rutkevicius, A. P. Richter, O. D. Velev, S. D. Stoyanov and V. N. Paunov, *ChemPhysChem.*, 2012, **13**, 4235.
98. H. Li, Y. H. Deng, H. S. Wu, Y. Ren, X. Q. Qiu, D. F. Zheng and C. L. Li, *Holzforschung*, 2016, **70**, 725.
99. L. Chen, J. Dou, Q. Ma, N. Li, R. Wu, H. Bian, D. J. Yelle, T. Vuorinen, S. Fu, X. Pan and J. J. Y. Zhu, *Sci. Adv.*, 2017, **3**, e1701735.

100. M. Norgren and H. Edlund, *Curr. Opin. Colloid Interface Sci.*, 2014, **19**(5), 409.
101. E. Adler, *Wood Sci. Technol.*, 1977, **11**(3), 169.
102. S. Beisl, A. Miltner and A. Friedl, *Int. J. Mol. Sci.*, 2017, **18**, 1244.
103. P. Figueiredo, K. Lintinen, A. Kiriazis, V. Hynninen, Z. Liu, T. Bauleth-Ramos, A. Rahikkala, A. Correia, T. Kohout, B. Sarmento, J. Yli-Kauhaluoma, J. Hirvonen, O. Ikkala, M. A. Kostiainen and H. A. Santos, *Biomaterials*, 2017, **121**, 97.
104. R. Kumar, A. Butreddy, N. Kommineni, P. G. Reddy, N. Bunekar, C. Sarkar, S. Dutt, V. K. Mishra, K. R. Aadil, Y. K. Mishra, D. Oupicky and A. Kaushiket, *Int. J. Nanomed.*, 2021, **16**, 2419.
105. L. Dai, R. Liu, L. Q. Hu, Z. F. Zou and C. L. Si, *ACS Sustainable Chem. Eng.*, 2017, **5**, 8241.
106. D. Ciolacu, A. M. Oprea, N. Anghel, G. Cazacu and M. Cazacu, *Mater. Sci. Eng., C*, 2012, **32**, 452.
107. E. Larraneta, M. Imizcoz, J. X. Toh, N. J. Irwin, A. Ripolin, A. Perminova, J. Dominguez-Robles, A. Rodriguez and R. F. Donnelly, *ACS Sustainable Chem. Eng.*, 2018, **6**(7), 9037.
108. M. S. Alqahtani, A. Alqahtani, A. Al-Thabit, M. Roni and R. Syed, *J. Mater. Chem. B*, 2019, **7**(28), 4461.
109. L. Siddiqui, J. Bag, Seetha, D. Mittal, A. Leekha, H. Mishra, M. Mishra, A. K. Verma, P. K. Mishra, A. Ekielski, Z. Iqbal and S. Talegaonkar, *Int. J. Biol. Macromol.*, 2020, **152**, 786.
110. L. Siddiqui, H. Mishra, P. K. Mishra, Z. Iqbal and S. Talegaonkar, *Med. Hypotheses*, 2018, **121**, 10.
111. S. M. Wahba, A. S. Darwish, I. H. Shehata and S. S. A. Elhalem, *Mater. Sci. Eng., C*, 2015, **48**, 599.
112. X. Wang, Z. Zhou, X. Guo, Q. He, C. Hao and C. Ge, *RSC Adv.*, 2016, **6**, 35550.
113. P. Figueiredo, C. Ferro, M. Kemell, Z. Liu, A. Kiriazis, K. Lintinen, H. F. Florindo, J. Yli-Kauhaluoma, J. Hirvonen, M. A. Kostiainen and H. A. Santos, *Nanomedicine*, 2017, **12**(21), 2581.
114. A. Tiwari, R. Kumar, O. Shefi and J. K. Randhawa, *ACS Appl. Bio. Mater.*, 2020, **3**(7), 4665.
115. M. Tortora, F. Cavalieri, P. Mosesso, F. Ciaffardini, F. Melone and C. Crestini, *Biomacromolecules*, 2014, **15**(5), 1634.
116. N. Chen, L. A. Dempere and Z. Tong, *ACS Sustainable Chem. Eng.*, 2016, **4**(10), 5204.
117. Y. Li, X. Qiu, Y. Qian, W. Xiong and D. Yang, *Chem. Eng. J.*, 2017, **327**, 1176.
118. J. Gil-Chávez, P. Gurikov, X. Hu, R. Meyer, W. Reynolds and I. Smirnova, *Biomass Convers. Biorefin.*, 2021, **11**, 2387.
119. Y. Hasegawa, Y. Kadota, C. Hasegawa and S. Kawaminami, *J. Nutr. Sci. Vitaminol.*, 2015, **61**, 449.
120. T. Norikura, Y. Mukai, S. Fujita, K. Mikame, M. Funaoka and S. Sato, *Basic Clin. Pharmacol. Toxicol.*, 2010, **107**, 813.

121. S. Elangovan, A. Afanasenko, J. Haupenthal, Z. Sun, Y. Liu, A. K. Hirsch and K. Barta, *ACS Cent. Sci.*, 2019, **5**, 1707.
122. M. Asanoma, K. Takahashi, M. Miyabe, K. Yamamoto, N. Yoshimi, H. Mori and Y. Kawazoe, *Carcinogenesis*, 1994, **15**, 2069.
123. H. Nakashima, T. Murakami, N. Yamamoto, T. Naoe, Y. Kawazoe, K. Konno and H. Sakagami, *Chem. Pharm. Bull.*, 1992, **40**, 2102.
124. D. Spasojević, M. Prokopijević, O. Prodanović, N. Zelenović, N. Polović, K. Radotić and R. Prodanović, *Macromol. Res.*, 2019, **27**, 764.
125. G. Kocak, C. Tuncer and V. Bütün, *Polym. Chem.*, 2017, **8**, 144.
126. D. Piccinino, E. Capecchi, L. Botta, P. Bollella, R. Antiochia, M. Crucianelli and R. Saladino, *Catal. Sci. Technol.*, 2019, **9**, 4125.
127. T. Zou, M. H. Sipponen and M. Osterberg, *Front. Chem.*, 2019, 7, 320.
128. M. H. Sipponen, H. Lange, M. Ago and C. Crestini, *ACS Sustainable Chem. Eng.*, 2018, **6**(7), 9342.
129. A. Rangan, M. V. Manchiganti, R. M. Thilaividankan, S. G. Kestur and R. Menon, *Ind. Crops Prod.*, 2017, **103**, 152.

CHAPTER 6

Antimicrobial and Antioxidant Properties of Lignin and Its Composites

SWARUP ROY,*[a,b,†] RUCHIR PRIYADARSHI,[a,†] SHIV DUTT PUROHIT[c] AND JONG-WHAN RHIM*[a]

[a] Department of Food and Nutrition, BioNanocomposite Research Center, Kyung Hee University, 26 Kyungheedae-ro, Dongdaemun-gu, Seoul 02447, Republic of Korea; [b] School of Bioengineering and Food Technology, Shoolini University, Solan, Himachal Pradesh 173229, India; [c] School of Chemical Engineering, Yeungnam University, 280 Daehak-Ro, Gyeongsan, Gyeongbuk 38541, Republic of Korea
*Emails: swaruproy2013@gmail.com; jwrhim@khu.ac.kr

6.1 Introduction

Biomass is a naturally renewable resource that forms the basis of sustainable development.[1] Forest biomass, such as wood, has been used by humanity since ancient times as a versatile material with applications ranging from construction material to a source of energy as a fuel. Lignin is a complex organic polymer found in abundance in plants as a structural component.[2] Besides polysaccharides cellulose, hemicellulose, and pectin, which are the major contributors to the plant cell wall mass, lignin is also a major component acting as their binding agent, which holds them together,

[†] These authors contributed equally.

Biomaterials Science Series No. 16
Lignin-based Materials: Health Care and Medical Applications
Edited by Kuruvilla Joseph, Runcy Wilson, Gejo George and Saritha Appukuttan

Published by the Royal Society of Chemistry, www.rsc.org

giving rise to an organized structure (Figure 6.1).[3] Chemically, lignin is an alkyl-aromatic heteropolymer composed of three phenylpropane units, *i.e.*, guaiacyl (G), syringyl (S), and *p*-hydroxyphenyl (H), which are cross-linked *via* aryl ether linkages, *viz.*, β-O-4′ and carbon–carbon linkages such as β-5′, β-β′, and 5-5′ (Figure 6.2).[2,4] Lignin is synthesized in plants *via* an enzymatic pathway involving the deaminase enzyme. This enzyme converts the precursor amino acids, *i.e.*, phenylalanine and tyrosine, into cinnamic acid and *p*-coumaric acid.[4] These carboxylic acids are reduced into *p*-coumaryl alcohol and a methoxylated aromatic ring, respectively, where the latter ultimately converts to coniferyl alcohol and sinapyl alcohol. The three hydroxycinnamyl alcohols – *p*-coumaryl alcohol, coniferyl alcohol, and sinapyl alcohol – are finally converted to derivatives *p*-hydroxyphenyl (H), guaiacyl (G), and syringyl (S) monomeric units that form an amorphous lignin structure through a combinatorial radical coupling reaction.[2] Apart from this basic assembly, the lignin structure is further complicated by several additional components such as hydroxycinnamic acids and flavonoids that adorn the aromatic heteropolymer with additional linkages and chemical functionality.[2–4]

Almost 70 million tons of lignin are produced annually as a byproduct of the paper industry during cellulose extraction.[5] Due to its complex structure and heterogeneity, only a small percentage of this amount is utilized for value-added applications such as an industrial stabilizer, dispersant, and surfactant.[5] Lignin is most commonly used in the form of a low-cost fuel material to produce heat and power in industrial incinerators.[6] Besides its complex structure, its brittle nature and incompatibility with other polymer systems were constraints on creating lignin-based high-performance materials.[5] Nevertheless, its structural complexity also presents several modifiable active sites in the form of phenolic hydroxyl groups which can be exploited to chemically modify this bioresource to alter its properties based on end applications.[4,7] Also, the economic aspect of lignin and its properties such as high carbon content, high thermal stability, biodegradability, and stiffness have attracted significant government and industrial funding worldwide in the past decade.[3,5] This has led to many discoveries and developments in utilizing lignin as a functional material for various

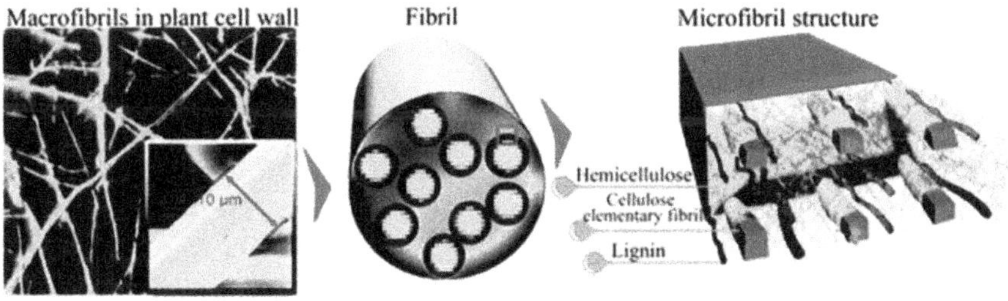

Figure 6.1 Schematic representation of lignin as a structural component within the lignocellulosic matrix of the plant cell wall. Adapted from ref. 5 with permission from the Royal Society of Chemistry.

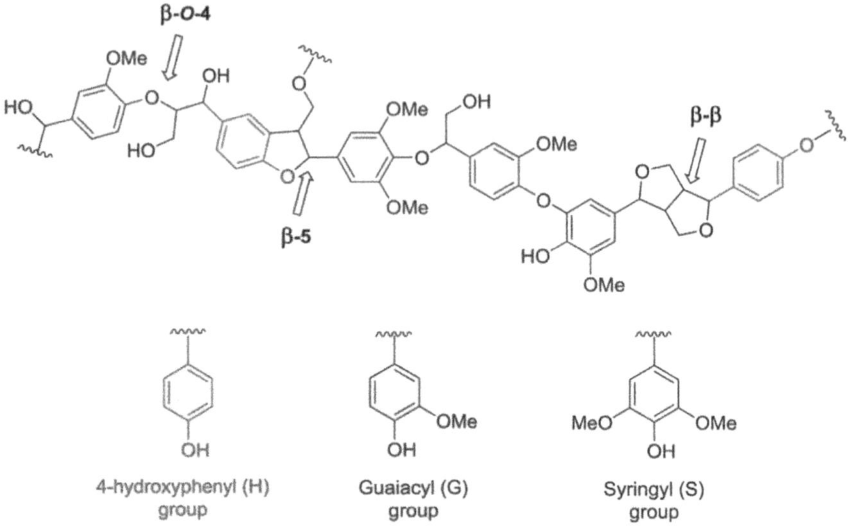

Figure 6.2 The lignin structure contains the most frequent bonds as well as the corresponding monomers that are part of its structure: 4-hydroxyphenyl (H), guaiacyl (G), and syringyl (S). Adapted from ref. 10, https://doi.org/10.1016/j.jbiotec.2016.08.011, under the terms of the CC BY 4.0 license https://creativecommons.org/licenses/by/4.0/.

applications. Its functionalities, such as antioxidant activity, have been utilized to develop biomedical and pharmaceutical materials.[4] Being structurally amphiphilic, lignin is also highly surface-active and has been applied to improve interfacial strength in the form of adhesives.[6,7] Its light absorption capacity in the UV-visible region has led to the development of anti-UV products and sunscreen agents.[4,8] The presence of various phenolic functional groups endows it with antimicrobial and antioxidant activity, which can be utilized for food packaging and preservation applications.[9] This chapter is dedicated to discussing the antioxidant and antimicrobial activities of lignin and their exploitation in food packaging and agricultural applications such as plant protection and growth.

6.2 Extraction and Isolation of Lignin

The efficient separation and/or extraction of high purity lignin from lignocellulose are critical for lignin valorization.[11,12] In nature, lignin is formed by the radical polymerization mechanism. Various studies suggested that the reactive components of lignin are attached to hemicellulose by covalent bonds, resulting in a lignin–carbohydrate complex (LCC); thus, lignin extraction is difficult.[13,14] Therefore, to extract lignin, splitting of LCC bonds and partial depolymerization of lignin are essential. Lignin breaks down into smaller fractions by the depolymerization process, which makes it easier to solubilize the lignin and separate it from the biomass. During lignin extraction, repolymerization reaction (new C–C bond formation) occurs due to the highly reactive functional groups on lignin and its fractions. This results in several lignin intermediates with different properties.[14] Generally, extracted lignin differs in structures owing to diverse extraction methods, feedstocks, and extraction severity.

At present, various industries such as paper and pulp and biorefinery processes (*e.g.*, kraft pulping, soda pulping, sulfite pulping, and hot water, alkaline, dilute acid, or enzymatic fractionation/hydrolysis) are large scale producers of lignin. The paper and pulp industry is a deep-rooted and vastly advanced industry with numerous established technologies. Subsequently, lignins (*e.g.*, soda lignin, kraft lignin, and lignosulfonates) extracted from these technologies have been extensively reviewed by researchers.[14,15] This section primarily emphasizes lignin extraction from biorefinery processes.

Biorefineries utilize biomass instead of fossil-based materials as a renewable resource to manufacture fuels and chemicals. Biorefineries rationally use all lignocellulose components, including lignin, hemicellulose, and cellulose. As specific lignocellulose constituents have apparent physical and chemical property variances, a separation/isolation process is typically required before selective conversion. The isolation of lignin, hemicellulose, and cellulose from the biomass matrix is usually carried out in a biomass pretreatment or fractionation stage.[14,16] Biomass pretreatment is the first step of biorefining directed at decreasing biomass recalcitrance and is critical for the downstream valorization. Several biomass pretreatment technologies have been devised, including alkaline hydrolysis, acid hydrolysis, oxidative/reductive catalytic fractionation, and solvolysis fractionation. Attributable to dissimilarities in fractionation approaches and severity, the same biomass feedstock could show variable lignin structure, lignin purity, and delignification efficiency.[13]

6.2.1 Acid Hydrolysis

Various concentrated acids (concentrated H_2SO_4, HCl, and HF) at room temperature (18–25 °C)[17] and dilute acids (0–5 wt% H_2SO_4, HCl, and H_3PO_4) at higher temperatures (120–240 °C)[18,19] have been used for acid hydrolysis of biomass. These methods aim to depolymerize cellulose and hemicellulose, whereas soluble lignin fragments could be obtained in low amounts. Under the acid hydrolysis conditions, the obtained lignin fragments are highly reactive, and due to this property, unwarranted condensation reactions occur. To stop the condensation reactions, researchers utilized a flow-through reactor. Owing to its design parameters, it can efficiently isolate the lignin from the heating zone. This approach aids in curtailing the condensation reactions.[20,21] The structural analysis of the lignin obtained from the flow-through reactor showed cleavage of β-O-4 linkages and occurrence of very limited lignin condensation. The major drawback of the flow-through reactor technique is high water consumption, which negatively affects the sugar concentration in downstream conversion. This drawback can be addressed by maneuvering the crucial aspects of the process, such as increasing mass and heat transfer and solid loading limitations. Various researchers reported the steps of lignin formation under dilute acid pretreatment conditions.

In the first step, droplets of lignin form at 120–200 °C, which is higher than lignin's phase temperature. In this step, lignin gets liquefied and amalgamated in the pretreatment mixture. Further, the droplets of lignin might transport from the cell wall matrix and deposit on the biomass surface.[22] Li *et al.* reported that droplets have a higher S/G ratio than original lignin.[23] In the second step, pseudo-lignin forms from the degraded products of lignin and other carbohydrates.[24] Here, concentrated acid hydrolysis (*e.g.*, 72 wt% H_2SO_4 hydrolysis at room temperatures) is used for converting carbohydrates into soluble sugars. Under concentrated acid hydrolysis conditions, lignin undergoes severe repolymerization reactions and degradation. The extracted lignin is extremely condensed and is known as Klason lignin. The calculation of Klason lignin is generally used to estimate the lignin content in biomass.[25] Due to its highly condensed microarchitecture, the selective conversion of Klason lignin to chemicals and/or fuels is challenging.

6.2.2 Alkaline Hydrolysis

The lignin extraction *via* alkaline hydrolysis involves the depolymerization of the polyphenolic macromolecule to separate it from the biomass matrix.[26] For this purpose, alkaline chemicals such as NaOH, $Ca(OH)_2$, and ammonia are used. Lignin oligomers, phenolic dimers, and monomers are formed due to partial hydrolysis of ether linkages. Several popular techniques are carried out for the pretreatment of biomass using ammonia or its aqueous solutions, including ammonia fiber expansion (AFEX),[27] anhydrous ammonia pretreatment (AAP),[28] and ammonia recycled percolation (ARP).[29] These pretreatments can alter the crystalline structure of cellulose, break the LCC bonds, and partially cleave lignin ether linkages, which facilitates the separation of lignin from the other biomass constituents. The recovery of ammonia due to its high volatility is an advantage of these technologies. The low treatment severity results in separation of lignin with a low condensation structure. However, the alkaline delignification process has relatively low efficiency.

6.2.3 Enzymatic Hydrolysis

The enzymatic hydrolysis process involves the treatment of biomass with cellulase and hemicellulase enzymes which convert polysaccharides into soluble oligo/monosaccharides, resulting in the generation of lignin-enriched solid residues.[28,30] Before this process, an additional physical or chemical pretreatment is usually required to deal with the recalcitrant structure of lignocellulosic biomass, which otherwise resists the enzymatic hydrolysis due to physical hindrance.[18] For this purpose, methods including ball-milling, hot water pretreatment, dilute acid or alkaline hydrolysis, AFEX, and mechanical refining have been used to increase the digestibility of plant polysaccharides by enzymes.[14] Due to the mild conditions during the enzymatic hydrolysis process, the type and severity of the pretreatment method dictate the purity and structural modification of lignin. Nevertheless, this process usually yields solid residues with relatively low lignin purity. Hence, post-treatment methods such as organic solvent extraction and mild acid hydrolysis are carried out to enhance the purity of the extracted lignin. These processes remove the residual carbohydrates, protein, and/or ash in the lignin-enriched solid residue.[31] Despite the complexity of this process, it is presumed to be crucial for future lignin valorization, given the rapidly increasing propensity towards developing biorefining processes.

6.2.4 Reductive or Oxidative Fractionation

Reductive or oxidative fractionation is a catalytic process involving cleaving LCC bonds and ether linkages in the lignocellulosic biomass using reducing (*e.g.*, H_2 or H_2-donating solvents) or oxidizing agents (*e.g.*, O_2, O_3, H_2O_2). The process is usually complemented with solvolytic extraction, *i.e.*, using organic solvents or their aqueous mixtures or acid/alkaline hydrolysis.[16,32–34] The process results in diverse lignin structures such as lignin oligomers, high molecular weight lignin, highly depolymerized and deoxygenated lignin oil, and lignin-derived acids. Solvolytic extraction is more common due to the high solubility of lignin in organic solvents, which results in high extraction efficiency. Lignin depolymerization and/or condensation occur when catalytic treatment is provided in organic solvents. Also, lignin structures can be modified in organic medium; for instance, lignin is etherified at the α-position of alkyl chains when it reacts with low molecular weight alcohol solvents.[35,36] Hence, the organic medium, type of catalyst, and pretreatment severity are crucial factors in determining the structure of the extracted lignin.

6.2.5 Combined Pretreatment Methods

Despite the development of various techniques, the efficient extraction of lignin remains a challenge. To deal with this issue, various research groups have tried combined methods.[11,19,37,38] Mechanocatalytic methods involve biomass pretreatment with acid catalysis and mechanical forces.[39,40] This combined treatment can efficiently convert the biomass into water-soluble oligomers of lignin and other polysaccharides, which after undergoing an additional catalyzed hydrolysis process yields monosaccharides along with lignin precipitates. Another combined method, known as the deacetylation and mechanical refining (DMR) method, involves removing acetyl groups by dilute alkaline treatment of biomass, followed by mechanical disc refining.[41,42] This process is followed by enzyme-catalyzed hydrolysis, which results in a high yield, usually over 80%. The isolated lignin is suitable for further conversion due to its less condensed structure. Other combined methods, such as those based on concentrated phosphoric acid plus hydrogen peroxide,[43] inorganic salts with calcium peroxide[38] or deep eutectic solvents,[34] deep eutectic solvent with microwave irradiation,[32] and electrolysis combined with ultra-sonication,[37] were also explored for lignin extraction.

6.3 Lignin Structure and Physicochemical Properties

Lignin is the important structural component of terrestrial plants, and to date, it has been considered to be the second most abundant polymer after cellulose.[44,45] In plant or wood lignocellulose, there are three primary components - cellulose, hemicellulose, and lignin (Figure 6.3a).[45] Lignin is the most significant and reliable renewable source of aromatic compounds.[45] Lignin is the smallest long-chain heterogeneous polymer. It is a complex amorphous polymer composed of phenylpropane linked with ether bonds.[14] The exact molecular weight of native lignin is yet to be explored. There is a concern owing to the structural variation that appeared during the isolation process of lignin. The approximate average molecular weight of native lignin is around 2500 to 10 000 g mol^{-1}.[46] Moreover, the lignin content and structure can vary based on the plant species. The three structural units of lignin are presented in Figure 6.3b. In general, lignin in softwood contains more coniferyl alcohol (~90%). In comparison, coniferyl alcohol (~25–50%) and sinapyl alcohol (~45–75%) are present in hardwood. On the other hand, in the case of grass, along with coniferyl and sinapyl alcohols, *p*-hydroxyphenyl (*p*-coumaryl alcohol) is also present, and its content is highest (~5–35%).[47–49] The monomer units of lignin are joined by many inter-lignin linkers (chemical bonds) (Figure 6.3c), which results in the formation of the phenylpyronoid polymer with strong resistance to its chemical and biochemical degradation.[44,50] Actually, several C–C and C–O bonds in the inter-lignin structure create difficulty in lignin degradation.[47] The key bond in lignin is β-O-4, which accounts for around 50–80% of all inter-unit bonds, and interestingly studies have shown that the β-O-4 bond can be more easily broken compared to the C–C bond.[51]

Lignin contains many phenolic hydroxyl groups, and as a result, it is soluble under alkaline conditions.[52] Increasing the proportion of phenolic -OH groups greatly increased the alkaline solubility of lignin.[53] The variation in the solubility of lignin comes from the various pulping methods. The condensation and degradation phenomena produce lignin with the soluble and insoluble parts. The soluble part is the amorphous component, while the non-soluble one is the raw material's fiber structure.[54]

Lignin is a thermoplastic amorphous polymer.[54] The glass transition temperature (T_g) of lignin varies from 90 to 170 °C depending on the type of lignin.[55] The T_g is highest for kraft lignin and lowest for organosolv lignin. The crystallinity, geometric factors, chain flexibility, molecular weight, cross-linking, *etc.*, are various factors that may affect the T_g.[52,56] Moreover,

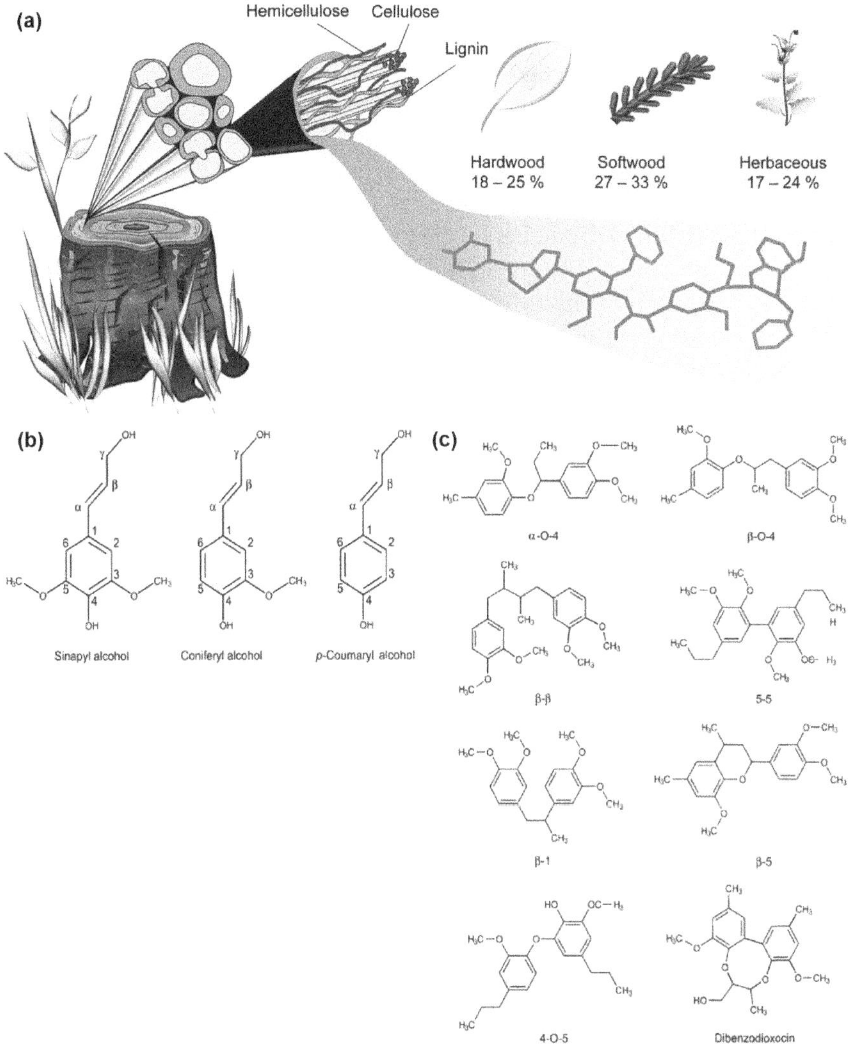

Figure 6.3 Distribution of lignin and its structure. (a) Lignocellulose contains cellulose, hemicellulose, and lignin, with variable lignin contents in herbs, hardwood, and softwood. Adapted from ref. 45 with permission from Elsevier, Copyright 2019. (b) The structural schematics of the three units in lignin are the syringyl group, guaiacyl group, and *p*-hydroxyphenyl group. (c) Type of inter-lignin bond consisting of three units. Adapted from ref. 44 with permission from Elsevier, Copyright 2016.

the thermal decomposition temperature of lignin also varies over a broad range from 200 to 500 °C owing to its complex structure, molecular weight, amorphousness, and non-uniformity.[57] Commonly, lignin breaks into many smaller fractions or shuffles its chain. As a result, there is a weight loss during thermal decomposition. It has been observed that softwood lignin is more stable than hardwood lignin.[52] Previously, it was shown that the main thermal degradation of lignin took place at a temperature of around 400 °C due to the degradation of phenolic hydroxyl groups.[52]

Lignin is a completely biodegradable polymer, and the degradation happens primarily under aerobic conditions with oxidative enzymes.[10,58] As lignin has a complex polymer structure with aromaticity, it is chemically resistant.[10] Even though many microbes can degrade plant cell walls and cellulose, not many microbes can degrade lignin. Few fungi and filamentous bacteria such as the *Basidiomycota* phylum, *Streptomyces*, and *Nocardia* can degrade lignin.[59] There are also lignin-degrading enzymes, like lignin peroxidase, which are found in the white rot fungus *Phanerochaete chrysosporium*.[60]

6.4 Functional Properties of Lignin

6.4.1 Antioxidant Properties

Lignin is a heteropolymeric aromatic compound containing various hydroxyl and methoxy groups. Due to these functional components, the oxidation reaction can be inhibited *via* hydrogen donation. Lignin is well known for its superb free-radical scavenging properties. Lignin can reduce oxygen radicals; moreover, it can prevent oxidation reactions.[61,62] It has been previously established that the antioxidant potential of lignin is very similar to commercial antioxidants such as butylated hydroxytoluene.[63] Alkali lignin contains hydroxyl-cinnamic acid derivatives, and is an excellent radical scavenger.

The antioxidant action of lignin primarily originates from the phenolic structures having oxygen-containing reactive radicals. Furthermore, studies have shown that the scavenging activity mainly comes from the non-etherified phenolic and aliphatic hydroxyl and methoxy groups that go through a proton-coupled electron transfer mechanism.[61] Studies have shown that for better antioxidant activity of lignin, lower molecular weight, low polydispersity, and low aliphatic hydroxyl groups are necessary factors.[62,64,65] The other minor factors affecting the antioxidant activity of lignin are various types of phenylpropane units and pi-conjugated systems. Due to the increase in the number of oxygen-containing groups in the side chains and pi-conjugated systems, the antioxidant activity of lignin gets reduced.[64] Recently, the strong antioxidant activity of lignin was used for making various polymer-based functional composite materials (Figure 6.4).[5] The incorporation of a small quantity of lignin (<10%) in polymer matrices tremendously improved its antioxidant activity, which could be very useful in food packaging applications. Lignin-based functional food packaging (coating or film) materials could be promising resources to prevent the loss of oxidation-prone foods' shelf-life and maintain the food quality. Although lignin is a very useful functional material, a recent report showed that higher quantities of lignin could show cytotoxic effects. As a result, it is prescribed that the content of lignin should be optimized before its use in food packaging.

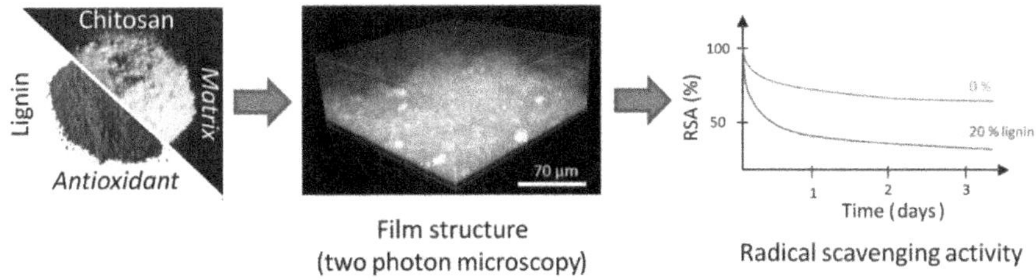

Figure 6.4 Antioxidant potential of lignin-incorporating chitosan-based composite films. Adapted from ref. 66 with permission from American Chemical Society, Copyright 2016.

6.4.2 Antimicrobial Activity

Lignin is a plant-derived polymer known for its good antimicrobial activity. The antimicrobial activity of lignin depends on the plant source and the extraction technique.[67] It was already shown that lignin derived from ten different sources and different isolation methods exhibited diverse antimicrobial activity towards Gram-negative and Gram-positive bacteria.[68] Basically, the antimicrobial activity of lignin comes from the phenolic compounds present in lignin.[5] It is already known that lignin is composed of monolignols like hydroxyphenyl, guaiacyl, and syringyl. These structures originate from precursor compounds: *p*-hydroxyphenyl from coumaryl alcohol, guaiacyl from coniferyl alcohol, and syringyl from sinapyl alcohol residues.[69] The various precursor compounds related to the antimicrobial activity of lignin are shown in Figure 6.2. The lignin functional components with oxygen-containing phenolic groups such as –OH, –CO, and –COOH in the side chain have less antimicrobial properties with respect to the isoeugenol phenolic fragment.[67,70] Lignin shows antimicrobial effects against food pathogenic microorganisms like *Escherichia coli* and *Listeria monocytogenes*.[71,72] The complete antimicrobial mechanism of lignin is yet to be elucidated. Still, it is believed that the bactericidal effects of lignin may originate from cell disruption. The schematic diagram proposed by Yang *et al.* (2018) is presented in Figure 6.5.[73] The antimicrobial action comes from phenolic compounds, and it has been earlier demonstrated that the phenolic types of compounds mainly disrupt the cell membrane, which results in the release of cell contents. Also, phenolic compounds can reduce adenosine triphosphate (ATP) production by infiltrating bacteria and reducing the pH.[74] The lignins extracted through kraft, sulfite, and organosolv techniques have shown variable antimicrobial activity, which has been extensively reviewed recently by Ndaba *et al.* (2020).[75] Recently, another lignin extraction technique, namely the soda process, is becoming popular due to the surface modification of lignin. Thus, as an ecofriendly and inexpensive polymer, lignin has a lot of scope as a potential antimicrobial agent.

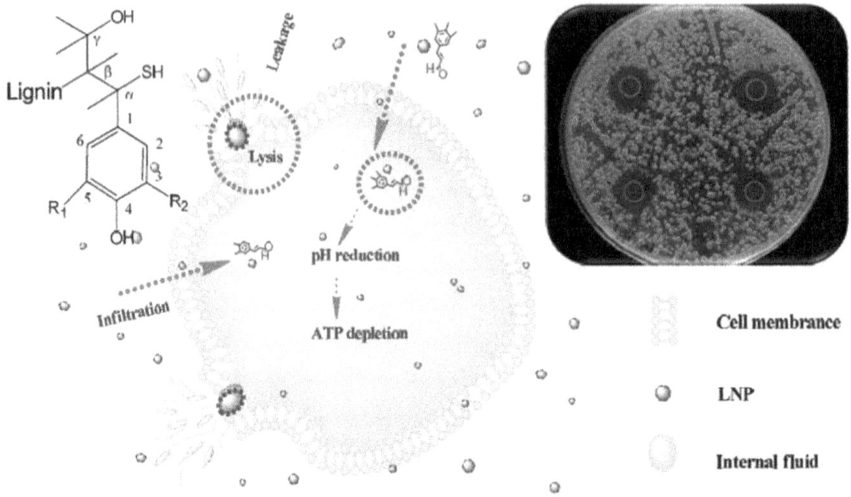

Figure 6.5 Proposed mechanism of action of lignin nanoparticles. Adapted from ref. 73 with permission from American Chemical Society, Copyright 2018.

6.5 Applications of Lignin and Lignin-based Composites

6.5.1 Food Packaging Applications

With the increasing trend for the consumption of packed food products, the demand for ready-to-eat and home meal replacement (HMR) food is continuously increasing. This calls for the design of new and more efficient food packaging systems that can preserve the food for a longer time and maintain its nutritional value.[7,76] Since microbial contamination and oxidation of food are the major factors that deteriorate the food quality,[77,78] the food packaging industry is increasingly focused on developing active packaging materials that involve the use of functional ingredients as filler materials or additives in packaging polymers.[78,79] However, the increasing consumer responsiveness towards synthetic functional materials is driving the research towards natural and safer alternatives, especially natural polyphenolic extracts.[76] As already discussed in the previous sections, lignin, a natural polyphenolic compound, exhibits several functional properties, making it a lucrative option to be used in this application regime.[7] Lignin has been used as an active filler in several polymeric films (Figure 6.6), including semi-synthetic polymers such as polyvinyl alcohol[80] and polylactic acid,[81] polysaccharides such as chitosan[66] and alginate,[82] proteins such as gelatin,[83] *etc.*

Both micro- and nanoforms of lignin have been used as active fillers in polymeric food packaging films.[84] Extrusion prepared PLA-based films

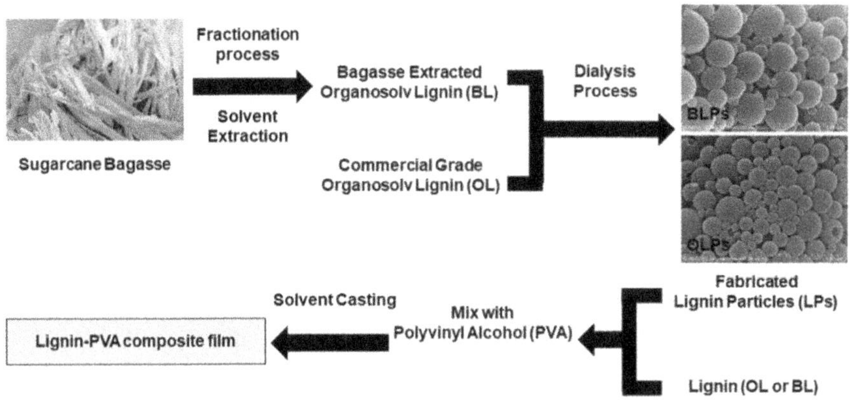

Figure 6.6 Lignin particles were prepared from sugarcane bagasse waste and used as functional filler materials in PVA-based food packaging films by Posoknistakul *et al.* Adapted from ref. 80 with permission from American Chemical Society, Copyright 2020.

incorporated with 1% lignin nanoparticles and 3% cellulose nanocrystals, by weight basis, effectively inhibited the growth of Gram-negative *Xanthomonas axonopodis* and *Xanthomonas arboricola*.[85] This control over plant pathogens was attained due to the antimicrobial action of nano-lignin. Lignin nanoparticles were also used with PVA/chitosan blend films to efficiently eradicate the bacterial contamination with Gram-negative plant/fruit pathogens.[86] Along with the antimicrobial effect, an appreciable antioxidant activity was also obtained for hydroxypropylmethylcellulose–chitosan–lignin films.[87] Also, adding grape seed lignin to polyhydroxybutyrate and polyhydroxyalkanoate films endowed them with antioxidant activity and enhanced their crystallinity and mechanical properties while reducing their gas permeability.[88]

Besides the use of nanoparticles of neat lignin, its combination with other polyphenolic active components has also been reported. Melanin is such a less-explored natural active polyphenolic component having significant antioxidant potential.[89–91] Core–shell nanoparticles of lignin–melanin were synthesized and integrated with PBAT-based films.[92] The nanocomposites not only displayed enhanced UV shielding, but also their mechanical performance was considerably ameliorated. Lignin was also used to synthesize hybrid nanoparticles with metal oxides such as zinc oxide (ZnO).[93] The use of ZnO/lignin hybrid nanoparticles in a polyurethane matrix resulted in an excellent UV-screening material with improved mechanical properties. Another approach that involved the synergistic enhancement of the antioxidant activity of lignin-based packaging was the grafting of metallic Au nanoparticles on lignocellulosic fibers in unbleached softwood pulp.[94] The fabricated nanocomposite paper displayed an antioxidant potential of above 98%, which is attributed to the synergistic effect provided by the catalytic activity of Au nanoparticles. Apart from these combinations, polyvinyl alcohol-based films incorporated with nano-lignin and boron nitride also displayed enhanced properties from a packaging perspective.[95] Besides using lignin as a filler material in the polymer matrix, there are reports where it is grafted in the polymer backbone, endowing it with antioxidant activity. The grafting of lignin in the chitosan chain was carried out, and the resulting lignin/chitosan grafted film was studied for its antioxidant potential.[96] The lignin/chitosan grafted film displayed considerable enhancement in the antioxidant potential. It was suggested to be a suitable material for active packaging to protect sensitive products, such as foods and cosmetics, against oxidation.

6.5.2 Applications in Plant Protection

It is known that the accumulation of protolignin increases in the cell walls of plants that are under biotic or abiotic stress. Cell walls play an important role in the primary defense system of plants against adverse conditions such as drought, temperature variation, pests, and salts.[4] When plants such as rhododendron are exposed to cold environments, the expression of the *p*-coumarate-3-hydroxylase (C3H) gene is increased, leading to increased lignin content, affecting the permeability and stiffness of the cell wall, thereby increasing cold resistance.[97] In a similar way, the Satsuma mandarin is also known for its increased lignin content under high temperature conditions.[98] Abiotic factors such as drought and salinity negatively affect plant growth and often result in death.[99] These two factors often affect the plant simultaneously and cause an osmotic imbalance in plant cells. Lignin regulates the water loss from plants by reducing the transpiration rate by decreasing cell wall permeability.[99] This increased lignin content under increased salinity and drought conditions has been studied in the stem of *Eucalyptus globulus* plants and *Leucaena* seedlings.[100,101]

Lignin also plays a paramount role in enhancing the pest resistance of plants. An increased expression of the *CmMYB19* gene, which is responsible for lignin production, was observed in chrysanthemum plants on aphid infection.[102] The increased lignin content limited the insect proliferation on the host. In another such observation, the insect-specific toxin peptide *LqhIT2* activated the phenylpropane pathway in rice, which enhanced the insect resistance in the plant by increasing the lignin content in cell walls.[103] Similarly, the resistance to root-knot nematodes in *Arabidopsis* increased with increased lignin content in roots.[104]

Besides the natural role of lignin in plant resistance and protection, the polyphenolic compound has also been used synthetically for this purpose. The antimicrobial activity of lignin nanoparticles against plant pathogens such as *Pseudomonas syringae* and *X. axonopodis* was observed at a low aqueous concentration of 8% w/v.[73] Besides, enzyme-responsive lignin nanocarriers (NCs) were fabricated and were loaded with fungicide pyraclostrobin with a high encapsulation efficiency.[105] These NCs were used to inject the trunk of *Vitis vinifera* grapevine as a curative and preventive

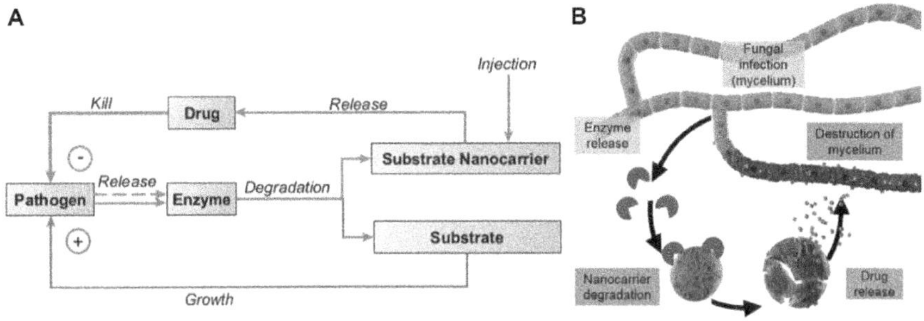

Figure 6.7 Concept of NC-mediated drug delivery: (A) top portion: degradation of a substrate leads to growth of the pathogen (infection, disease); and bottom portion: a negative feedback loop is initiated by masking a drug that fights the pathogen in NCs based on the food source of the pathogen. (B) Concept shown in (A) used to fight the fungal grapevine trunk disease "Esca": fungicide-loaded lignin NCs release the drug load only when the Esca fungi release lignin-degrading enzymes. Adapted from ref. 105, https://doi.org/10.1002/advs.201802315, under the terms of the CC BY 4.0 license https://creativecommons.org/licenses/by/4.0/.

treatment against Esca-related pathogens such as *Phaeoacremonium minimum* and *Phaeomoniella chlamydospora.* During the infection, the pathogens released the lignin-degrading enzymes, which also degraded the lignin-based encapsulation, thereby releasing the fungicides, hence killing the pathogens (Figure 6.7).[105] Since the fungicides were not released until the infection occurred, this strategy was found to be efficient for a prolonged period of 5 years. Several other researchers also carried out studies on enzyme-responsive lignin NCs, reporting similar long-term efficiency.[106–108] This strategy had the potential to reduce the environmental pollution induced by the overuse of chemical pesticides. Another reported strategy was using lignin–Cu organic–inorganic hybrid material as an agronomic antimicrobial agent.[109] This lignin-based antimicrobial material was tested on greenhouse-grown tomato crops. Its efficacy against *Rhizoctonia solani* was better than commercial pesticides containing copper(II) hydroxides.

6.5.3 Applications in Plant Growth

Lignin is an essential structural macromolecule in plants and plays an important role in normal plant growth and development. Abnormal development in plants, such as xylem collapse, stem bending, and growth retardation, have been observed on the inhibition of the lignin biosynthesis gene.[110] For instance, a considerable reduction in the seed germination rate was observed in mutant *Arabidopsis thaliana* with lower lignin content in the seed coat.[111] Further, a loss of apical dominance and retarded growth were observed in *Arabidopsis* mutants with lower lignin content.[112] The close association of lignin with cell expansion and growth in plants was also demonstrated on the decreased lignin content due to the overexpression and downregulation of the *OsPEX1* gene in rice.[113,114] Additionally, the proto-lignins in plant cells were found to have biological activities similar to humic substances that help in plant growth promotion.[115,116] Humic acid (HA) is a phytohormone that helps regulate plant growth by controlling its physiological metabolism. The biological activities of HA have a close relationship to its structure and functional groups,[117,118] and it is known that similar functional groups are present in the lignin structure.[4] The growth promotional effect of lignin has already been studied on tomato and maize seedlings.[119,120] The HA-like lignin isolated from giant reeds also displayed a similar positive effect on the development of tomato, watercress, and chicory seeds as displayed by phytohormone gibberellin.[121] The discovery of the HA-like bioactivity of lignin has generated interest in recovering lignocellulosic residues as plant growth regulators.

This discovery of HA-like lignin could increase the economic benefits of non-food and energy crops and play an important role in enhancing sustainable agriculture where natural substances may replace synthetic seed germination promoters and biological stimulants for plant growth.[4] The natural humification of plant lignin occurs due to its decomposition resulting from non-specific oxidation by fungi. Commercially, lignin sulfate can be subjected to Fenton oxidation for converting it into an HA-like plant fertilizer.[122] Such a commercially produced HA-like lignin variant could effectively promote the sprouting of *Arabidopsis* seeds and elevate the abiotic stress tolerance in plants (Figure 6.8). The stimulation in plant growth was similar to commercial HA or even better.[122] Besides affecting the endogenous hormonal balance, lignin affects the plant growth and development *via* other routes such as regulating photosynthesis, altering the meristematic ability of plant tissue, and varying nutrient absorption by roots.[4] A significant increase in the photosynthetic ability of nano-lignin treated maize plants was observed, which was attributed to elevated chlorophyll content.[123] Also, the effect of lignin on the mitotic activity of seeds was observed, which promoted seed germination.[124] Lignin also promoted phosphorus absorption by plant roots, thereby directly promoting plant growth and development.[125]

Besides lignin and lignin nanoparticles, lignin-based nanocomposites were also used for the delivery and sustained release applications in agriculture. Lignin nanocapsules (LNCs) were used to deliver bioactive compounds or plant hormones, such as gibberellin, to modulate plant growth and development.[126] Lignin-based materials have also been used for the controlled release of fertilizers.[127,128] An enzymatic polymerization approach was used to prepare lignosulfonate granules for sustained and controlled release of fertilizers.[129] The system was successfully used for the sustained release of potassium nitrate and potassium phosphate for over 10 days. Further, adding alginates into the lignosulfonate granules prolonged the sustained release for over 25 days.[129] These results unveiled the possibility of synthesizing completely biodegradable and sustainable

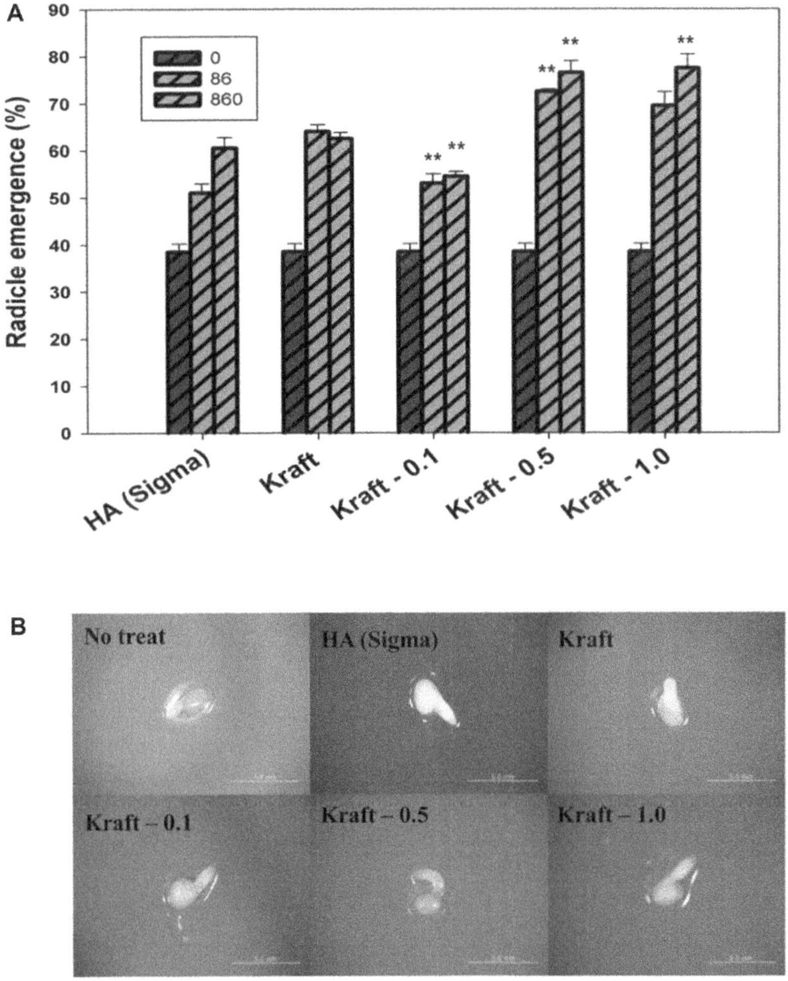

Figure 6.8 (A) Accelerated rates and (B) representative photographic images of *A. thaliana* germination in the presence of commercial HAs, kraft lignins, and their Fenton variants (86 or 860 mg L^{-1}). Data are means ± standard error (SE) ($n=3$). Significant differences are shown as asterisks (*$P<0.05$; **$P<0.01$; ***$P<0.001$). Abbreviations: HA, commercial HAs; kraft, untreated kraft lignin; kraft-0.1, lignin treated with 0.05 g of $FeSO_4 \cdot 7H_2O$; kraft-0.5, lignin treated with 0.25 g of $FeSO_4 \cdot 7H_2O$; kraft-1.0, lignin treated with 0.50 g of $FeSO_4 \cdot 7H_2O$. Adapted from ref. 122 with permission from American Chemical Society, Copyright 2018.

systems for fertilizer delivery. Another study in this regard reported the synthesis of a lignin/clay porous composite material for sustained release of urea.[7] The release studies demonstrated best results when the lignin/clay/urea concentration was 2 : 1 : 4. Furthermore, the composite material was capable of absorbing moisture from air, thereby facilitating the retention of water by dry soil.

6.6 Conclusions, Challenges, and Future Perspectives

In nature, lignin is present as a structural biopolymer in plant cells, providing them with strength and rigidity. The potential of this polyphenolic macromolecule has gone unnoticed and unexplored for a long time as the focus was always on the commercialization of paper as a bioresource product. Nevertheless, the high calorific value of lignin led to its utilization as a low-cost industrial fuel. With the gradual research progress in the past few years, other lucrative properties of lignin such as antioxidant and antimicrobial activity were discovered. These properties made this abundant biomass an economically active additive for various functional and high-performance materials, such as reinforcement fillers in food packaging polymers, amphiphilic surfactants in adhesives, and nanocarriers in biomedical and agricultural applications. Despite the emerging research interest in lignin's functional properties and biological activities, its complex nature and undefined chemical composition largely limit its application potential and scale-up possibilities. Depending on the biological origins and processing methods, there is a variation in the obtained lignin's heterogeneity, complexity, macromolecular structure, and molar mass. The structure differs in terms of the monomeric units and the interunit linkages. Besides, there is a diversity in the distribution of the functional groups that directly affect its functional properties, including its solubility and reactivity. To address this limitation, extensive research is required on the structure–property relationship of this polyphenolic macromolecule. Also, the extraction and pretreatment processes, which largely determine the lignin structure and hence its properties, should be developed and standardized.

Certainly, lignin has continued to emerge as a sustainable prospect for various applications from the industrial perspective. Successful trials have been conducted on integrating lignin with other materials in high-performance applications such as adhesives and thermoplastic materials. Extensive research is underway in energy storage applications where lignin is utilized as an electrolyte separator in batteries. Due to its high carbon content, it has also been considered as a sustainable precursor for carbon-based nanomaterials such as carbon nanotubes and nanodots. Its hydrophobic nature has been utilized as a waterproofing agent in paints and coatings. However, the accurate determination of the environmental impacts of lignin production and utilization processes is also necessary. The quantification of the environmental impact of a product, starting from its extraction to its end-of-life, is carried out by the life cycle assessment (LCA) process. While conducting LCA, factors such as eutrophication, water use, land use, and toxicity are most commonly considered. Yet, the main focus remains on determining the global warming potential (GWP). For instance, the reported GWP for the production of 1 kg lignin nanoparticles from wheat straw in the form of aqueous suspension ranges from 36.6 to 199 kg CO_2, which depends on virgin ethanol and thermal energy requirements during recovery. This GWP is much lower than 800 kg CO_2 produced during 1 kg nano-fibrillated cellulose production by the TEMPO oxidation process. Additionally, lignin can substantially reduce the industrial carbon footprint given its biodegradable nature, unlike petroleum-based plastic wastes, which are generally incinerated or landfilled, and give rise to additional CO_2 and CH_4 emissions even when recycled. Hence, compared to fossil-based products, lignin offers improved environmental performances.

Overall, lignin-centered biorefineries have a multi-output characteristic; *i.e.*, they give out many different product types, which makes the exact assessment relatively difficult. The continuously evolving technology and non-standardized methods make the assessment even more challenging. Nevertheless, in the era where the focus is on innovations in plant-derived materials, lignin-based materials are assumed to have immense potential for competing with fossil-based materials and mitigating the negative environmental impacts of high-performance functional materials. While lignin-based technologies are still in their very early stages of development, they are prospects with enormous potential.

Abbreviations

AAP	Anhydrous ammonia pretreatment
AFEX	Ammonia fiber expansion
ARP	Ammonia recycled percolation
ATP	Adenosine triphosphate
C3H	*p*-Coumarate-3-hydroxylase
DMR	Deacetylation and mechanical refining
GWP	Global warming potential
HA	Humic acid
HMR	Home meal replacement
LCA	Life cycle assessment
LCC	Lignin–carbohydrate complex
LNC	Lignin nanocapsule
NC	Lignin nanocarrier
PBAT	Poly(butylene adipate-*co*-terephthalate)
PLA	Polylactic acid
PVA	Polyvinyl alcohol
TEMPO	(2,2,6,6-Tetramethylpiperidin-1-yl)oxyl
T_g	Glass transition temperature

References

1. F. Bilgili, E. Koçak, Ü. Bulut and S. Kuşkaya, *Renewable Sustainable Energy Rev.*, 2017, **71**, 830–845.
2. R. Vanholme, B. De Meester, J. Ralph and W. Boerjan, *Curr. Opin. Biotechnol.*, 2019, **56**, 230–239.

3. R. Katahira, T. J. Elder and G. T. Beckham, in *RSC Energy and Environment Series*, Royal Society of Chemistry, 2018, vol. 2018, pp. 1–20.
4. F. Shu, B. Jiang, Y. Yuan, M. Li, W. Wu, Y. Jin and H. Xiao, *Biomacromolecules*, 2021, **22**, 4905–4918.
5. D. Kai, M. J. Tan, P. L. Chee, Y. K. Chua, Y. L. Yap and X. J. Loh, *Green Chem.*, 2016, **18**, 1175–1200.
6. D. Stewart, *Ind. Crops Prod.*, 2008, **27**, 202–207.
7. E. Lizundia, M. H. Sipponen, L. G. Greca, M. Balakshin, B. L. Tardy, O. J. Rojas and D. Puglia, *Green Chem.*, 2021, **23**, 6698–6760.
8. P. Widsten, *Cosmetics*, 2020, **7**, 85.
9. B. Rukmanikrishnan, S. Ramalingam, S. K. Rajasekharan, J. Lee and J. Lee, *Int. J. Biol. Macromol.*, 2020, **153**, 55–62.
10. G. de Gonzalo, D. I. Colpa, M. H. M. Habib and M. W. Fraaije, *J. Biotechnol.*, 2016, **236**, 110–119.
11. S. S. Hassan, G. A. Williams and A. K. Jaiswal, *Bioresour. Technol.*, 2018, **262**, 310–318.
12. D. Wang, F. Shen, G. Yang, Y. Zhang, S. Deng, J. Zhang, Y. Zeng, T. Luo and Z. Mei, *Bioresour. Technol.*, 2018, **249**, 117–124.
13. A. J. Ragauskas, G. T. Beckham, M. J. Biddy, R. Chandra, F. Chen, M. F. Davis, B. H. Davison, R. A. Dixon, P. Gilna, M. Keller, P. Langan, A. K. Naskar, J. N. Saddler, T. J. Tschaplinski, G. A. Tuskan and C. E. Wyman, *Science*, 2014, **344**, 1246843.
14. W. Schutyser, T. Renders, S. Van Den Bosch, S. F. Koelewijn, G. T. Beckham and B. F. Sels, *Chem. Soc. Rev.*, 2018, **47**, 852–908.
15. J. Zakzeski, P. C. A. Bruijnincx, A. L. Jongerius and B. M. Weckhuysen, *Chem. Rev.*, 2010, **110**, 3552–3599.
16. K. H. Kim, T. Dutta, J. Sun, B. Simmons and S. Singh, *Green Chem.*, 2018, **20**, 809–815.
17. A. Shahbazi and B. Zhang, in *Bioalcohol Production: Biochemical Conversion of Lignocellulosic Biomass*, Woodhead Publishing, 2010, pp. 143–158.
18. V. A. Thomas, B. S. Donohoe, M. Li, Y. Pu, A. J. Ragauskas, R. Kumar, T. Y. Nguyen, C. M. Cai and C. E. Wyman, *Biotechnol. Biofuels*, 2017, **10**, 1–13.
19. W. Liu, W. Chen, Q. Hou, S. Wang and F. Liu, *RSC Adv.*, 2018, **8**, 10207–10214.
20. D. D. Laskar, J. Zeng, L. Yan, S. Chen and B. Yang, *Ind. Crops Prod.*, 2013, **50**, 391–399.
21. S. Bhagia, H. Li, X. Gao, R. Kumar and C. E. Wyman, *Biotechnol. Biofuels*, 2016, **9**, 1–15.
22. M. J. Selig, S. Viamajala, S. R. Decker, M. P. Tucker, M. E. Himmel and T. B. Vinzant, *Biotechnol. Prog.*, 2007, **23**, 1333–1339.
23. H. Li, Y. Pu, R. Kumar, A. J. Ragauskas and C. E. Wyman, *Biotechnol. Bioeng.*, 2014, **111**, 485–492.
24. P. Sannigrahi, D. H. Kim, S. Jung and A. Ragauskas, *Energy Environ. Sci.*, 2011, **4**, 1306–1310.

25. A. B. Ibáñez and S. Bauer, *Biomass Bioenergy*, 2014, **68**, 75–81.
26. J. Kim, M. Sunagawa, S. Kobayashi, T. Shin and C. Takayama, *Neurosci. Res.*, 2016, **105**, 42–48.
27. M. W. Lau, C. Gunawan and B. E. Dale, *Biotechnol. Biofuels*, 2009, **2**, 1–11.
28. C. G. Yoo, C. W. Lee and T. H. Kim, *Appl. Biochem. Biotechnol.*, 2014, **174**, 2639–2651.
29. T. H. Kim, R. Gupta and Y. Y. Lee, *Methods Mol. Biol.*, 2009, **581**, 79–91.
30. B. Jiang, T. Cao, F. Gu, W. Wu and Y. Jin, *ACS Sustainable Chem. Eng.*, 2017, **5**, 342–349.
31. A. Tolbert, H. Akinosho, R. Khunsupat, A. K. Naskar and A. J. Ragauskas, *Biofuels, Bioprod. Biorefin.*, 2014, **8**, 836–856.
32. Z. Chen and C. Wan, *Bioresour. Technol.*, 2018, **250**, 532–537.
33. M. V. Galkin and J. S. M. Samec, *ChemSusChem*, 2016, **9**, 1544–1558.
34. Y. L. Loow, T. Y. Wu, G. H. Yang, L. Y. Ang, E. K. New, L. F. Siow, J. Md. Jahim, A. W. Mohammad and W. H. Teoh, *Bioresour. Technol.*, 2018, **249**, 818–825.
35. C. S. Lancefield, I. Panovic, P. J. Deuss, K. Barta and N. J. Westwood, *Green Chem.*, 2017, **19**, 202–214.
36. T. Renders, S. Van Den Bosch, S. F. Koelewijn, W. Schutyser and B. F. Sels, *Energy Environ. Sci.*, 2017, **10**, 1551–1557.
37. G. Kumar, P. Sivagurunathan, G. Zhen, T. Kobayashi, S. H. Kim and K. Xu, *Bioresour. Technol.*, 2017, **245**, 196–200.
38. R. G. Saratale, H. S. Shin, G. S. Ghodake, G. Kumar, M. K. Oh and G. D. Saratale, *Bioresour. Technol.*, 2018, **258**, 26–32.
39. G. Calvaruso, M. T. Clough and R. Rinaldi, *Green Chem.*, 2017, **19**, 2803–2811.
40. M. Kessler, R. T. Woodward, N. Wong and R. Rinaldi, *ChemSusChem*, 2018, **11**, 552–561.
41. X. Chen, J. Shekiro, T. Pschorn, M. Sabourin, L. Tao, R. Elander, S. Park, E. Jennings, R. Nelson, O. Trass, K. Flanegan, W. Wang, M. E. Himmel, D. Johnson and M. P. Tucker, *Biotechnol. Biofuels*, 2014, **7**, 1–12.
42. X. Chen, W. Wang, P. Ciesielski, O. Trass, S. Park, L. Tao and M. P. Tucker, *ACS Sustainable Chem. Eng.*, 2016, **4**, 324–333.
43. Q. Wang, Z. Wang, F. Shen, J. Hu, F. Sun, L. Lin, G. Yang, Y. Zhang and S. Deng, *Bioresour. Technol.*, 2014, **166**, 420–428.
44. O. Y. Abdelaziz, D. P. Brink, J. Prothmann, K. Ravi, M. Sun, J. García-Hidalgo, M. Sandahl, C. P. Hulteberg, C. Turner, G. Lidén and M. F. Gorwa-Grauslund, *Biotechnol. Adv.*, 2016, **34**, 1318–1346.
45. J. Becker and C. Wittmann, *Biotechnol. Adv.*, 2019, **37**, 107360.
46. Z. Anwar, M. Gulfraz and M. Irshad, *J. Radiat. Res. Appl. Sci.*, 2014, 7, 163–173.
47. A. Duval and M. Lawoko, *React. Funct. Polym.*, 2014, **85**, 78–96.
48. M. Lei, S. Wu, J. Liang and C. Liu, *J. Anal. Appl. Pyrolysis*, 2019, **138**, 249–260.

49. G. L. F. Gellerstedt and E. G. Henriksson, in *Monomers, Polymers and Composites from Renewable Resources*, Elsevier, 2008, pp. 201–224.
50. Z. Chen and C. Wan, *Renewable Sustainable Energy Rev.*, 2017, **73**, 610–621.
51. S. Van den Bosch, S. F. Koelewijn, T. Renders, G. Van den Bossche, T. Vangeel, W. Schutyser and B. F. Sels, *Top. Curr. Chem.*, 2018, **376**, 1–40.
52. O. Faruk and M. Sain, *Lignin in Polymer Composites*, Elsevier Inc., Waltham, MA, USA, 1st edn, 2016, vol. 1.
53. E. I. Evstigneev, *Russ. J. Appl. Chem.*, 2011, **84**, 1040–1045.
54. H. Chen, *Biotechnology of lignocellulose: Theory and practice*, Springer Netherlands, 2014.
55. W. G. Glasser, *ACS Symp. Ser.*, 1999, **742**, 216–238.
56. H. Hatakeyama and T. Hatakeyama, in *Advances in Polymer Science*, Springer, Berlin, Heidelberg, 2009, vol. 232, pp. 1–63.
57. M. Brebu and C. Vasile, *Cellul. Chem. Technol.*, 2010, **44**, 353–363.
58. T. Furukawa, F. O. Bello and L. Horsfall, *Front. Biol.*, 2014, **9**, 448–471.
59. A. M. Abdel-Hamid, J. O. Solbiati and I. K. O. Cann, in *Advances in Applied Microbiology*, Academic Press, 2013, vol. 82, pp. 1–28.
60. K. Piontek, A. T. Smith and W. Blodig, *Biochem. Soc. Trans.*, 2001, **29**, 111–116.
61. F. J. Lu, L. H. Chu and R. J. Gau, *Nutr. Cancer*, 1998, **30**, 31–38.
62. T. Dizhbite, G. Telysheva, V. Jurkjane and U. Viesturs, *Bioresour. Technol.*, 2004, **95**, 309–317.
63. M. Azadfar, A. H. Gao and S. Chen, *Int. J. Biol. Macromol.*, 2015, **75**, 58–66.
64. J. Ponomarenko, T. Dizhbite, M. Lauberts, A. Volperts, G. Dobele and G. Telysheva, *J. Anal. Appl. Pyrolysis*, 2015, **113**, 360–369.
65. V. Ugartondo, M. Mitjans and M. P. Vinardell, *Bioresour. Technol.*, 2008, **99**, 6683–6687.
66. K. Crouvisier-Urion, P. R. Bodart, P. Winckler, J. Raya, R. D. Gougeon, P. Cayot, S. Domenek, F. Debeaufort and T. Karbowiak, *ACS Sustainable Chem. Eng.*, 2016, **4**, 6371–6381.
67. K. Sriroth and J. Sunthornvarabhas, *Biochem. Pharmacol.*, 2018, 7, 501–2167.
68. Ł. Klapiszewski, T. Rzemieniecki, M. Krawczyk, D. Malina, M. Norman, J. Zdarta, I. Majchrzak, A. Dobrowolska, K. Czaczyk and T. Jesionowski, *Colloids Surf., B*, 2015, **134**, 220–228.
69. B. Ndaba, I. Chiyanzu, S. Marx and G. Obiero, *Biomass Bioenergy*, 2014, **71**, 350–356.
70. M. Hyldgaard, T. Mygind, R. Piotrowska, M. Foss and R. L. Meyer, *Front. Microbiol.*, 2015, **6**, 754.
71. S. Nandi, J. J. Maurer, C. Hofacre and A. O. Summers, *Proc. Natl. Acad. Sci.*, 2004, **101**, 7118–7122.
72. M. Guo, T. Jin, N. P. Nghiem, X. Fan, P. X. Qi, C. H. Jang, L. Shao and C. Wu, *Appl. Biochem. Biotechnol.*, 2018, **184**, 350–365.

73. W. Yang, E. Fortunati, D. Gao, G. M. Balestra, G. Giovanale, X. He, L. Torre, J. M. Kenny and D. Puglia, *ACS Sustainable Chem. Eng.*, 2018, **6**, 3502–3514.
74. S. Dumitriu and V. Popa, *Polymeric biomaterials: Structure and function*, CRC Press, 2013, vol. 1.
75. B. Ndaba, A. Roopnarain, M. O. Daramola and R. Adeleke, *Sustainable Chem. Pharm.*, 2020, **18**, 100342.
76. R. Priyadarshi, S. Roy, T. Ghosh, D. Biswas and J.-W. Rhim, *Sustainable Mater. Technol.*, 2021, e00353.
77. S. Roy, R. Priyadarshi and J.-W. Rhim, *Foods*, 2021, **10**, 2789.
78. R. Priyadarshi, S.-M. Kim and J.-W. Rhim, *Sustainable Mater. Technol.*, 2021, **29**, e00325.
79. R. Priyadarshi, Z. Riahi, J.-W. Rhim, S. Han and S.-G. Lee, *ACS Appl. Nano Mater.*, 2021, **4**, 14292–14302.
80. P. Posoknistakul, C. Tangkrakul, P. Chaosuanphae, S. Deepentham, W. Techasawong, N. Phonphirunrot, S. Bairak, C. Sakdaronnarong and N. Laosiripojana, *ACS Omega*, 2020, **5**, 20976–20982.
81. S. Shankar, J. W. Rhim and K. Won, *Int. J. Biol. Macromol.*, 2018, **107**, 1724–1731.
82. K. R. Aadil, D. Prajapati and H. Jha, *Food Packag. Shelf Life*, 2016, **10**, 25–33.
83. R. Núñez-Flores, B. Giménez, F. Fernández-Martín, M. E. López-Caballero, M. P. Montero and M. C. Gómez-Guillén, *Food Hydrocolloids*, 2013, **30**, 163–172.
84. S. Beisl, A. Friedl and A. Miltner, *Int. J. Mol. Sci.*, 2017, **18**, 2367.
85. W. Yang, E. Fortunati, F. Dominici, G. Giovanale, A. Mazzaglia, G. M. Balestra, J. M. Kenny and D. Puglia, *Int. J. Biol. Macromol.*, 2016, **89**, 360–368.
86. W. Yang, J. S. Owczarek, E. Fortunati, M. Kozanecki, A. Mazzaglia, G. M. Balestra, J. M. Kenny, L. Torre and D. Puglia, *Ind. Crops Prod.*, 2016, **94**, 800–811.
87. A. Alzagameem, S. E. Klein, M. Bergs, X. T. Do, I. Korte, S. Dohlen, C. Hüwe, J. Kreyenschmidt, B. Kamm, M. Larkins and M. Schulze, *Polymers*, 2019, **11**, 670.
88. P. Vostrejs, D. Adamcová, M. D. Vaverková, V. Enev, M. Kalina, M. Machovsky, M. Šourková, I. Marova and A. Kovalcik, *RSC Adv.*, 2020, **10**, 29202–29213.
89. S. Roy, L. Van Hai, H. C. Kim, L. Zhai and J. Kim, *Carbohydr. Polym.*, 2020, **231**, 115729.
90. S. Roy and J. W. Rhim, *J. Nanomater.*, 2019, **2019**, 2840517.
91. S. Roy and J.-W. Rhim, *Food Hydrocolloids*, 2021, **118**, 106823.
92. Q. Xing, P. Buono, D. Ruch, P. Dubois, L. Wu and W. J. Wang, *ACS Sustainable Chem. Eng.*, 2019, **7**, 4147–4157.
93. H. Wang, W. Lin, X. Qiu, F. Fu, R. Zhong, W. Liu and D. Yang, *ACS Sustainable Chem. Eng.*, 2018, **6**, 3696–3705.
94. N. Bumbudsanpharoke and S. Ko, *Int. J. Biol. Macromol.*, 2018, **107**, 1782–1791.

95. X. Wang, S. L. Ji, X. Q. Wang, H. Y. Bian, L. R. Lin, H. Q. Dai and H. Xiao, *J. Mater. Chem. C*, 2019, 7, 14159–14169.
96. K. Crouvisier-Urion, F. Regina Da Silva Farias, S. Arunatat, D. Griffin, M. Gerometta, J. R. Rocca-Smith, G. Weber, N. Sok and T. Karbowiak, *Green Chem.*, 2019, **21**, 4633–4641.
97. H. Wei, A. L. Dhanaraj, R. Arora, L. J. Rowland, Y. Fu and L. Sun, *Plant, Cell Environ.*, 2006, **29**, 558–570.
98. Z. Yun, H. Gao, P. Liu, S. Liu, T. Luo, S. Jin, Q. Xu, J. Xu, Y. Cheng and X. Deng, *BMC Plant Biol.*, 2013, **13**, 1–16.
99. P. K. Agarwal, P. S. Shukla, K. Gupta and B. Jha, *Mol. Biotechnol.*, 2012, **54**, 102–123.
100. J. Moura-Sobczak, U. Souza and P. Mazzafera, *BMC Proc.*, 2011, **5**, 1.
101. S. Srivastava, R. K. Vishwakarma, Y. A. Arafat, S. K. Gupta and B. M. Khan, *Physiol. Mol. Biol. Plants*, 2015, **21**, 197–205.
102. Y. Wang, L. Sheng, H. Zhang, X. Du, C. An, X. Xia, F. Chen, J. Jiang and S. Chen, *Int. J. Mol. Sci.*, 2017, **18**, 619.
103. X. Tianpei, D. Li, P. Qiu, J. Luo, Y. Zhu and S. Li, *Plant Sci.*, 2015, **230**, 1–11.
104. T. Fujimoto, T. Mizukubo, H. Abe and S. Seo, *Mol. Plant-Microbe Interact.*, 2015, **28**, 398–407.
105. J. Fischer, S. J. Beckers, D. Yiamsawas, E. Thines, K. Landfester, F. R. Wurm, J. Fischer, E. Thines, S. J. Beckers, D. Yiamsawas, K. Landfester and F. R. Wurm, *Adv. Sci.*, 2019, **6**, 1802315.
106. T. O. MacHado, S. J. Beckers, J. Fischer, B. Müller, C. Sayer, P. H. H. De Araújo, K. Landfester and F. R. Wurm, *Biomacromolecules*, 2020, **21**, 2755–2763.
107. S. Peil, S. J. Beckers, J. Fischer and F. R. Wurm, *Mater. Today Bio*, 2020, 7, 100061.
108. M. H. Sipponen, H. Lange, C. Crestini, A. Henn and M. Österberg, *ChemSusChem*, 2019, **12**, 2039–2054.
109. V. Sinisi, P. Pelagatti, M. Carcelli, A. Migliori, L. Mantovani, L. Righi, G. Leonardi, S. Pietarinen, C. Hubsch and D. Rogolino, *ACS Sustainable Chem. Eng.*, 2019, 7, 3213–3221.
110. J. Yoon, H. Choi and G. An, *J. Integr. Plant Biol.*, 2015, **57**, 902–912.
111. M. Liang, E. Davis, D. Gardner, X. Cai and Y. Wu, *Planta*, 2006, **224**, 1185–1196.
112. A. L. Schilmiller, J. Stout, J. K. Weng, J. Humphreys, M. O. Ruegger and C. Chapple, *Plant J.*, 2009, **60**, 771–782.
113. L. A. Rogers and M. M. Campbell, *New Phytol.*, 2004, **164**, 17–30.
114. S. Ke, X. Luan, J. Liang, Y. H. Hung, T. F. Hsieh and X. Q. Zhang, *Plant Mol. Biol.*, 2019, **100**, 151–161.
115. C. Xiao, R. Bolton and W. L. Pan, *Bioresour. Technol.*, 2007, **98**, 1482–1488.
116. D. Savy, V. Cozzolino, A. Nebbioso, M. Drosos, A. Nuzzo, P. Mazzei and A. Piccolo, *Plant Soil*, 2016, **402**, 221–233.
117. P. Calvo, L. Nelson and J. W. Kloepper, *Plant Soil*, 2014, **383**, 3–41.

118. L. P. Canellas, R. Spaccini, A. Piccolo, L. B. Dobbss, A. L. Okorokova-Façanha, G. D. A. Santos, F. L. Olivares and A. R. Façanha, *Soil Sci.*, 2009, **174**, 611–620.
119. A. Balas and V. I. Popa, *BioResources*, 2007, **2**, 363–370.
120. D. Savy, V. Cozzolino, G. Vinci, A. Nebbioso and A. Piccolo, *Molecules*, 2015, **20**, 19958–19970.
121. D. Savy, L. Canellas, G. Vinci, V. Cozzolino and A. Piccolo, *J. Plant Growth Regul.*, 2017, **36**, 995–1001.
122. H. J. Jeong, J. Y. Cha, J. H. Choi, K. S. Jang, J. Lim, W. Y. Kim, D. C. Seo and J. R. Jeon, *ACS Omega*, 2018, **3**, 7441–7453.
123. D. Del Buono, F. Luzi and D. Puglia, *Nanomaterials*, 2021, **11**, 846.
124. V. I. Popa, M. Dumitru, I. Volf and N. Anghel, *Ind. Crops Prod.*, 2008, **27**, 144–149.
125. T. Lv, H. Yang, R. Zhang, W. Fan, Y. Xu, H. Cao, L. Ning, C. Zhou, L. Wang, T. Lv, H. Yang, R. Zhang, W. Fan, Y. Xu, H. Cao, L. Ning, C. Zhou and L. Wang, *Agric. Sci.*, 2017, **8**, 341–347.
126. S. Falsini, I. Clemente, A. Papini, C. Tani, S. Schiff, M. C. Salvatici, R. Petruccelli, C. Benelli, C. Giordano, C. Gonnelli and S. Ristori, *ACS Sustainable Chem. Eng.*, 2019, 7, 19935–19942.
127. R. Weiss, E. Ghitti, M. Sumetzberger-Hasinger, G. M. Guebitz and G. S. Nyanhongo, *ACS Omega*, 2020, **5**, 4322–4329.
128. J. Chen, X. Fan, L. Zhang, X. Chen, S. Sun and R. C. Sun, *ChemSusChem*, 2020, **13**, 4356–4366.
129. D. Legras-Lecarpentier, K. Stadler, R. Weiss, G. M. Guebitz and G. S. Nyanhongo, *ACS Sustainable Chem. Eng.*, 2019, 7, 12621–12628.

CHAPTER 7

Lignin and Its Composites for Drug Delivery Applications

LUC ZONGO[a] AND HEIKO LANGE*[b]

[a] Department of Pharmacy, University of Saint Dominic of West Africa (USDAO), Doulougou, Burkina Faso; [b] Department of Environmental and Earth Sciences, University of Milano-Bicocca, Milan, Italy
*Email: heiko.lange@unimib.it

7.1 Biomass-based Polyphenols and Their Potential for Pharmaceutical Applications

Polyphenolic secondary metabolites represent an interesting renewable pool of a variety of structurally different molecules, oligomers, and polymers with specific intrinsic characteristics that are interesting for the biomedical and pharmacological sectors; these polyphenols comprise lignins, lignans, and tannins.[1,2] Traditionally lignans and tannins have already their applications in the biomedical and pharmacological sectors, often in the form of the active ingredient in traditional treatments.[2–4] Lignin, the truly polymeric representative, has yet to make its dedicated entry into these fields. The idea of using lignin for pharmaceutical and biomedical applications remains fascinating,[5–7] and the following paragraphs will discuss some important characteristics of lignins that impact their use, studies regarding the uses of lignins in various forms in the biomedical and pharmacological sectors, and the remaining and emerging challenges.

Biomaterials Science Series No. 16
Lignin-based Materials: Health Care and Medical Applications
Edited by Kuruvilla Joseph, Runcy Wilson, Gejo George and Saritha Appukuttan

Published by the Royal Society of Chemistry, www.rsc.org

7.2 Lignin – Occurrence, Isolation, Structural Features, and Refinements

7.2.1 Occurrence and Isolation

Although lignin is becoming ever more widely available in course of the establishment of different biorefinery schemes alongside the traditional pulp and paper industries,[8,9] it is still not exploited to its fullest potential when it comes to designing a holistic valorisation of biomass within the claimed green transition of modern societies. Lignin is the second most abundant biopolymer produced in land-based plants; it appears as a comparably complex natural polymer. Its straightforward use is generally hampered by difficulties that reach back to its structural features, which compared to cellulose and other biopolymers are more complex.[1,10] Lignin, with its varying peculiar polymer characteristics, remains a challenge with respect to standard valorisations that target bulk properties in simpler materials applications.

In order to deduce applications in higher value areas beyond the materials field, an initial effort is necessary to structurally map lignin samples intended for use. The state-of-the-art methods as listed in the following allow deep structural insights, but would still need further refinement, correlation, and independent validation also by advanced *in silico* methods, in order to render the process of acquiring structural data of lignin easier and more applicable also for non-lignin experts. Reliable and understood structural information is necessary for being able to understand then the reactivity of the analysed isolated lignin, and to delineate which chemistries are suitable for eventual derivatisations of lignins with respect to a specific application.

The structural characterisation and partial structural control during isolation of lignin from biomasses is still an active field of research, and refined measurements and the development of novel techniques allow regularly for structural refinements that improve the structural picture of this natural polymer and the emerging application potential. Several reviews and monographs can be consulted to get a head start at least on a theoretical level.[11–16]

Lignin as a hydrophobic polymer, summing up to 15–35% of the dry mass of lignocellulosic material,[9,17–19] is present mainly in plant cell walls, and chemically and physically interacts with hemicelluloses, thereby giving rise to so-called lignin carbohydrate complexes (LCCs)[20] and cellulose,[18] respectively.

In order to fulfil its manifold functions for the plant, lignin *in planta* is distributed within the different parts of the plant, *i.e.*, between the stem, branching points, branches, and leaves, and between the different walls of the plant cells themselves.[9,21]

Lignin does not display a singular regular repetition motif, or more or less regular sequences of motifs connecting the various monomers (*vide infra*).

This is still a surprising and difficult to cope with aspect, especially when considering the close interaction between lignin and the two other more regularly structured main structural plant polymers, *i.e.*, cellulose and hemicelluloses. Furthermore, its structure changes not only between different plants, but also within the same plant depending on where it fulfils its function for growth, stability, protection, *etc.*[1,21]

The difficulty in understanding the complex structure of lignin emerges not only from the fact that the physical and chemical means used within isolation processes can give rise to structural modification of lignins, as discussed later, but also from the yet poorly understood biosynthetic pathway of lignin formation. It is yet to be elucidated whether there is really no enzymatic control that determines the structure of lignin in plants, or whether lignin is allowed to form according microenvironmental conditions at the site of chain propagation. Momentarily, lignin is believed within the community to display by and large random sequences of different interunit bonding motifs (see Figure 7.1, *vide infra*),[22] exhibiting thus very different polymer aspects and characteristics with respect to carbohydrate-based biopolymers.

Lignin formation in cells has been proposed to be a post-mortem process in plant cells.[23,24] The three monolignols *p*-coumaryl alcohol (H-type lignin), coniferyl alcohol (G-type lignin), and sinapyl alcohol (S-type lignin), which are synthesised by plants starting from L-tyrosine or L-phenylalanine,[25–28] are consumed in radical polymerisations, which in turn are initiated by enzymatic generation of phenoxy radicals. Different natural sources in combination with different biomass fractionation and isolation methods will result in varying ratios of the different monomer types incorporated into the lignin, as well as the nature and the distribution of the interunit bonding motifs *via* which the three phenol types are linked.[17,25,26,28,29]

Three different general types of lignin are distinguished: (i) softwood lignin, which contains around 95% and more G-type monomers and only a small percentage of H-type monomers; (ii) hardwood lignin, which shows all three types of monomers in the typical ratio H/G/S = 1/1/3; and (iii) herbaceous lignin, which contains as well all three monomers, but with enrichment in H-type monomers.[18]

Oxidative coupling of the monomers, initiated, as stated, *via* phenoxy radicals and proceeding *via* direct couplings and/or couplings after relocalisation of the radical across the resonance system of the monomers, leads to various binding motifs. Positions O-4, C-1, C-3, C-5, and C-β (IUPAC numbering) are primarily involved in natural polymerisation events. Based on theoretically possible coupling modes, it is still controversial whether lignin *in planta* is a branched and thus a three-dimensional polymer, or whether it consists of linear chains that are rather oligomeric than polymeric (see Figure 7.2).[30]

Despite the different asymmetric carbon atoms that are formed during its biosynthesis, lignin appears as an overall racemic mixture with respect to the various chiral centres – another fact that suggests that the *in vivo* lignin

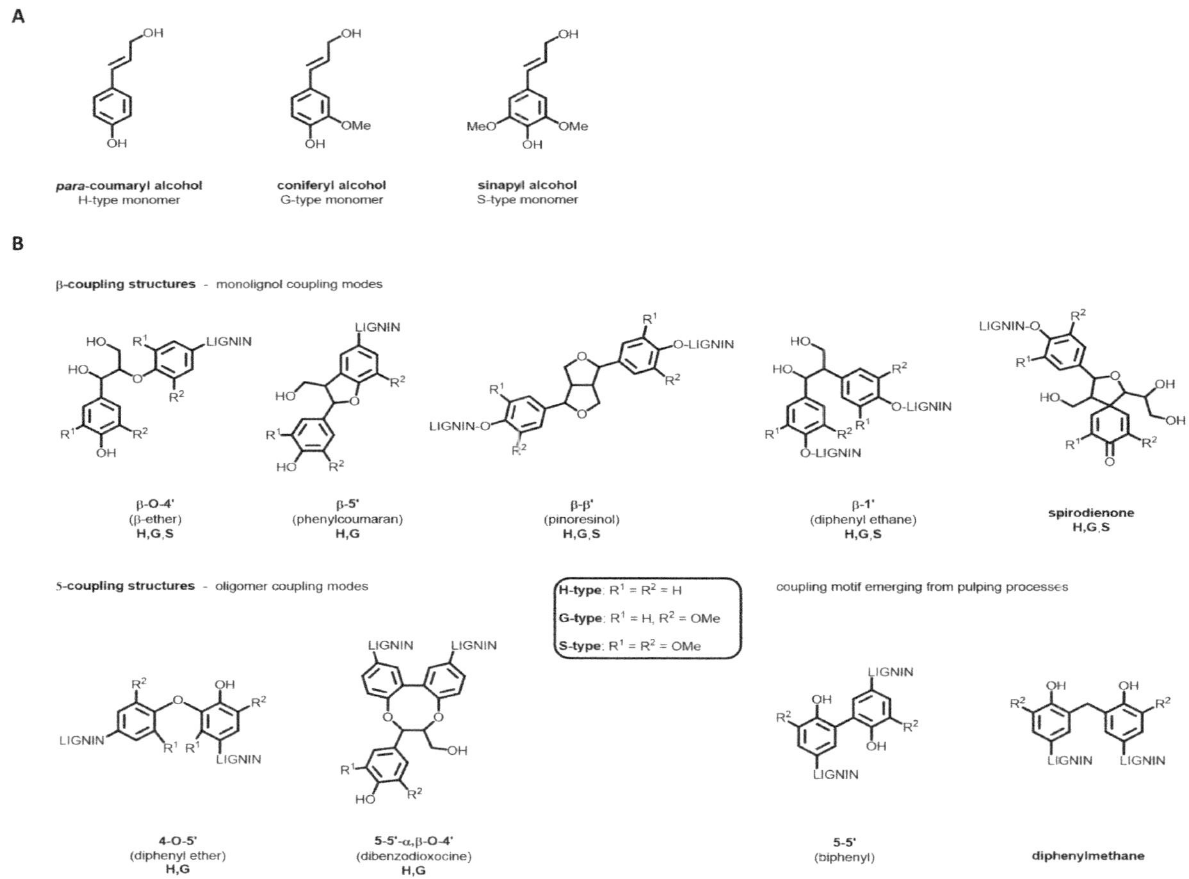

Figure 7.1 (A) Monolignol building blocks of lignins and (B) lignin interunit bonding motifs.

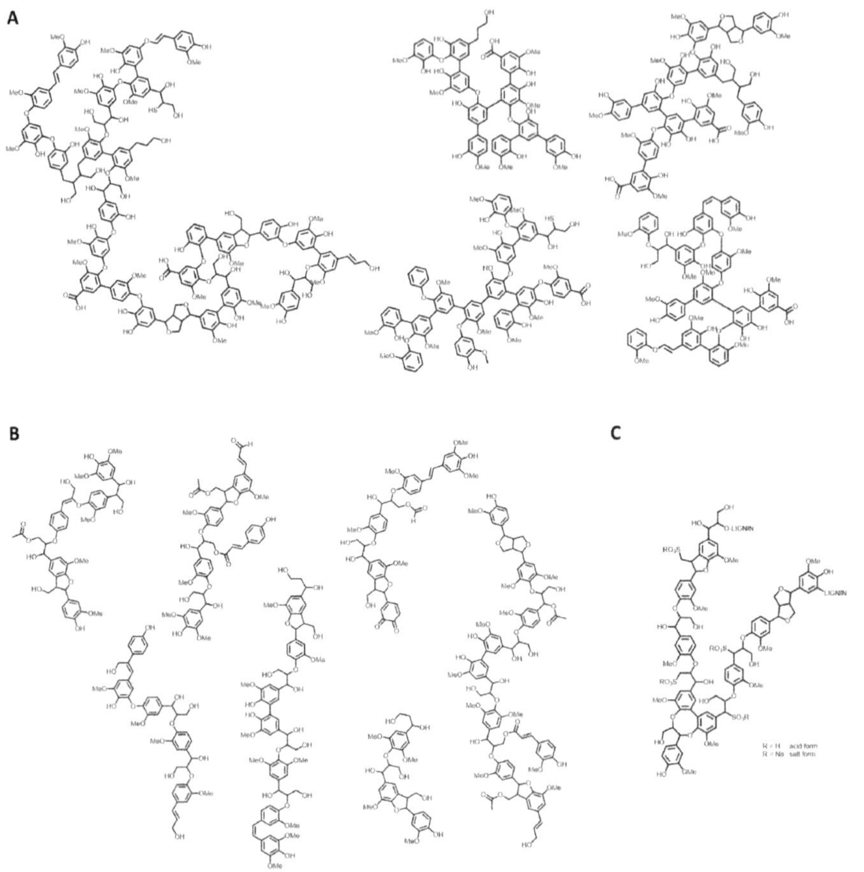

Figure 7.2 Lignin structures showing characteristic interunit bonding motifs and functional groups for different types of lignin: (A) softwood kraft lignin; (B) wheat straw organosolv lignin; and (C) (generic) lignosulfonate.

formation is probably not controlled in detail. It has been shown, however, that the ratio of the different aromatic units does correlate with the ratio between *threo-* and *erythro-*configured aliphatic parts in lignin.[31,32] This point is obviously another interesting difference between lignin and carbohydrate-based polymers in plants. With respect to the use in a pharmaceutical context, this in the macroscopic average 'hidden' chirality displays an interesting aspect yet to be elucidated in its molecular consequences.

Since any isolation process can affect the natural structure of lignin, it is important to acknowledge the source of any eventually exploited lignin in the form of the biomass it comes from and at least the three different processes that are established for the isolation of lignin on meaningful scales: (i) organosolv lignin;[33–35] (ii) kraft lignin;[36–38] and (iii) lignosulfonates.[39–41]

Figure 7.2 shows the structures of the three main lignin types. Organosolv lignins contain rather linear oligomeric chains. Kraft lignins, the most abundantly available technical lignins, are composed of a fraction of truly polymeric chains, slightly branched, and a fraction of small, branched, and presumably oxidatively coupled oligomers. Lignosulfonates are structurally different from the other two in that they contain sulfonate groups, allowing them to exist in either a salt or an acid form. The latter aspect caused lignosulfonates to be the lignin type that is most abundantly exploited.

7.2.2 Structural Features and Selected Physico-chemical Properties of Isolated Lignins

A sound structural understanding of lignin is essential for designing valorisation strategies, understanding and elucidating potential and promising mechanisms for the formation of derivatives for targeted applications, and, last but not least, eventually managing regulatory issues. Classical wet-chemical and spectroscopic techniques can provide qualitative and quantitative information about the functional groups and linkages of constituents in lignin, as well as the derivatisation and degradation products that are formed in valorisation efforts. Modern nuclear magnetic resonance (NMR)-based techniques allow most easily analysis of isolated lignins without touching the oligo- or polymeric structure, and give more detailed insights than infrared or UV-vis spectroscopy.[11–16]

7.2.2.1 NMR Spectroscopy-based Analysis Methods

Free hydroxyl groups can be simultaneously identified and quantified by quantitative ^{31}P NMR spectroscopy after phosphitylation of these hydroxyl groups in lignins. Quantification is achieved using an internal standard, such as cholesterol, cyclohexanol, or *N*-hydroxy-5-norbornene-2,3-dicarboxylic acid imide.[15,42–45] Quantitative ^{31}P NMR can be seen as one of the standard analytical tools in lignin characterisation. Advantages in the use of ^{31}P NMR lie in the low amount of sample needed, the possibility of using ionic liquids for samples that dissolve only scarcely, reduced analysis time, easy sample preparation, and high reproducibility. The different chemical environments around the phosphorus atoms lead to characteristic shifts, which in turn allows for the differentiation and quantification.

Interunit bonding motifs in lignin are most conveniently qualitatively elucidated nowadays using heteronuclear single quantum coherence (HSQC) spectroscopy. For obtaining a directly quantifiable HSQC spectrum, quick-quantitative heteronuclear single quantum coherence (QQ-HSQC) spectroscopy is a powerful tool, but a high-field (400 MHz or more) NMR machine equipped with a cryoprobe is necessary for running the demanding QQ-HSQC pulse sequence.[46] An alternative approach that is realisable without cryoprobes consists of a series of HSQC measurements with incremented repetition times; integral values for the signals of interest of at least three consecutive measurements of the same sample are then plotted and mathematically backward-extrapolated, to arrive at integration values for the hypothetical HSQC0 spectrum, which gives the name to the procedure.[47] In both QQ-HSQC and HSQC0, quantification can be achieved based on the intrinsic standard represented by the distinct shifts of the aromatic hydrogen atoms in the 2-position, which can be easily detected, summed up, and used as reference. As a third, easier, and universally applicable option, quantification of a high quality standard ^{1}H–^{13}C HSQC spectrum can be achieved on the basis of a quantitative ^{13}C NMR analysis of the very same sample used for acquisition of the HSQC, given that the sample exhibits the necessary structural stability in the solvent of choice in the time frames needed for the separate measurements.[48]

7.2.2.2 *Fourier Transform Infrared Spectroscopy and Raman Spectroscopy*

FT-IR spectroscopy remains an important tool for structural characterisation also in the field of lignin and its derivatives due to its facile applicability, nowadays in form of the ATR technique. Careful application of band deconvolution[49,50] and/or band fitting[51,52] allows for more detailed insights within a certain context, but does normally not allow for insights as detailed as obtainable using NMR techniques.

In 1984, the first reports were published regarding the use of Raman spectroscopy for characterising structural features in lignin.[53,54] Since then, numerous efforts have been made to overcome the initial low sensitivity stemming from the interfering natural fluorescence of lignin, and from the visible light absorption by lignin, which made it necessary to develop tailor-made irradiation sources[55] and to assist analysis by *in silico* approaches.[56]

7.2.2.3 *Size Exclusion and Gel Permeation Chromatographic Methods*

The mean average molecular weight (M_n), weight average molecular weight (M_w), and polydispersity (PD) of lignins can be determined using size exclusion chromatography (SEC) or gel permeation chromatography (GPC).[57–63] Calibration is normally achieved based on commercially available polystyrene standards, working with the acetylated form of technical lignins in tetrahydrofuran (THF) as master solvent. When dimethyl sulfoxide (DMSO) and dimethylformamide (DMF) are used as solvents, polystyrene sulfonates in acid form are used as standards, and minimum amounts of inorganic salts such as lithium chloride are normally added for breaking aggregates. UV-based detectors, best in the form of flexible PDA detectors, are most commonly used, exploiting the aromatic moieties in lignin; refractive index detectors, which sense the aliphatic part in lignin, have been shown to perform comparably poorly for lignin, while being naturally more suitable to highlight potential carbohydrate-stemming impurities, including LCCs. Molecular weight sensitive detectors based on viscosimetry[64] or laser light scattering,[65–69] which do not need to be calibrated and which are thus free from potential biasing due to the inappropriate choice of standards, represent the state of the art in terms of detection in the GPC and SEC fields.

7.2.2.4 Antioxidant Activity Assays

Antioxidant activity is one of the most important properties of natural polyphenols when it comes to applications in the pharmaceutical and biomedical sectors, since it allows for various exploitations, aiming eventually even at synergistic effects. Measuring antioxidant activity is eventually tricky, and in the case of lignins additionally hampered by their low solubility in some cases. Quantitative data for antioxidant activity can be obtained most reliably by averaging across two different methods, which can be chosen between (i) the oxygen-radical absorbance capacity (ORAC) test,[70] (ii) 2,2-diphenyl-1-picrylhydrazyl (DPPH) method,[71] and (iii) the 2,2′-azino-bis(3-ethylbenzothiazoline-6-sulfonic acid) (ABTS) radical cation assay.[72,73]

Most of the methods express the activity of the sample as standard antioxidant equivalents (E) per gram of freshly weighed samples against suitable calibration systems. Gallic acid (GAE), tannic acid (TAE), trolox, ascorbic acid (AAE), and others can be used as standard antioxidants for means of calibration and as reference points. Existing standard methods might eventually have to be modified in terms of solvents, so as to ensure the complete solubility of the lignin sample.

7.2.3 Refinement Methods

Fractionation of technical lignins is currently the only scalable procedure to approach the issue of structural inhomogeneity. Various fractionation techniques have been developed for refining for structural and/or molecular size features.[74,75] Fractionation can be achieved using: (i) sequential precipitation out of alkaline solutions;[76] (ii) fractional precipitation of re-dissolved kraft lignin and a wheat straw lignin in a gradually changed binary solvent system;[77,78] (iii) (sequential) extractions using different solvents;[76,79–87] and (iv) fractionation by ultrafiltration of black liquor using ceramic membranes.[88–90] Efforts have recently been summarised.[91,92] With respect to regulatory aspects important in the pharmaceutical and biomedical fields, fractionation can be seen as a more practical means to arrive at more describable, but eventually homogenised structural features. With respect to the practical issue of an eventual toxicity, fractionation might allow selectively taking out the problematic type of structure.

7.3 Biodegradability and Toxicity of Lignins

Biodegradation of both natural and synthetic materials is needed to eventually convert back complex structures into smaller motifs and molecules that are then eventually able, after eventual mineralisation, to enter the natural distribution cycles for the respective elements.[93,94] While biological degradation by microorganisms plays an important role especially with respect to biomineralisation and elemental valorisation, upstream physical degradation, either by natural friction-generating events and/or weathering or by industrial mechanical treatments, is necessary, since physical pre-treatments generate larger surface areas that facilitate subsequent degradation by microorganisms. Photolytic bond cleavages, oxidative bond cleavages, and simple hydrolytic cleavages represent more chemical approaches that can, facultatively in close interplay with the mentioned physical methods, cause the initial degradation of polymeric structures.[95] Transition from polymeric to oligomeric structures is also achieved by some organisms, which produce an ensemble of enzymes that are able to work on even larger polymeric structures, exploiting even mediator systems for augmenting efficacy (*vide infra*). Biomineralisation represents then the last step. Due to the complexity of the biodegradation as such, and the interplay of physical and chemical factors, it is still very often difficult or even impossible to obtain sufficient experimental data on the basis of which a rational design of a biodegradable polymer for a specific application would be possible. Going for natural polymers with fitting natural characteristics during the design of novel materials, for example, is thus a winning alternative from the 'green-by-design' concept.[96] Also in this context, the numerous efforts to establish potentially the use of lignin in the pharmaceutical sector are justified.[5,97,98]

7.3.1 Biodegradation

The construction of lignin involves mainly alkyl aryl ethers, and the redox potential needed for effective bond cleavages is too high as plant-based oxidoreductases and laccases, which are involved in the construction of lignin, could cleave them. Lignin degradation can thus only be effectuated by organisms outside the plant that host so-called lignocellulolytic enzymes. This group of enzymes includes cellulases,[99] hemicellulases,[100] and ligninases.[101] Lignocellulolytic enzymes are typically found in fungi, *e.g.*, *Trichoderma reesei* (ascomycetes)[102] and *Phanerochaete chrysosporium* (white-rot fungus)[103] and *Fomitopsis palustris* (brown-rot fungus)[104] (the latter two from the Basidiomycota phylum), and include both cellulose and lignin degrading enzymes. Some organisms have specialised in lignin degradation as carbon sources, for example non-pathogenic *Ceriporiopsis subvermispora*[105] or pathogenic *Fusarium solani*.[106] In research efforts dedicated also to the valorisation of lignin by degradation, commonly applied enzymes are laccases, also called phenol oxidases (EC 1.10.3.x), lignin peroxidase (LiP, EC 1.11.1.14), and manganese peroxidase (MnP, EC 1.11.1.13); these can be summarised as ligninases, since they target more specifically lignins.

7.3.2 Lignin Toxicity

Lignin toxicity is a difficult aspect, and only a few studies exist that give reliable insights. Due to the structural heterogeneity stemming from the biomass source and the isolation procedure as outlined earlier, comparison is difficult, and generalisation is essentially impossible. Estimating cytotoxicity and toxicity on the basis of certain structural features is possible, but does obviously have pitfalls and the impossibility to consider synergistic effects, *etc.* It has been found in one study, for example, that the cytotoxicity increases with the antioxidant activity of lignins,[107] but such simple correlations cannot always be made. In the following paragraphs, toxicity studies done using technical lignins or their fractions are discussed. The archival literature knows several studies in which lignin degradation products, structurally only scarcely defined, have been used, often in combination with other substances, in toxicity screens. Any results of such lignin-derived materials do not classify, however, for judging on lignin toxicity, and have therefore been excluded from the lignin-centred discussion. Also chemically altered lignins for which no details regarding the alteration process are available shall be excluded here since eventually poor chemical derivatisation and purification techniques, which are unfortunately quite frequent, increase the risk that results are biased.

An ecotoxicity profile of lignin was also mapped *versus* the marine alga *Phaeodactylum tricornutum* (Bohlin). The results indicated a 50% reduction in growth rate (ErC50) at 114 $mg\,L^{-1}$ of Sigma-Aldrich alkali lignin. In addition, the no observed effect concentration (NOEC) and lowest observed effect concentration (LOEC) values were found at $<0.1\ mg\,L^{-1}$ and $0.1\ mg\,L^{-1}$, respectively, thus indicating that the lignin is in interesting contrast to a simultaneously tested tannin.[108]

In other respects, the mutagenicity and developmental toxicity of the two primary monolignols, guaiacol (G) and syringol (S), were investigated.[109] The results showed a sign of mutagenicity for S-type monomers, whereas G-type monomers did not trigger mutations. However, both S- and G-type monomers displayed developmental toxicity. Comparable developmental toxicity levels were detected for S- and G-type mixtures, as well as for three preparations of depolymerised lignins, comprising two softwoods and one hardwood. The study is especially interesting since the monomeric species could represent products formed during the biodegradation of lignin.

The toxicological responses of engineered lignin nanoparticles (LNPs), consisting of a highly purified lignin core and a chitosan coating, were reported using embryonic zebrafish toxicity assays.[110] Zebrafish embryos were exposed to the nanoparticles with an intact chorionic membrane and with the chorion enzymatically removed to allow for direct contact of the particles with the developing embryo. The results showed that the mortality rates were significantly greater for chitosan-coated LNPs, *i.e.*, Ch-LNPs, compared to plain LNPs and control groups. The toxicity of the lignin in non-particulate form as control was unfortunately not reported.

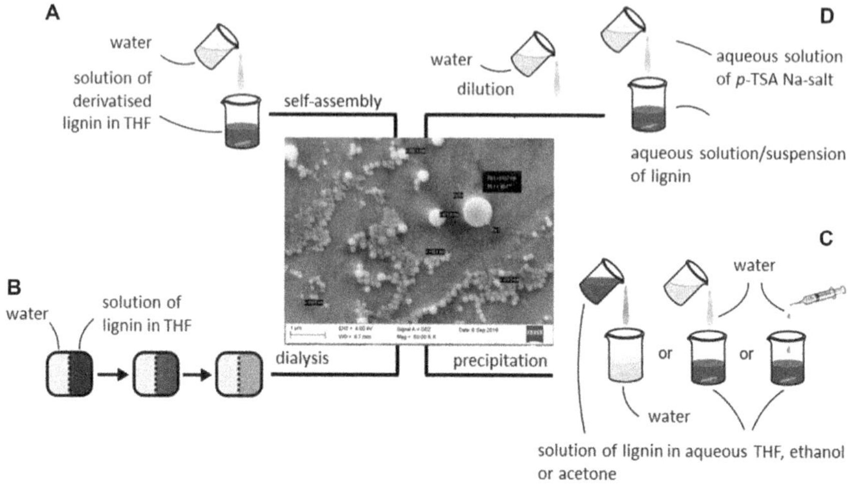

Figure 7.3 Fabrication modes for generating nano- and microscale lignin particles: (A) adding water into a solution from an organic solvent like THF or ethanol of suitable or suitably modified lignin; (B) dialysing a solution of suitable lignin in organic solvent against deionised water; (C) nanoprecipitation by adding water into a solution of a suitable lignin in a merely organic solvent; and (D) formation of LNPs using a hydrotropic system. Refer to main text for leading references.

The scattered nature of these very few studies on materials tested with scarce control tests clearly indicates that further studies are highly needed to better understand the eventual toxicity of certain lignins, and the underlying mechanisms, in order to truly exploit the pharmaceutically interesting properties of lignin in a synergistic fashion. So far, it has been obvious that there are only very few indications, but eventually no true trends, which could be delineated for designing lignin-based materials in the pharmaceutical field with a surely predictable toxicity profile.

7.4 Lignin in Pharmaceutical and Biomedical Applications

7.4.1 Lignin for Drug Delivery Purposes

Isolated lignins display generally a series of intrinsic properties that are beneficial with respect to the development of drug delivery vehicles; some of these properties can even give rise to the emergence of synergistic effects in pharmaceutical and biomedical applications. The antioxidant, UV-shielding, and metal complexing nature and generally stacking propensity render the polymer very suitable for self-assembly-driven construction of materials of various complexity that might even benefit from synergistic effects. Various forms of drug delivery systems are possible based on lignin properties, namely particulate systems, core–shell structures like capsules, and matrices from spun fibers.

7.4.1.1 Particulate Systems

The generation of particulate systems in the area of lignin valorisation for pharmaceutical and or biomedical application emerged roughly ten years ago, and has since then seen remarkable dynamics. This dynamics, however, especially in recent years, has seen rather a horizontal than a vertical development, as reflected in numerous recent review articles.[111–118] Three main routes for generating LNPs have been established. Most often applied within the archival literature for the preparation of LNPs are solvent exchange concepts, based on the strategic use of an anti-solvent that is added to a solution of lignin, or *vice versa*, resulting in the formation of solid spheres in both cases (see Figure 7.3A–C).[119] Using pure water as non-solvent, reported systems most commonly use aqueous and non-aqueous THF,[120,121] dioxane,[122] DMSO,[123,124] acetone,[124–126] or ethanol.[127,128]

Autotrophs with sodium *para*-toluenesulfonate (Na-PTSA) can be exploited to generate LNPs from various lignins (see Figure 7.3D).[129,130] The LNPs precipitate upon rapid dilution with water of an acidic solution containing lignin and Na-PTSA. With respect to the other three methods presented, this approach bears the risk that the rapid precipitation in the presence of the aromatic surfactant leads to notable incorporation of surfactant molecules into the LNPs. In the case of LNPs to be used in the pharmaceutical sector, generated *via* this approach; the chance that certain unwanted ingredients (like surfactants) remain in the final product is a hurdle during regulatory processes.

Aerosol technology[131] and ultrasonication-based[132] approaches represent other possibilities to generate LNPs and lignin microparticles. In these processes, solvent is vaporised from a solution of lignin, forming particles at the hydrophobic solvent–air interface, or forcing intermolecular interactions in the violent collapse of microbubbles generated upon sonication, respectively. Particles formed do not necessarily exhibit a spherical shape, and exhibit eventually larger PDs.

It is commonly found that LNPs exhibit antioxidant and UV-shielding properties, reflecting as such the intrinsic characteristics of the starting lignins.[133] This finding goes together with elucidations that confirm that the lignins remain chemically, *i.e.*, structurally essentially unaltered,[130,134] unless they have been chemically transformed prior to the formation of nanosystems. Chemical modifications, eventually necessary in the form of attachment of cross-linkers for inducing stability during a curing process, or necessary for attaching polyethylene glycol motifs for rendering LNPs invisible to metabolic degradation, or in the form of target-sensing motifs in the case of targeted delivery applications, must ideally not interfere with the desired biodegradation or toxicity profiles. The latter aspects are especially important with respect to regulatory issues and an eventually guaranteed biodegradability of the lignin part of any pharmaceutical product on the basis of lignins in after-life assessments.

7.4.1.1.1 Micro- and Nanoscaled Core–Shell Structures. Core–shell structures naturally exhibit different structural characteristics with respect to solid particle systems. Such differences exist in various forms: (i) construction approach; (ii) maturing behaviours; (iii) morphology; (iv) loading techniques; (v) release characteristics, *etc.* Core–shell structures can exist in different forms, with solid or liquid cores, and with more than one shell, which gives as such a wider scope of tuning possibilities, also in the case of lignins as shell materials or components.

Oil-in-water emulsions have been used as templates for the preparation of lignin microcapsules (0.3–1.4 μm) by ultrasonication (see Figure 7.4).[135–139] By applying the same approach, generation of lignin–metal frameworks as shell materials was demonstrated.[140] Ultrastirring of lignin-containing emulsions was demonstrated to be applicable to lignin micro- and nanoparticle formation.[141]

Lignin nanotubes (LNTs) could be realised using aluminium with interpenetrating holes as sacrificial templates.[142] Once the template was removed, the relaxation of structures occurred, and the length and diameter of the nanotubes with aspect ratios between 55 and 95 were influenced only slightly by the different types of lignin raw materials used.

As in the case of the formation of most of the discussed lignin-based particles, no chemical derivatisation of lignins is necessary to achieve capsule structures with decent stabilities and eventually tuneable decomposition characteristics, allowing once more for generating materials that exhibit unaltered biodegradation profiles. As in the case of lignin particles, functionalisation of lignins with the aim of introducing cross-linkers or sensing motifs for targeted delivery aspects can alter the intrinsic biodegradability, especially in case these lignin-alien motifs are introduced with simple chemistries *via* the phenolic OH groups. The latter does, of course, also represent a reduction of some other intrinsic benefits of polyphenols, namely the antioxidant activity.

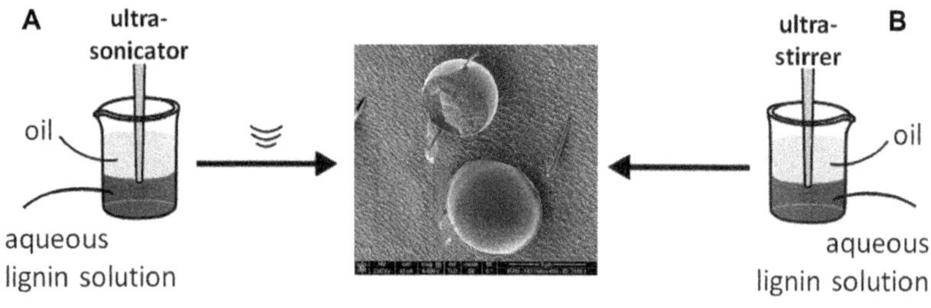

Figure 7.4 Modes for generating of nano- and microscale lignin capsules: (A) ultrasonication of an emulsion of an oil and an aqueous solution of a suitable lignin and (B) ultrastirring of an emulsion of an oil and an aqueous solution of a suitable lignin. Refer to main text and Table 7.1 for references.

7.4.1.1.2 Incorporation of Actives in Lignin Particles and Lignin Capsule Structures. Lignin particles and capsules can exhibit trigger-related slow or fast release of entrapped or encapsulated actives, which are protected from external, eventually degrading factors such as UV radiation, redox-active species, and metabolic activity. *Vice versa*, biological and/or environmental matrices in which particles and capsules will be eventually immersed are protected from the active ingredients, unless matrix characteristics change to trigger slow or fast particle and capsule decomposition. Entrapment, adsorption, and covalent binding are common methods for incorporating actives into LNPs; encapsulation is the technique of choice for LNC systems. In case loading of actives *via* entrapment or encapsulation is realised during the formation of LNPs or LNCs, suitable solubility profiles of the actives, which in the case of generations of lignin nanomaterials from biphasic systems can become challenging given the ultimately necessary hydrophilicity of most actives, are a key factor to consider. The various general methods for loading and releasing actives into/from a variety of different carriers are shown in Figure 7.5.

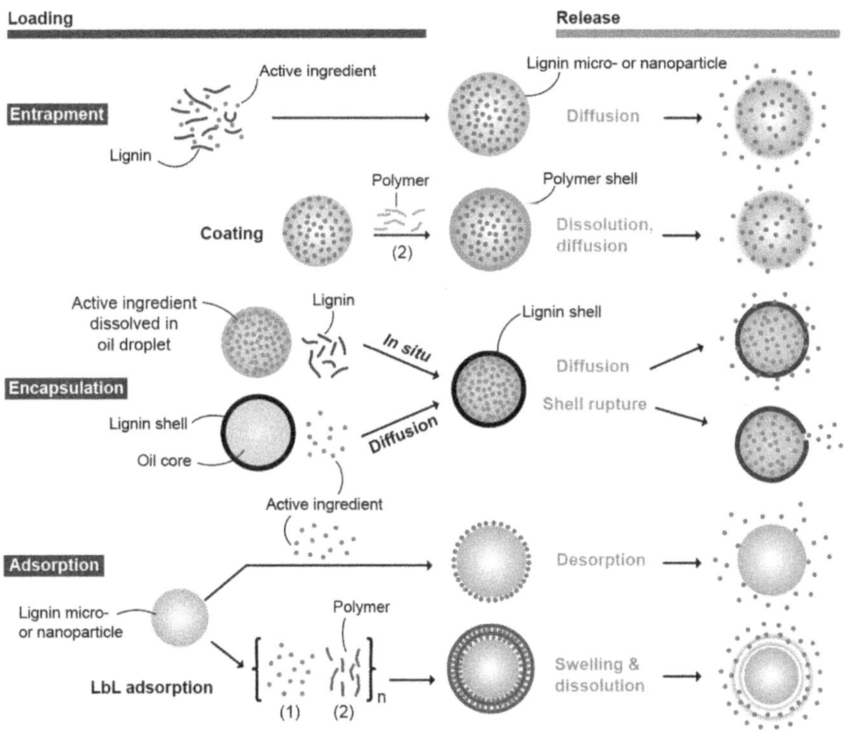

Figure 7.5 General methods for loading and releasing actives from a variety of carriers. Reproduced from ref. 97 with permission from John Wiley & Sons, Copyright © 2019 The Authors. Published by Wiley-VCH Verlag GmbH & Co. KGaA.

7.4.1.1.3 Entrapment *Vs.* Encapsulation *Vs.* Adsorption. Entrapment is the most common method for incorporating small active molecules into solid lignin particles. LNPs are normally found to form following supramolecular assembly strategies in which poorly soluble molecules start, *via* electrical interactions involving mainly the aromatic rings,[120,134,143] to aggregate into initial nucleation points on which consequently technically ever more soluble lignin molecules are deposited. In accordance with such a mechanism, most of the substances entrapped in LNPs are rather insoluble in water. A couple of pioneering reports describe the loading of pharmaceutically relevant small molecules and biological macromolecules such as antibodies and enzymes. Some selected works are discussed in the form of Table 7.1.

Encapsulation can be used to incorporate lipophilic drugs in the water-immiscible cores of capsules during the formation of LNCs from biphasic media at oil–water interfaces (see Figure 7.4). Correspondingly, hydrophilic drugs can be incorporated in a reverse system in which the liquid core is formed from the aqueous phase. In both cases, the amphiphilic lignin can act as a shell material. The surface area-to-volume ratio, equal to diameter/6, renders microcapsules more efficient than nanocapsules, when considering for efficiency only the volume available for loading of the active.

The entrapment/encapsulation efficiencies (EE%) are commonly determined in order to describe the success in loading efforts. The reported values are found to range from 4% to >95%, depending on the system type and formation process, as well as on the type of active. Due to this triple dependence, a comparison of entrapment and encapsulation values is rather difficult. Some substances like curcumin and coumarin derivatives have been used more often especially in proof-of-concept works, but these dummy actives present on their own some intrinsic difficulties, such as solubility in various solvents, that render analysis difficult to compare.

Instead of entrapment and encapsulation, immobilisation of macromolecules, including oligopeptides and enzymes, is normally achieved *via* covalent surface-linking or simple electrostatic adsorption. In pharmaceutically relevant applications, this approach is mainly used for attaching enzymes to the surfaces; for example, lipase was adsorbed onto composite materials consisting of chitin and oxidised kraft lignin,[162] and cutinases and porcine pancreatic α-amylase were adsorbed onto organosolv lignin.[165]

Table 7.1 lists a selection of recent works in which lignin-based particulate or capsule systems were employed for supporting actives relevant to the pharmaceutical sector. Listed are systems that were used entrapping, encapsulating, and adsorbing actives. Systems used for hosting substances relevant to the agricultural sector, such as pesticides, herbicides, and insecticides, have not been considered here, but should, of course, be taken into account for the valuable insights they deliver on a general technical ground. The relatively small number of entries compared to the overall relatively high number of publications, as well as patents detailing the synthesis and potential use of lignin microcapsules, is once more indicative of the misbalance in terms of envisaged fields of application.

Table 7.1 Micro- and nanoscaled particle and capsule systems based on lignin for use as carrier systems. Selected examples report incorporation of pharmacologically active small molecules and macromolecules like enyzmes.

Nanoscaled system (lignin source)	Active	Active incorporation	Entrapment/encapsulation/adsorption efficiency (%)	Loading capacity (wt%)	Year/ref.
LMCs (poplar alkali lignin)	Avermectin	ET	61	17	2016/144
SLS-CTAB microspheres	Avermectin	EC	63	71	2017/145
LNPs and LNCs (**KL**)	2-Propyl-pyridine	EC	n.a.	n.a.	2017/136
Complex micelles (**SKL**)	Ibuprofen	EC	74	46	2017/146
LNPs (**SKL**)	BZL, SFN	ET	77 BZL, 68 SFN	8 BZL, 7 SFN	2017/147
LNPs (alkali lignin from HT-pre-treated corn cobs)	Resveratrol	ET	71–95	19–26	2017/148
LNPs (succinylated **SKL**)	Benzazulene	ET	50–57	9–11	2017/149
LNPs (**WS** soda lignin)	Budesonide	ET	35	3.5	2018/134
LNPs (**KL**, Sigma-Aldrich)	DOX, GFLX	ET	43 DOX, 5–37 GFLX	n.a.	2018/150
LNC (**EHL**)	Avobenzone, octinoxate	EC	98	53	2018/151
LNCs (enzymatically hydrolysable lignin (**EHL**))	Doxorubicin	EC	45–65	6–14	2019/152
Chitosan-coated LNPs (**SKL**)	Ciprofloxacin	EC	n.a.	n.a.	2019/153
LMCs (alkali lignin)	Avermectin	ET	76–85	4.4–4.7	2020/141
LNCs (**SLS**)	Thymol derivatives	EC	10–76	n.a.	2020/154
Lignin micelles (methacrylated alkali lignin)	Ibuprofen	EC	84	n.a.	2020/155
LNPs (**OSL**)	Curcumin	ET	91	n.a.	2020/156
NPs(poly(lactic-*co*-glycolic) acid conjugated alkaline lignin)	GDC-0623	ET	n.a.	n.a.	2020/157

Table 7.1 (*Continued*)

Nanoscaled system (lignin source)	Active	Active incorporation	Entrapment/encapsulation/adsorption efficiency (%)	Loading capacity (wt%)	Year/ref.
LNPs (**SKL** fractions)	Folic acid	ET	67	n.a.	2022/158
LNPs (**KL**, Sigma-Aldrich)	Biomimetic thymine photo-adduct	ET	37–75	5.6–14	2022/159
LNPs (**SKL**)	Doxorubicin	ET	73	n.a.	2022/160
LNCs (**SLS**)	Sulfo-rhodamine	EC	n.a.	n.a.	2014/161
Chitin–**KL** composite	Lipase	AD	1.0–2.0	33–11	2015/162
Chitosan-LNPs (calcium-**LS**)	RNase A	ET	61–40	6.6–43	2013/163
Silica–lignin composite (alkali lignin)	Lipase	AD	1.7	42	2017/164
LNPs (acetic acid lignin from bamboo shoot shells)	α-Amylase	AD	n.a.	1.9	2017/165
LNPs (**OSL**)	Tyrosinase	ET	69	12	2018/166
LNPs (**OSL**)	Tyrosinase	AD	71–90	12–15	2018/166
Cationic LNPs (**SKL**)	Lipase, cutinase	AD	96	5.5	2018/167

7.4.1.1.4 Functionalisation of Particle and Capsule Surfaces. As already mentioned in the preceding paragraphs, functionalisation of the surfaces of particles and capsules might be necessary in order to improve their applicability and/or performance in an application setting. Yet, only a few reports in the archival literature describe such efforts. Caicedo *et al.* functionalised LNTs with avidin for specific immobilisation onto desthiobiotin-grafted glass surfaces.[142] Carboxylated LNPs (CLNPs) were realised starting from succinylated softwood kraft lignin, and were further modified by chemically attaching a block copolymer comprising poly(ethylene glycol) (PEG), poly(histidine), and a cell-penetrating peptide.[149]

Anionic LNPs have been modified by adopting a layer-by-layer (LbL) technique generating a layer of cationic polymers such as poly(diallyldimethylammonium chloride) on their surfaces.[120,168] Using a variation of the LbL approach, adsorption of chitosan on LNPs was shown to form colloidally stable cationic particles that fully stabilised olive oil-in-water emulsions.[153]

A matrix comprised of poly(vinyl alcohol) (PVA) and softwood kraft lignin was used to generate silver nanoparticle composites with boosted antibacterial activity.[153] Commercial silica material was modified with *N*-(2-aminoethyl)-3-aminopropyltrimethoxysilane to increase the affinity to an oxidised kraft lignin. Silver NPs grafted onto the resulting silica/lignin hybrids were stable and active against *Pseudomonas aeruginosa*.[169] Silver-ion-containing LNPs, generated exploiting the strong metal-complexing characteristic of the phenolic groups in lignin, have been realised as an alternative to silver nanoparticle-based applications in antimicrobial applications. This system design is potentially suitable to circumvent expectable regulatory issues in the nearer future in the field of metal nanoparticles (*vide infra*).[170]

7.4.1.2 Fibrous Systems

Fibrous systems based entirely on, or at least containing lignin can eventually represent interesting matrices for topical applications in which a combination of sustained effects by the matrix, synergistic effects exerted by the lignin component, and a tuneable release of actives from the matrix is desired. Such fibrous matrices are best to be realised, at least on a small scale, using electrospinning.[171] Scale-up would be easier using a melt-spinning approach or needleless electrospinning;[172] yet the latter represents often too severe conditions as actives could withstand in the desired form. In electrospinning applications, lignin is normally used together with another polymer, in order to endow the matrix with antioxidant and anti-inflammatory characteristics. PLA–lignin composites were combined with poly(L-lactide) (PLLA) and electrospun into nanofibrous mats.[173] The mats were intended as tissue engineering scaffolds, just as in a study in which electrospun fibers containing lignin and PVA and poly(glycerol sebacate) (PGS) resulted in a matrix applicable for nerve tissue engineering.[174] Additionally, different nanofibers containing lignin have been successfully produced by electrospinning and present high potential for application in wound dressings, especially when realised with a chitin/chitosan component, due to the antimicrobial properties exerted by the polyphenolic lignin component.[175–177] Abudula *et al.* reported on poly(caprolactone)-coated chitin–lignin gel fibrous scaffolds for drug delivery,[178] using a wheat straw organosolv lignin and PGS as an active. More recently, nanofibers from a mixture of cellulose acetate, softwood lignosulfonate, and *N*-vanillidene-phenylthiazole copper(II) complexes have been reported, for potential diaper dermatitis control.[179]

7.4.1.3 *Lignin-containing Hydrogels for the Pharmaceutical Sector*

Hydrogels are three-dimensional, highly porous networks of cross-linked polymers that can absorb quantities of water or biological fluids that exceed their dry weight in magnitude.[180,181] Always with the same idea of exploiting the natural benefits of lignins in the resulting materials, lignin-based hydrogels have been designed in the biomedical context for drug delivery, wound dressing, or tissue engineering.[182] Hydrogels for the controlled release of hydrophobic compounds have been realised in the form of lignin–xanthan mixtures for the controlled release of bisoprolol fumarate.[183] Lignin has been combined with chitosan, PVA, alginate, or cellulose to produce hydrogels for wound healing applications[184,185] and tissue engineering.[107,186]

7.4.2 Other Emerging Uses

Lignin has been successfully combined with poly(lactic acid) (PLA) and poly(hydroxybutyrate) (PHB) to produce biocomposite filaments for 3D printing,[187–190] to make lignin benefits enter into the trend to use 3D printing technology involving biopolymers in biomedical applications,[191] exploiting the inherent antioxidant and antimicrobial properties of lignin alongside other (bio)polymers for specified wound dressing and tissue engineering, as well as for drug delivery.

Lignin has more recently been shown to be a versatile pharmaceutical excipient, rendering drugs more bioavailable, and improving as such eventually the oral dosage conditions:[192–194] it was found that lignin in combination with other classic excipients enhanced the aspirin release rates from tablets, and/or altered the dissolution profiles of such tablets.

A couple of reports exist that detail the beneficial contribution of lignins in sunscreen preparations, aiming especially at a synergistic effect that emerges from the intrinsic sunscreening characteristic of the lignin molecules.[195–198] Due to its broad-spectrum UV-absorbing and antioxidant properties, lignin can be seen as a sustainable alternative to synthetic sunscreen actives that are known to be problematic in terms of long-term environmental effects. In the UV-blocking context, poly(propylene carbonate) composite films filled with titanium dioxide decorated lignin were able to absorb about 90% of UV light in the full UV band (200–400 nm).[199,200] Composite films of gellan gum, hydroxyethyl cellulose, and lignin have been realised, exhibiting almost 100% protection against UVB and 90% protection against UVA.[201]

7.5 Regulatory Aspects Concerning Use of Natural Substances in Pharmaceutical Applications

The exploitation of lignin characteristics in pharmaceutical and biomedical applications is an emerging field that has seen so far mainly proof-of-concept works. In most cases, regulatory aspects are not considered; yet such approval of the regulatory bodies, *i.e.*, the US Food and Drug Administration (FDA) and the European Medicines Agency (EMA), for use of lignin in pharmaceutical and biomedical applications is needed. Much more efforts would be necessary involving standard animal models in the context of concise toxicological evaluations, eventually also on various model organisms, so as to evaluate in a much less fragmented way (*vide supra*) the various means of exploitation of lignin in the biomedical sector. Specific guidelines were developed by the interested regulatory bodies of the leading drug manufacturing countries of the Western Hemisphere and its allies, *i.e.*, the FDA, the EMA, and the authorised Ministry of Health, Labour, and Welfare in Japan, in conjunction with The International Council for Harmonisation (ICH). This *vade mecum*, known as the common technical document (CTD) from 2003, requires a detailed report on the function and conditions of excipients, covering production, physico-chemical characterisation, control tests and methods, and toxicological information. As proposed by the Committee for Medicinal Products for Human Use (CHMP) in 2007, introduction of novel excipients must be well motivated; in the case of lignin, this can be done on the basis of its intrinsic characteristics, the potential for synergistic effects including an enhancement of activity of the active, its eventually retained biodegradability, and socio-economic advantages with respect to sustainability.

Nanopreparations from lignin require standardised toxicological assessments of novel nanomaterials, and prerequisite steps during nanoparticle generation and optimisation are suggested to improve the ecotoxicity analysis of nanomaterials, which will play an additional role when it comes to meeting regulatory aspects.[202,203] Engineered nanomaterials (ENMs) pose five major challenges:[204] (i) variable physico-chemical characteristics; (ii) unknown and difficult-to-monitor end-of-life processes in the environment; (iii) lack of reliable techniques for quantifying exposures to ENMs; (iv) bioavailability of ENMs on wildlife and the environment practically being non-predictable; and (v) knowledge gaps in the toxicity and toxic mechanisms of ENMs. Risk assessment of nanomaterials and toxicological impacts of nanomaterials on agriculture, food, and the environment are still in discussion in the EU, in search for meaningful regulatory guidelines for the safer use of nanomaterials, and an eventual complete prohibition in the future.

7.6 Summary and Outlook

The natural polyphenol lignin is available in industrial quantities from the traditional pulp and paper industries in the form of kraft lignin or lignosulfonates. More modern biorefinery industries are generating meaningful quantities of structurally different organosolv lignins.[205–207] Yet, the most convincing application is use of lignins as solid fuels in the pulp industry itself. Two aspects are responsible for this, namely the structural complexity and varying physico-chemical characteristics of the various isolated lignins available in the market, and the yet *de facto* prohibitive costs connected with their use as substituents for standard fossil-based resources.[8,208] The first aspect can be overcome by ongoing research regarding isolation, characterisation, and proper/correct structural description of available lignins. The second aspect can be tackled only by arriving at robust methods to produce either value-added chemicals or advanced functional materials from lignin on market scales, to arrive at an economically viable, holistic biorefinery.[8–10,190,206,207,209] The development of materials on the basis of lignin for pharmaceutical and biomedical use is a principally interesting way to arrive at high value products from lignin to improve returns within larger biorefinery efforts. However, the field sees mainly proof-of-concept studies regarding the exploitation of eventual beneficial physico-chemical properties of lignins in pharmaceutical or biomedical applications; an application in real markets does not exist yet. While the proof-of-concept studies are good for highlighting potential, they reflect, most probably unwillingly, the difficulties in general availability of 'standard lignins' with a set of physico-chemical properties that would allow for direct comparisons between various application studies, and the difficulty within the interested community to fully structurally understand the irregular natural polyphenol lignin. As soon as these two interconnected problems are solved, significant hurdles in terms of regulatory aspects in the pharmaceutical and biomedical fields, as well as in terms of the targeted holistic and economically viable biorefinery, become conquerable. It would thus be necessary to focus research efforts, and work in a transdisciplinary manner across the various sciences and sectors touched by the pharmaceutical and biomedical fields, so as to make a necessary next significant step forward.

References

1. C. Heitner, D. Dimmel and J. Schmidt, *Lignin and Lignans: Advances in Chemistry*, CRC Press, Boca Raton, 2010.
2. A. Pizzi, in *Monomers, Polymers and Composites from Renewable Resources*, ed. M. N. Belgacem and A. Gandini, Elsevier, Amsterdam, 2008, pp. 179–199.
3. T. Deyama and S. Nishibe, in *Lignin and Lignans*, ed. C. Heitner, D. Dimmel and J. Schmidt, CRC Press, Boca Raton, 2010, pp. 585–629.

4. J. Serrano, R. Puupponen-Pimiä, A. Dauer, A.-M. Aura and F. Saura-Calixto, *Mol. Nutr. Food Res.*, 2009, **53**, S310.
5. J. Domínguez-Robles, Á. Cárcamo-Martínez, S. A. Stewart, R. F. Donnelly, E. Larrañeta and M. Borrega, *Sustainable Chem. Pharm.*, 2020, **18**, 100320.
6. I. Ullah, Z. Chen, Y. Xie, S. S. Khan, S. Singh, C. Yu and G. Cheng, *Int. J. Biol. Macromol.*, 2022, **208**, 819.
7. N. Nan, W. Hu and J. Wang, *Biomedicines*, 2022, **10**, 747.
8. G. T. Beckham, Lignin Valorization: Emerging Approaches, in *Energy and Environment Series*, Royal Society of Chemistry, Cambridge, 2018.
9. M. Aresta, A. Dibenedetto and F. Dumeignil, *Biorefineries, An Introduction*, De Gruyter, Berlin, Boston, 2015.
10. K. Wagemann and N. Tippkötter, Biorefineries, in *Advances in Biochemical Engineering/Biotechnology*, Springer Nature Switzerland, Cham, 2019.
11. H. Lange, P. Giannì and C. Crestini, Lignin Valorization: Emerging Approaches, in *Energy and Environment Series*, ed. G. T. Beckham, Royal Society of Chemistry, Cambridge, 2018, pp. 413–476.
12. J. S. Lupoi, S. Singh, R. Parthasarathi, B. A. Simmons and R. J. Henry, *Renewable Sustainable Energy Rev.*, 2015, **49**, 871.
13. C. Lapierre, in *Lignin and Lignans*, ed. C. Heitner, D. Dimmel and J. Schmidt, CRC Press, Boca Raton, 2010, pp. 11–48.
14. J. Schmidt, in *Lignin and Lignans*, ed. C. Heitner, D. Dimmel and J. Schmidt, CRC Press, Boca Raton, 2010, pp. 49–102.
15. D. Argyropoulos, in *Lignin and Lignans*, ed. C. Heitner, D. Dimmel and J. Schmidt, CRC Press, Boca Raton, 2010, pp. 245–265.
16. H. Hatakeyama and T. Hatakeyama, in *Lignin and Lignans*, ed. C. Heitner, D. Dimmel and J. Schmidt, CRC Press, Boca Raton, 2010, pp. 301–319.
17. N. G. Lewis and E. Yamamoto, *Annu. Rev. Plant Physiol. Plant Mol. Biol.*, 1990, **41**, 455.
18. K. V. Sarkanen and C. H. Ludwig, *Lignins: occurrence, formation, structure and reactions*, John Wiley & Sons, Inc., New York, 1971.
19. E. Adler, *Wood Sci. Technol.*, 1977, **11**, 169.
20. Y. Zhao, U. Shakeel, M. Saif Ur Rehman, H. Li, X. Xu and J. Xu, *J. Cleaner Prod.*, 2020, **253**, 120076.
21. W. G. Glasser and S. Sarkanen, *Lignin: Properties and Materials*, American Chemical Society, Washington, DC, 1989.
22. G. Brunow and K. Lundquist, in *Lignin and Lignans*, ed. C. Heitner, D. Dimmel and J. Schmidt, CRC Press, Boca Raton, 2010, pp. 267–299.
23. E. Pesquet, B. Zhang, A. Gorzsás, T. Puhakainen, H. Serk, S. Escamez, O. Barbier, L. Gerber, C. Courtois-Moreau, E. Alatalo, L. Paulin, J. Kangasjärvi, B. Sundberg, D. Goffner and H. Tuominen, *Plant Cell*, 2013, **25**, 1314–1328.

24. R. Decou, H. Serk, D. Ménard, E. Pesquet, in *Xylem*, ed. M. de Lucas, J. P. Etchhells and M. de Lucas, Springer New York, 2017, pp. 233–247.
25. R. Vanholme, B. Demedts, K. Morreel, J. Ralph and W. Boerjan, *Plant Physiol.*, 2010, **153**, 895.
26. W. Boerjan, J. Ralph and M. Baucher, *Annu. Rev. Plant Biol.*, 2003, **54**, 519.
27. L. A. Donaldson, *Phytochemistry*, 2001, **57**, 859.
28. R. Whetten and R. Sederoff, *Plant Cell*, 1995, **7**, 1001.
29. J. M. Humphreys and C. Chapple, *Curr. Opin. Plant Biol.*, 2002, **5**, 224.
30. R. Hatfield and R. S. Fukushima, *Crop Sci.*, 2005, **45**, 832.
31. T. Akiyama, H. Goto, D. S. Nawawi, W. Syafii, Y. Matsumoto and G. Meshitsuka, *Holzforschung*, 2005, **59**, 276.
32. T. Akiyama, T. Sugimoto, Y. Matsumoto and G. Meshitsuka, *J. Wood Sci.*, 2002, **48**, 210.
33. A. Guerra, I. Filpponen, L. A. Lucia and D. S. Argyropoulos, *J. Agric. Food Chem.*, 2006, **54**, 9696.
34. T. J. McDonough, *Tappi J.*, 1993, **76**, 186.
35. E. K. Pye and J. H. Lora, *Tappi J.*, 1991, **74**, 113.
36. F. S. Chakar and A. J. Ragauskas, *Ind. Crops Prod.*, 2004, **20**, 131.
37. J. Gierer, *Wood Sci. Technol.*, 1980, **14**, 241.
38. P. Tomani, *Cellul. Chem. Technol.*, 2010, **44**, 53.
39. G. C. Howard, *Ind. Eng. Chem.*, 1934, **26**, 614.
40. J. Ruwoldt, *Surfaces*, 2020, **3**, 622.
41. M. Kienberger, S. Maitz, T. Pichler and P. Demmelmayer, *Processes*, 2021, **9**, 804.
42. X. Meng, C. Crestini, H. Ben, N. Hao, Y. Pu, A. J. Ragauskas and D. S. Argyropoulos, *Nat. Protoc.*, 2019, **14**, 2627.
43. B. Saake, D. S. Argyropoulos, O. Beinhoff and O. Faix, *Phytochemistry*, 1996, **43**, 499.
44. A. Granata and D. S. Argyropoulos, *J. Agric. Food Chem.*, 1995, **43**, 1538.
45. A. E. Wroblewski, C. Lensink, R. Markuszewski and J. G. Verkade, *Energy Fuels*, 1988, **2**, 765.
46. D. J. Peterson and N. M. Loening, *Magn. Reson. Chem.*, 2007, **45**, 937.
47. K. Hu, W. M. Westler and J. L. Markley, *J. Am. Chem. Soc.*, 2011, **133**, 1662.
48. L. Zhang and G. Gellerstedt, *Magn. Reson. Chem.*, 2007, **45**, 37.
49. W.-J. Yang, P. R. Griffiths, D. M. Byler and H. Susi, *Appl. Spectrosc.*, 1985, **39**, 282–287.
50. J. K. Kauppinen, D. J. Moffatt, H. H. Mantsch and D. G. Cameron, *Appl. Spectrosc.*, 1981, **35**, 271.
51. W. F. Maddams, *Appl. Spectrosc.*, 1980, **34**, 245.
52. J. A. Pierce, R. S. Jackson, K. W. Van Every, P. R. Griffiths and H. Gao, *Anal. Chem.*, 1990, **62**, 477.

53. H. L. Hergert, PhD thesis, Georgia Institute of Technology, 1984.
54. U. Agarwal and R. Atalla, in *Lignin and Lignans*, ed. C. Heitner, D. Dimmel and J. Schmidt, CRC Press, Boca Raton, 2010, pp. 103–136.
55. M. W. Meyer, J. S. Lupoi and E. A. Smith, *Anal. Chim. Acta*, 2011, **706**, 164.
56. K. L. Larsen and S. Barsberg, *J. Phys. Chem. B*, 2010, **114**, 8009.
57. T. I. Obiaga and M. Wayman, *J. Appl. Polym. Sci.*, 1974, **18**, 1943.
58. F. Chen and J. Li, *J. Wood Chem. Technol.*, 2000, **20**, 265.
59. A. Salanti, L. Zoia, E.-L. Tolppa and M. Orlandi, *Biomacromolecules*, 2012, **13**, 445.
60. K. K. Y. Wong and E. de Jong, *J. Chromatogr. A*, 1996, **737**, 193.
61. J. Pellinen and M. Salkinoja-Salonen, *J. Chromatogr. A*, 1985, **322**, 129.
62. H. L. Chum, D. K. Johnson, M. P. Tucker and M. E. Himmel, *Holzforschung*, 1987, **41**, 97.
63. T. Bikova, A. Treimanis, G. Rossinska and G. Telysheva, *Holzforschung*, 2004, **58**, 489.
64. O. Ringena, S. Lebioda, R. Lehnen and B. Saake, *J. Chromatogr. A*, 2006, **1102**, 154.
65. A. V. Gidh, S. R. Decker, T. B. Vinzant, M. E. Himmel and C. Williford, *J. Chromatogr. A*, 2006, **1114**, 102.
66. A. V. Gidh, S. R. Decker, C. H. See, M. E. Himmel and C. W. Williford, *Anal. Chim. Acta*, 2006, **555**, 250.
67. S. M. Braaten, P. B. E. Christensen and G. E. Fredheim, *J. Wood Chem. Technol.*, 2003, **23**, 197.
68. B. Cathala, B. Saake, O. Faix and B. Monties, *J. Chromatogr. A*, 2003, **1020**, 229.
69. G. E. Fredheim, S. M. Braaten and B. E. Christensen, *J. Chromatogr. A*, 2002, **942**, 191.
70. B. Ou, D. Huang, M. Hampsch-Woodill, J. A. Flanagan and E. K. Deemer, *J. Agric. Food Chem.*, 2002, **50**, 3122.
71. M. R. Szabo, C. Idiţoiu, D. Chambre and A. X. Lupea, *Chem. Pap.*, 2007, **61**, 214.
72. A. Serpen, E. Capuano, V. Fogliano and V. Gökmen, *J. Agric. Food Chem.*, 2007, **55**, 7676.
73. R. Re, N. Pellegrini, A. Proteggente, A. Pannala, M. Yang and C. Rice-Evans, *Free Radicals Biol. Med.*, 1999, **26**, 1231.
74. K. Lundquist and T. K. Kirk, *Tappi J*, 1980, **63**, 80.
75. C. Schuerch, *J. Am. Chem. Soc.*, 1952, **74**, 5061.
76. P. Mousavioun and W. O. S. Doherty, *Ind. Crops Prod.*, 2010, **31**, 52.
77. C. Cui, R. Sun and D. S. Argyropoulos, *ACS Sustainable Chem. Eng.*, 2014, **2**, 959.
78. H. Lange, P. Schiffels, M. Sette, O. Sevastyanova and C. Crestini, *ACS Sustainable Chem. Eng.*, 2016, **4**, 5136.
79. R. W. Thring, M. N. Vanderlaan and S. L. Griffin, *J. Wood Chem. Technol.*, 1996, **16**, 139.

80. N.-Y. Teng, I. Dallmeyer and J. F. Kadla, *Ind. Eng. Chem. Res.*, 2013, **52**, 6311–6317.
81. R. J. A. Gosselink, D. J. E. G. van, J. E. de, E. L. Scott, J. P. M. Sanders, J. Li and G. Gellerstedt, *Holzforschung*, 2010, **64**, 193.
82. F. Zikeli, T. Ters, K. Fackler, E. Srebotnik and J. Li, *Ind. Crops Prod.*, 2014, **61**, 249.
83. A. Duval, F. Vilaplana, C. Crestini and M. Lawoko, *Holzforschung*, 2015, **70**, 11.
84. G. Griffini, V. Passoni, R. Suriano, M. Levi and S. Turri, *ACS Sustainable Chem. Eng.*, 2015, **3**, 1145.
85. V. Passoni, C. Scarica, M. Levi, S. Turri and G. Griffini, *ACS Sustainable Chem. Eng.*, 2016, **4**, 2232.
86. R. E. Majdar, C. Crestini and H. Lange, *ChemSusChem*, 2020, **13**, 4735.
87. R. J. A. Gosselink, J. C. van der Putten and D. S. van Es, *Fractionation of Technical Lignins*, WO2015178771A1, 2015.
88. A. Toledano, A. García, I. Mondragon and J. Labidi, *Sep. Purif. Technol.*, 2010, **71**, 38.
89. O. Sevastyanova, M. Helander, S. Chowdhury, H. Lange, H. Wedin, L. Zhang, M. Ek, J. F. Kadla, C. Crestini and M. E. Lindström, *J. Appl. Polym. Sci.*, 2014, **131**, e40799.
90. A. Duval, S. Molina-Boisseau and C. Chirat, *Holzforschung*, 2014, **69**, 127.
91. H. Sadeghifar and A. Ragauskas, *ACS Sustainable Chem. Eng.*, 2020, **8**, 8086.
92. M. Gigli and C. Crestini, *Green Chem.*, 2020, **22**, 4722.
93. P. M. Vitousek, L. O. Hedin, P. A. Matson, J. H. Fownes and J. Neff, in *Success, Limitations, and Frontiers in Ecosystem Science*, ed. M. L. Pace and P. M. Groffman, Springer, New York, 1998, pp. 432–451.
94. M. Vert, Y. Doi, K.-H. Hellwich, M. Hess, P. Hodge, P. Kubisa, M. Rinaudo and F. Schué, *Pure Appl. Chem.*, 2012, **84**, 377.
95. S. Lambert, C. Sinclair and A. Boxall, in *Reviews of Environmental Contamination and Toxicology*, ed. D. M. Whitacre, Springer, Cham, Heidelberg, New York, Dordrecht, London, 2014, vol. 227, pp. 1–53.
96. *Lignin Biodegradation: Microbiology, Chemistry and Potential Applications*, ed. T. Higuchi, H. Chang and T. K. Kirk, CRC Press, Boca Raton, 1980.
97. M. H. Sipponen, H. Lange, C. Crestini, A. Henn and M. Österberg, *ChemSusChem*, 2019, **12**, 2039.
98. S. Sugiarto, Y. Leow, C. L. Tan, G. Wang and D. Kai, *Bioact. Mater.*, 2022, **8**, 71.
99. C. M. Payne, B. C. Knott, H. B. Mayes, H. Hansson, M. E. Himmel, M. Sandgren, J. Ståhlberg and G. T. Beckham, *Chem. Rev.*, 2015, **115**, 1308.
100. D. Shallom and Y. Shoham, *Curr. Opin. Microbiol.*, 2003, **6**, 219.
101. D. W. S. Wong, *Appl. Biochem. Biotechnol.*, 2009, **157**, 174.

102. A. Levasseur, M. Saloheimo, D. Navarro, M. Andberg, P. Pontarotti, K. Kruus and E. Record, *BMC Biochem.*, 2010, **11**(1), 32.
103. T. K. Kirk, S. Croan, M. Tien, K. E. Murtagh and R. L. Farrell, *Enzyme Microb. Technol.*, 1986, **8**, 27.
104. E. Munir, J. J. Yoon, T. Tokimatsu, T. Hattori and M. Shimada, *Proc. Natl. Acad. Sci.*, 2001, **98**, 11126.
105. H. Nguyen, K. Kondo, Y. Yagi, Y. Iseki, N. Okuoka, T. Watanabe, B. Mikami, T. Nagata and M. Katahira, *ACS Sustainable Chem. Eng.*, 2022, **10**, 923.
106. J. J. Coleman, *Mol. Plant Pathol.*, 2016, **17**, 146.
107. S. Quraishi, M. Martins, A. A. Barros, P. Gurikov, S. P. Raman, I. Smirnova, A. R. C. Duarte and R. L. Reis, *J. Supercrit. Fluids*, 2015, **105**, 1.
108. G. Libralato, F. Avezzù and A. Volpi Ghirardini, *J. Hazard. Mater.*, 2011, **194**, 435.
109. X. Zhang, D. F. Levia, E. O. Ebikade, J. Chang, D. G. Vlachos and C. Wu, *J. Appl. Toxicol.*, 2022, **42**, 423.
110. J. S. Stine, B. J. Harper, C. G. Conner, O. D. Velev and S. L. Harper, *Nanomaterials*, 2021, **11**, 111.
111. N. Nan, W. Hu and J. Wang, *Biomedicines*, 2022, **10**, 747.
112. K. Liu, Y. Zhuang, J. Chen, G. Yang and L. Dai, *Int. J. Mol. Sci.*, 2022, **23**, 7254.
113. V. K. Yadav, N. Gupta, P. Kumar, M. G. Dashti, V. Tirth, S. H. Khan, K. K. Yadav, S. Islam, N. Choudhary, A. Algahtani, S. P. Bera, D.-H. Kim and B.-H. Jeon, *Materials*, 2022, **15**, 953.
114. J. Garg, M. Nee Chiu, S. Krishnan, L. Kumar Tripathi, S. Pandit, B. Farasati Far, N. Kumar Jha, K. Kumar Kesari, V. Tripathi, S. Pandey and P. Kumar Gupta, *Mater. Lett.*, 2022, **311**, 131573.
115. K. Chen, S. Wang, Y. Qi, H. Guo, Y. Guo and H. Li, *ChemSusChem*, 2021, **14**, 1284.
116. L. E. Low, K. C. Teh, S. P. Siva, I. M. L. Chew, W. W. Mwangi, C. L. Chew, B.-H. Goh, E. S. Chan and B. T. Tey, *Environ. Nanotechnol., Monit. Manage.*, 2021, **15**, 100398.
117. Z. Zhang, V. Terrasson and E. Guénin, *Nanomaterials*, 2021, **11**, 1336.
118. M. Österberg, M. H. Sipponen, B. D. Mattos and O. J. Rojas, *Green Chem.*, 2020, **22**, 2712.
119. F. Xiong, Y. Han, S. Wang, G. Li, T. Qin, Y. Chen and F. Chu, *Ind. Crops Prod.*, 2017, **100**, 146.
120. M. Lievonen, J. J. Valle-Delgado, M.-L. Mattinen, E.-L. Hult, K. Lintinen, M. A. Kostiainen, A. Paananen, G. R. Szilvay, H. Setälä and M. Österberg, *Green Chem.*, 2016, **18**, 1416.
121. M. H. Sipponen, M. Smyth, T. Leskinen, L.-S. Johansson and M. Österberg, *Green Chem.*, 2017, **19**, 5831.
122. H. Li, Y. Deng, H. Wu, Y. Ren, X. Qiu, D. Zheng and C. Li, *Holzforschung*, 2016, **70**, 725.
123. D. Tian, J. Hu, R. P. Chandra, J. N. Saddler and C. Lu, *ACS Sustainable Chem. Eng.*, 2017, **5**, 2702.

124. D. Tian, J. Hu, J. Bao, R. P. Chandra, J. N. Saddler and C. Lu, *Biotechnol. Biofuels*, 2017, **10**, 192.
125. S. R. Yearla and K. Padmasree, *Environ. Sci. Pollut. Res.*, 2016, **23**, 18085.
126. M. Farooq, T. Zou, G. Riviere, M. H. Sipponen and M. Österberg, *Biomacromolecules*, 2019, **20**, 693.
127. T. Leskinen, M. Smyth, Y. Xiao, K. Lintinen, M. Mattinen, M. A. Kostiainen and P. Oinas, *Nord. Pulp Pap. Res. J.*, 2017, **32**, 586.
128. H. Li, Y. Deng, B. Liu, Y. Ren, J. Liang, Y. Qian, X. Qiu, C. Li and D. Zheng, *ACS Sustainable Chem. Eng.*, 2016, **4**, 1946.
129. L. Chen, X. Zhou, Y. Shi, B. Gao, J. Wu, T. B. Kirk, J. Xu and W. Xue, *Chem. Eng. J.*, 2018, **346**, 217.
130. S. Cailotto, M. Gigli, M. Bonini, F. Rigoni and C. Crestini, *ChemSusChem*, 2020, **13**, 4759.
131. M. Ago, S. Huan, M. Borghei, J. Raula, E. I. Kauppinen and O. J. Rojas, *ACS Appl. Mater. Interfaces*, 2016, **8**, 23302.
132. I. A. Gilca, V. I. Popa and C. Crestini, *Ultrason. Sonochem.*, 2015, **23**, 369–375.
133. S. R. Yearla and K. Padmasree, *J. Exp. Nanosci.*, 2016, **11**, 289.
134. M. H. Sipponen, H. Lange, M. Ago and C. Crestini, *ACS Sustainable Chem. Eng.*, 2018, **6**, 9342.
135. M. Tortora, F. Cavalieri, P. Mosesso, F. Ciaffardini, F. Melone and C. Crestini, *Biomacromolecules*, 2014, **15**, 1634.
136. D. Yiamsawas, S. Beckers, H. Lu, K. Landfester and F. R. Wurm, *ACS Biomater. Sci. Eng.*, 2017, **3**, 2375.
137. N. Chen, L. A. Dempere and Z. Tong, *ACS Sustainable Chem. Eng.*, 2016, **4**, 5204.
138. J. Pan, Y. Yin, M. Gan, M. Meng, X. Dai, R. Wu, W. Shi and Y. Yan, *Chem. Eng. J.*, 2015, **266**, 299.
139. Y. Yang, Z. Wei, C. Wang and Z. Tong, *Chem. Commun.*, 2013, **49**, 7144.
140. E. D. Bartzoka, H. Lange, K. Thiel and C. Crestini, *ACS Sustainable Chem. Eng.*, 2016, **4**, 5194.
141. Y. Pang, Z. Qin, S. Wang, C. Yi, M. Zhou, H. Lou and X. Qiu, *Colloid Polym. Sci.*, 2020, **298**, 1001.
142. H. M. Caicedo, L. A. Dempere and W. Vermerris, *Nanotechnology*, 2012, **23**, 105605.
143. F. Xiong, Y. Han, S. Wang, G. Li, T. Qin, Y. Chen and F. Chu, *Ind. Crops Prod.*, 2017, **100**, 146.
144. Y. Deng, H. Zhao, Y. Qian, L. Lü, B. Wang and X. Qiu, *Ind. Crops Prod.*, 2016, **87**, 191.
145. Y. Li, M. Zhou, Y. Pang and X. Qiu, *ACS Sustainable Chem. Eng.*, 2017, **5**, 3321.
146. Y. Li, X. Qiu, Y. Qian, W. Xiong and D. Yang, *Chem. Eng. J.*, 2017, **327**, 1176.

147. P. Figueiredo, K. Lintinen, A. Kiriazis, V. Hynninen, Z. Liu, T. Bauleth-Ramos, A. Rahikkala, A. Correia, T. Kohout, B. Sarmento, J. Yli-Kauhaluoma, J. Hirvonen, O. Ikkala, M. A. Kostiainen and H. A. Santos, *Biomaterials*, 2017, **121**, 97.
148. L. Dai, R. Liu, L. Hu, Z. Zou and C. Si, *ACS Sustainable Chem. Eng.*, 2017, **5**, 8241.
149. P. Figueiredo, C. Ferro, M. Kemell, Z. Liu, A. Kiriazis, K. Lintinen, H. F. Florindo, J. Yli-Kauhaluoma, J. Hirvonen, M. A. Kostiainen and H. A. Santos, *Nanomedicine*, 2017, **12**, 2581.
150. L. Chen, X. Zhou, Y. Shi, B. Gao, J. Wu, T. B. Kirk, J. Xu and W. Xue, *Chem. Eng. J.*, 2018, **346**, 217.
151. X. Qiu, Y. Li, Y. Qian, J. Wang and S. Zhu, *ACS Appl. Bio Mater.*, 2018, **1**, 1276.
152. Y. Zhou, Y. Han, G. Li and F. Chu, *Materials*, 2019, **12**, 1694.
153. T. Zou, M. H. Sipponen and M. Österberg, *Front. Chem.*, 2019, **7**, 370.
154. C. Piombino, H. Lange, F. Sabuzi, P. Galloni, V. Conte and C. Crestini, *Molecules*, 2020, **25**, 866.
155. L. Cheng, B. Deng, W. Luo, S. Nie, X. Liu, Y. Yin, S. Liu, Z. Wu, P. Zhan, L. Zhang and J. Chen, *J. Agric. Food Chem.*, 2020, **68**, 5249.
156. M. S. Alqahtani, A. Alqahtani, M. Kazi, M. Z. Ahmad, A. Alahmari, M. A. Alsenaidy and R. Syed, *J. Drug Delivery Sci. Technol.*, 2020, **60**, 102020.
157. C. E. Byrne, C. E. Astete, M. Vaithiyanathan, A. T. Melvin, M. Moradipour, S. E. Rankin, B. L. Knutson, C. M. Sabliov and E. C. Martin, *Nanomedicine*, 2020, **15**, 10.
158. Z. Pourmoazzen, H. Sadeghifar, G. Yang and L. Lucia, *Colloids Surf., B*, 2020, **186**, 110685.
159. S. Gabellone, D. Piccinino, S. Filippi, T. Castrignanò, C. Zippilli, D. Del Buono and R. Saladino, *Int. J. Mol. Sci.*, 2022, **23**, 915.
160. J. H. Lee, J. S. Im, X. Jin, T. M. Kim and J. W. Choi, *ACS Sustainable Chem. Eng.*, 2022, **10**, 5792.
161. D. Yiamsawas, G. Baier, E. Thines, K. Landfester and F. R. Wurm, *RSC Adv.*, 2014, **4**, 11661.
162. J. Zdarta, Ł. Klapiszewski, M. Wysokowski, M. Norman, A. Kołodziejczak-Radzimska, D. Moszyński, H. Ehrlich, H. Maciejewski, A. L. Stelling and T. Jesionowski, *Mar. Drugs*, 2015, **13**, 2424.
163. S. Kim, M. M. Fernandes, T. Matamá, A. Loureiro, A. C. Gomes and A. Cavaco-Paulo, *Colloids Surf., B*, 2013, **103**, 1.
164. J. Zdarta, L. Klapiszewski, A. Jedrzak, M. Nowicki, D. Moszynski and T. Jesionowski, *Catalysts*, 2017, 7(1).
165. W. Gong, Z. Ran, F. Ye and G. Zhao, *Food Chem.*, 2017, **228**, 455.
166. E. Capecchi, D. Piccinino, I. Delfino, P. Bollella, R. Antiochia and R. Saladino, *Nanomaterials*, 2018, **8**, 438.
167. M. H. Sipponen, M. Farooq, J. Koivisto, A. Pellis, J. Seitsonen and M. Österberg, *Nat. Commun.*, 2018, **9**, 2300.

168. A. P. Richter, B. Bharti, H. B. Armstrong, J. S. Brown, D. Plemmons, V. N. Paunov, S. D. Stoyanov and O. D. Velev, *Langmuir*, 2016, **32**, 6468.
169. Ł. Klapiszewski, T. Rzemieniecki, M. Krawczyk, D. Malina, M. Norman, J. Zdarta, I. Majchrzak, A. Dobrowolska, K. Czaczyk and T. Jesionowski, *Colloids Surf., B*, 2015, **134**, 220.
170. A. P. Richter, J. S. Brown, B. Bharti, A. Wang, S. Gangwal, K. Houck, E. A. Cohen Hubal, V. N. Paunov, S. D. Stoyanov and O. D. Velev, *Nat. Nanotechnol.*, 2015, **10**, 817.
171. S. K. Singh, A. Singh and S. Mishra, in *Electrospun Nanofibers: Fabrication, Functionalisation and Applications*, ed. S. K. Tiwari, K. Sharma, V. Sharma and V. Kumar, Springer International Publishing, Cham, 2021, pp. 171–197.
172. P. Mikeš, D. A. Baker, A. Uhlin, D. Lukáš, E. Kuželová-Košťáková, A. Vidrich, J. Valtera, B. Kopřivová, N. Asatiani, L. Salmén and P. Tomani, *J. Polym. Environ.*, 2021, **29**, 2164.
173. D. Kai, W. Ren, L. Tian, P. L. Chee, Y. Liu, S. Ramakrishna and X. J. Loh, *ACS Sustainable Chem. Eng.*, 2016, **4**, 5268.
174. A. Saudi, S. Amini, N. Amirpour, M. Kazemi, A. Zargar Kharazi, H. Salehi and M. Rafienia, *Mater. Sci. Eng., C*, 2019, **104**, 110005.
175. K. R. Aadil, S. I. Mussatto and H. Jha, *Int. J. Biol. Macromol.*, 2018, **120**, 763.
176. E.-S. Lee, Y.-O. Kim, Y.-M. Ha, D. Lim, J. Y. Hwang, J. Kim, M. Park, J. W. Cho and Y. C. Jung, *Eur. Polym. J.*, 2018, **105**, 79.
177. A. Massey-Brooker, M. Vaccaro, S. Scialla, S. Walker, P. Morganti, F. Carezzi, B. Benjelloun-Mlayah, C. Crestini, H. Lange and E. Bartzoka, *A consumer goods product comprising chitin, lignin and a polymer or co-polymer*, CA2960127A1, 2016.
178. T. Abudula, K. Gauthaman, A. Mostafavi, A. Alshahrie, N. Salah, P. Morganti, A. Chianese, A. Tamayol and A. Memic, *Sci. Rep.*, 2020, **10**, 20428.
179. D. A. Elsherbiny, A. M. Abdelgawad, M. E. El-Naggar, B. A. Hemdan, S. Ghazanfari, S. Jockenhövel and O. J. Rojas, *Int. J. Biol. Macromol.*, 2022, **205**, 703.
180. E. M. Ahmed, *J. Adv. Res.*, 2015, **6**, 105.
181. A. Mahmood, D. Patel, B. Hickson, J. DesRochers and X. Hu, *Int. J. Mol. Sci.*, 2022, **23**, 1415.
182. R. Mohammadinejad, H. Maleki, E. Larrañeta, A. R. Fajardo, A. B. Nik, A. Shavandi, A. Sheikhi, M. Ghorbanpour, M. Farokhi, P. Govindh, E. Cabane, S. Azizi, A. R. Aref, M. Mozafari, M. Mehrali, S. Thomas, J. F. Mano, Y. K. Mishra and V. K. Thakur, *Appl. Mater. Today*, 2019, **16**, 213.
183. I. E. Raschip, A. D. Panainte, D. Pamfil, L. Profire and C. Vasile, *Med. Surg. J.*, 2015, **119**, 1189.
184. K. Ravishankar, M. Venkatesan, R. P. Desingh, A. Mahalingam, B. Sadhasivam, R. Subramaniyam and R. Dhamodharan, *Mater. Sci. Eng., C*, 2019, **102**, 447.

185. Y. Zhang, M. Jiang, Y. Zhang, Q. Cao, X. Wang, Y. Han, G. Sun, Y. Li and J. Zhou, *Mater. Sci. Eng., C*, 2019, **104**, 110002.
186. L. Berglund, F. Forsberg, M. Jonoobi and K. Oksman, *RSC Adv.*, 2018, **8**, 38219.
187. J. Domínguez-Robles, C. Mancinelli, E. Mancuso, I. García-Romero, B. F. Gilmore, L. Casettari, E. Larrañeta and D. A. Lamprou, *Pharmaceutics*, 2020, **12**, 63.
188. A. A. Vaidya, C. Collet, M. Gaugler and G. Lloyd-Jones, *Mater. Today Commun.*, 2019, **19**, 286.
189. J. Domínguez-Robles, N. K. Martin, M. L. Fong, S. A. Stewart, N. J. Irwin, M. I. Rial-Hermida, R. F. Donnelly and E. Larrañeta, *Pharmaceutics*, 2019, **11**, 165.
190. M. Tanase-Opedal, E. Espinosa, A. Rodríguez and G. Chinga-Carrasco, *Materials*, 2019, **12**, 3006.
191. A. J. Capel, R. P. Rimington, M. P. Lewis and S. D. R. Christie, *Nat. Rev. Chem.*, 2018, **2**, 422.
192. M. Pishnamazi, J. Iqbal, S. Shirazian, G. M. Walker and M. N. Collins, *Int. J. Biol. Macromol.*, 2019, **124**, 354.
193. M. Pishnamazi, H. Hafizi, S. Shirazian, M. Culebras, G. M. Walker and M. N. Collins, *Polymers*, 2019, **11**, 1059.
194. J. Domínguez-Robles, S. A. Stewart, A. Rendl, Z. González, R. F. Donnelly and E. Larrañeta, *Biomolecules*, 2019, **9**, 423.
195. S. C. Lee, E. Yoo, S. H. Lee and K. Won, *Polymers*, 2020, **12**, 699.
196. Y. Qian, X. Zhong, Y. Li and X. Qiu, *Ind. Crops Prod.*, 2017, **101**, 54.
197. Y. Qian, X. Qiu and S. Zhu, *ACS Sustainable Chem. Eng.*, 2016, **4**, 4029.
198. Y. Qian, X. Qiu and S. Zhu, *Green Chem.*, 2014, **17**, 320.
199. B. Rukmanikrishnan, S. Ramalingam, S. K. Rajasekharan, J. Lee and J. Lee, *Int. J. Biol. Macromol.*, 2020, **153**, 55.
200. W. Wu, T. Liu, X. Deng, Q. Sun, X. Cao, Y. Feng, B. Wang, V. A. L. Roy and R. K. Y. Li, *Int. J. Biol. Macromol.*, 2019, **126**, 1030.
201. B. Rukmanikrishnan, S. Ramalingam, S. K. Rajasekharan, J. Lee and J. Lee, *Int. J. Biol. Macromol.*, 2020, **153**, 55.
202. D. B. Warheit, P. J. Borm, C. Hennes and J. Lademann, *Inhalation Toxicol.*, 2007, **19**, 631.
203. H. Rauscher, K. Rasmussen and B. Sokull-Klüttgen, *Chem. Ing. Tech.*, 2017, **89**, 224.
204. R. W. S. Lai, K. W. Y. Yeung, M. M. N. Yung, A. B. Djurišić, J. P. Giesy and K. M. Y. Leung, *Environ. Sci. Pollut. Res.*, 2018, **25**, 3060.
205. A.-S. Jääskeläinen, T. Liitiä, A. Mikkelson and T. Tamminen, *Ind. Crops Prod.*, 2017, **103**, 51.
206. A. J. Ragauskas, G. T. Beckham, M. J. Biddy, R. Chandra, F. Chen, M. F. Davis, B. H. Davison, R. A. Dixon, P. Gilna, M. Keller, P. Langan, A. K. Naskar, J. N. Saddler, T. J. Tschaplinski, G. A. Tuskan and C. E. Wyman, *Science*, 2014, **344**, 1246843.

207. H. Wang, Y. Pu, A. Ragauskas and B. Yang, *Bioresour. Technol.*, 2019, **271**, 449.
208. D. S. Bajwa, G. Pourhashem, A. H. Ullah and S. G. Bajwa, *Ind. Crops Prod.*, 2019, **139**, 111526.
209. F. H. M. Graichen, W. J. Grigsby, S. J. Hill, L. G. Raymond, M. Sanglard, D. A. Smith, G. J. Thorlby, K. M. Torr and J. M. Warnes, *Ind. Crops Prod.*, 2017, **106**, 74.

CHAPTER 8

Lignin and Its Composites for Tissue Engineering

SARAH HANI SHOUSHRAH,[a] ABLA ALZAGAMEEM,[a,b] JONAS BERGRATH,[a] EDDA TOBIASCH[a] AND MARGIT SCHULZE*[a]

[a] Department of Natural Sciences, Bonn-Rhein-Sieg University of Applied Sciences, von-Liebig-Strasse 20, 53359 Rheinbach, Germany; [b] Chemistry Department, Tafila Technical University, PO Box 179, 66110 Tafila, Jordan
*Email: margit.schulze@h-brs.de

8.1 Introduction to Components of Tissue Engineering

Tissue engineering (TE) is a branch of regenerative medicine that aims to restore and/or improve tissue function. It is an interdisciplinary field that combines biological, chemical, and engineering properties.[1] In a tissue-engineered construct, cells (progenitor/stem cells or mature differentiated cells) are provided with a supporting structure or a template (scaffold or matrix) to achieve a three-dimensional (3D) construct with suitable stimuli (biomolecules and physical signals) that will provoke the cells to behave as desired and the whole construct to reproduce or replace the damaged tissue (see Figure 8.1). The interaction between these components and the surrounding tissue is what determines the success of the construct. The scaffold provides mechanical support for the cells, allowing a structured 3D cell mesh. In comparison to cellular therapy where only cells are injected, using a scaffold can modulate the behaviour of the cells since their development

Biomaterials Science Series No. 16
Lignin-based Materials: Health Care and Medical Applications
Edited by Kuruvilla Joseph, Runcy Wilson, Gejo George and Saritha Appukuttan

Published by the Royal Society of Chemistry, www.rsc.org

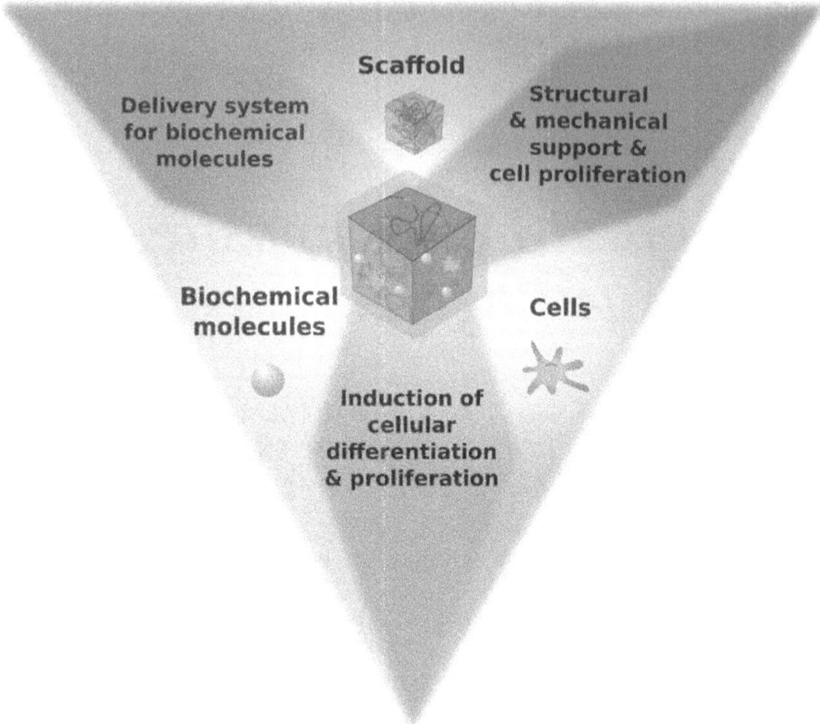

Figure 8.1 Components involved in tissue engineering. Tissue engineering approaches can be based on treatment with cells, biochemical molecules, or scaffolds separately. However, combination of these components can provide a better therapeutic benefit of constructs for the respective target tissue.

and function are affected by their microenvironment.[2–5] Here we also have discussed the importance of the biochemical molecules that can be loaded onto the scaffold. In this context, the term "biochemical molecules" is used to describe substances that can be combined with the scaffold to stimulate cells to regenerate the tissue, whether they be drugs, growth factors, cytokines, nucleic acids, *etc.* The scaffold acts as a tool for the delivery of the biochemical molecules whether to the cells seeded into the construct or to the tissues surrounding the construct to promote regeneration. It is not necessary to have all three elements in one construct, as there are approaches that depend on only one or two of the components, such as cellular/scaffold-free approaches.

TE constructs offer an alternative means for current treatment standards such as autologous grafts, allografts, and xenografts. These grafts despite being the go-to treatment option have their limitations. Examples include the risks of immune rejection and disease transmission, a limited donor pool, and insufficient vascularization.[6–8] In this chapter, TE of five major tissues will be discussed: cardiovascular TE (CVTE), bone TE (BTE), cartilage TE (CTE), skin TE (STE), and neural tissue TE (NTE).

CVTE constructs are used to treat cardiovascular diseases (CVDs) such as ischemic heart disease and myocardial infarction. Since the heart tissue has a limited regenerative ability, prolonged heart damage may cause irreversible acute or chronic heart failure. Current treatment choices include symptomatic treatment and transplantation in the end stages of CVDs. Possible applications for CVTE constructs include grafts/patches that could replace or restore the damaged heart tissue, heart valve replacements, and vessel grafting.[9] One promising emerging approach for production of CVTE constructs is the 3D bioprinting technology. 3D bioprinting entails the deposition of biomaterials and cells in a pattern to produce grafts in a certain spatial shape. However, it is still in its early stages of development, since the choices of bioinks and cells still need to be optimized.[10]

For neural tissue engineering, the aim is to repair nerve damage caused by injury in the peripheral nervous system (PNS) or the central nervous system (CNS).[11] In the PNS, nerve damage occurs mainly as a result of trauma which causes stretching and tearing of peripheral nerves that could result in the formation of a gap within the nerve. When the gap is significant in size, meaning more than 1 cm and up to 5 cm, nerve autografts are used to surgically treat the damage as it is not possible to repair such injury with microscopic surgery.[12] However, nerve autografts have several drawbacks such as neuroma formation, donor site morbidity, and limited sources.[13] An alternative to nerve grafts is a 3D synthetic device called a nerve guidance channel (catheter or conduit) (NGC). The NGC is a tubular device with a single lumen that bridges the severed nerve stumps, reducing the growth of scar tissues and protecting the nerve from compression exerted by the surrounding tissue. This helps in guiding the growth of nerve axons. In addition, NGCs can contain neurotrophic factors that improve axon regeneration.[14,15] Nerve regeneration within the CNS is even more challenging. Damage to the brain and spine results in cell death and an inflammation cascade. Current treatments focus on handling the symptoms and limiting the extent of injury. However, such treatments do not induce nerve regeneration. One reason why treating CNS injuries is more challenging is that access to the brain is hindered by the protective tissue surrounding it, namely the bony structure of the skull and the blood–brain barrier/blood–spinal cord barrier. Establishing direct access for treatment delivery is risky and can cause further tissue damage. In the case of drugs that can bypass the barrier, often high doses are required which can have adverse side effects also in other body parts.[13]

In contrast to nerve injuries, most bone injuries can usually be repaired with medical interventions such as screws and a cask supporting the body's own repair mechanism. However, there are injuries that require more than surgical interventions in order to heal, such as critical size bone defects.[16,17] Like nerve tissues, these gaps are too large to heal and thus require the use of a graft. Nowadays, the gold standard is still using autografts, if possible.

As an alternative, there are various synthetic scaffold materials available for synthetic bone grafts. However, they still have their drawbacks such as inefficient integration and vascularization.[18]

As for cartilage, degenerative musculoskeletal disorders such as osteoarthritis cause a need for regenerative options. Since it is a tissue that lacks vascularization and neural tissues, it has limited regenerative capability. Besides palliative treatments, reparative approaches are needed which include a method that provides access for mesenchymal stem cells (MSCs), located in the subchondral bone, to migrate into the cartilage, which is called a bone marrow stimulation technique.[19] Osteochondral transplantations using autografts are used for treatment of severe defects. Yet they suffer the same drawbacks that come with using auto- and allografts. Therefore, efforts are made towards developing synthetic grafts that can repair the damaged tissue by delivering the appropriate cells with an artificial scaffold, and signalling factors to the defect site. The scaffold should then provide a structural and functional extracellular matrix (ECM) that imitates the native cartilage tissue.[20,21]

Each of the tissues mentioned has different cellular, physical, and chemical compositions. Scaffolds designed to interact with these tissues should have similar compositions, suitable to regenerate or replace the respective tissue. Also, there are essential safety criteria that the scaffold must meet before it can be applied in clinical settings. In the next sub-section these criteria will be discussed.

8.1.1 Optimal Scaffold Design Criteria for TE

When designing scaffolds, there are multiple criteria that need to be taken into consideration. Some aspects are related to safety for clinical applications, while others are tissue specific. In this chapter these criteria will be discussed (see Table 8.1). The first criterion is biocompatibility; it is an essential requirement for the scaffold and the material it is made of. Biocompatibility has been defined as the ability of the scaffold to support cellular behaviour and functions appropriately in an intended application without inducing an undesirable response in host tissues locally or systemically,[22,23] meaning that a biocompatible scaffold will not induce undesirable reactions such as hemolysis, coagulation, toxic side effects, inflammation, and immune rejection in the host system, while serving its purpose.[13] The second criterion, closely related to biocompatibility, is biodegradability. Most scaffolds used for tissue regeneration purposes are supposed to be degradable since their purpose is to support the tissue to regrow and thus be replaced in the end. Therefore, it is essential that when the scaffold is degraded, its degradation products also cause no harm to the host system, meaning mainly that they cause no toxicity and can be safely eliminated *via* and from the host system. In this sense, biodegradability can be subsumed under the term biocompatibility, though the term biodegradability can also encompass other considerations besides biosafety.

Table 8.1 Necessary criteria to consider when designing a tissue-specific scaffold.

Criteria	Definition and techniques
Biocompatibility	The ability of the scaffold to support cellular behaviour and functions appropriately in an intended application without inducing an undesirable response in host tissues locally or systemically. ISO 10993-1 provides guidelines for determining the biocompatibility of medical devices. To ensure the safety of the fabricated scaffolds, they are tested *in vitro* and *in vivo*. *In vitro* assays include cell viability assays such as MTT assay, detection of apoptotic markers, and cell morphology analysis using microscopy. *In vivo*, animal experiments are used to examine the scaffold biocompatibility.[210]
Biodegradability	Scaffolds used for tissue regeneration purposes should be degradable and successively be replaced by newly formed tissue. The by-product of degradation should not be toxic to the host and must be safely eliminated from the host system. Biodegradability can be determined *in vitro* by hydrolytic degradation studies and Fourier transform infrared spectroscopy.[211]
Mechanical properties	The mechanical properties of the scaffold should be similar to those of the target tissue. The mechanical properties of scaffolds can encompass tensile strength, elasticity, and compressive strength. Mechanical properties can be tested with an axial compression test, *via* elastic modulus calculations, and with an electromechanical testing machine.[212,213]
Pores	Porosity is crucial for vascularization. It allows for transport of nutrients and oxygen to the cells inside the scaffold. Different cells require different pore sizes. Porosity and pore size affect the degradability and mechanical stability of the scaffold. Porosity and pore size can be measured with techniques such as the gravimetric method, mercury porosimetry, the liquid displacement method, SEM analysis, microcomputed tomography imaging, capillary flow porometry, and a permeability-based method.[30]
Electroconductivity	Electroconductive scaffolds contain materials that can transfer electrical signals. Such materials can improve the function of scaffolds for cardiovascular and neural tissue engineering. Electroconductivity can be measured with a digital multimeter.[214]
Vascularization	Vascularization provides nutrients and oxygen to the cells in the scaffold and is vital for TE constructs. Successful vascularization is highly dependent on interconnective pores within the scaffold. Certain cellular and biochemical features can also improve vascularization. There are several *in vitro* and *in vivo* approaches for testing vascularization of scaffolds. *In vitro* assays include migration assays and tube formation assay. *In vivo* the chick chorioallantoic membrane assay is frequently used.[215,216]

Table 8.1 (*Continued*)

Criteria	Definition and techniques
Osteoconductivity and osteoinductivity	Osteoinductive scaffolds can induce the differentiation of stem cells into an osteogenic lineage. An osteoconductive scaffold allows bone growth on its surface or inside its pores, meaning that it supports osteoblasts' bone forming function. There are several osteoinductive and conductive materials such as hydroxyapatite. Using biomolecules in combination with scaffolds could increase these properties. Osteoconductivity and osteoinductivity can be tested *in vitro* by examining the proliferation and osteogenic differentiation of cells on the scaffolds. Techniques used include alkaline phosphatase quantification, Alizarin Red S staining, and specific protein marker expression determination.[217]
Antioxidant properties	Scaffolds that exhibit antioxidant properties can improve tissue regeneration by preventing tissue damage caused by oxidative stress and reactive oxygen species (ROS). *In vitro* assays such as the DPPH assay, the hydroxyl radical scavenging assay, and the ferric reducing/antioxidant power (FRAP) assay can be used to determine the antioxidant properties of scaffolds.[218]
Antimicrobial properties	Scaffolds with antimicrobial properties reduce the risk of infections in the tissues they are transplanted in and complications that can arise with them. *In vitro* bacterial cultures are usually used for testing antimicrobial properties such as a zone of inhibition test and Gram staining.[219]

The degradation rate of a biodegradable scaffold should match the rate of tissue regeneration. Therefore, the degradation rate will vary depending on the target tissue as some tissues heal faster than others. On the other hand, materials used as long-term tissue replacements should be undegradable as the alternative is detrimental to the host.[13,22]

The mechanical properties of scaffolds and their source materials are another important criterion to consider. The mechanical properties of the scaffold should resemble those of the target tissue in order to support its function and regeneration properly. For instance, a scaffold designed to repair damage to a load bearing bone will have to stand different pressures applied to it than that used for skin wound healing. The mechanical properties of scaffolds can encompass tensile strength, elastic modulus, and compressive strength.[24] As an example, the experimental Young's modulus reading of the femur can go up to 500 MPa, while for skin tissues it varies from 2.9 to 150 MPa.[25,26] Another criterion related to mechanical scaffold properties is pores. Pore size and porosity affect the scaffold's mechanical stability and biodegradability.[27] For example, scaffolds designed for CVTE should have a thickness and mechanical strength that will tolerate the strain of transplantation and later the contraction of the heart until the tissue is regenerated.[28,29]

Another important feature of pores is that they are necessary for proper vascularization which will be discussed later.[30] The pore size must be considered carefully. An example for this are NGCs. While a large pore size supports vascularization, if the size is too large, leakage can happen that blocks the nerve duct with fibrous tissue. On the other hand, a pore size that is too small will lead to a decrease in nerve regeneration.[27] Although there are several methods available to fabricate NGCs, each has its own limitations such as the formation of irregularly shaped pores and low porosity. Vijayavenkataraman and colleagues fabricated a poly(caprolactone) (PCL)-based 3D porous NGC using electrohydrodynamic jet 3D printing (EHD-jetting) which provides better control in the fabrication process as other methods. It allows for the fine tuning of properties such as fiber diameter, pore size, and porosity. The authors investigated different pore sizes (125 to 550 μm) and porosities (65 to 88%) and reported that a pore size of 125 ± 15 μm and a porosity of greater than 60% provided an optimal degradation rate to match the nerve regeneration rate.[31] An optimal porosity will of course be different for other tissues. For STE, scaffolds are sometimes designed multi-layered to mimic layers of the skin with their respective cellular and tissue composition. In STE scaffolds, a pore size of 5 μm was reported to support neovascularization, but other processes such as fibroblast growth and dermal repair would require larger pores.[32] In comparison to bone tissue, the optimal porosity for skin tissue vascularization and growth is between 100 and 400 μm.[33]

There are also tissue-specific criteria. Electroconductivity is an example of a criterion that is important for neural and cardiovascular tissues. Biomaterials with electroconductive abilities include poly(3,4-ethylenedioxythiophene):poly(4-styrenesulfonate) (PEDOT:PSS), poly(pyrrole) (PPy), polyaniline, and polythiophene. The addition of such materials can also increase the biodegradability of the scaffold.[34–36] Roshanbinfar and colleagues designed an electroconductive biohybrid hydrogel for the treatment of heart failure. They intended to solve the arrhythmia issue caused by inefficient electrical coupling between graft and host tissue. The scaffold was composed of collagen, alginate, and the electroconductive PEDOT:PSS. The function of rat cardiomyocytes and human induced pluripotent stem cell (iPSC)-derived cardiomyocytes was tested. The authors reported enhanced electrical conductivity, increased beating frequencies, and improved tissue maturation.[37] In another study electroconductive NGCs were produced by 3D printing using a combination of PPy and PCL (PPy/PCL). The addition of PPy/PCL resulted in improved conductivity and softer scaffolds. Vijayavenkataraman and colleagues also reported that scaffolds with a concentration of 1% v/v of PPy/PCL supported the growth and maturation of human embryonic stem cell (ESC)-derived neural crest stem cells into neuronal cells.[38]

Vascularization is a vital criterion for successful integration of a TE construct, and it is one of the major challenges that must be faced before translation of TE into clinical settings.[39–41] Vascularization allows the transport of nutrients and oxygen to the cells within the scaffold, thus improving regeneration. There are several vascularization strategies. They can be divided into strategies that stimulate vascularization inside the scaffold by using an optimized combination of biomaterials, biomolecules, and cells, and pre-vascularization strategies that hinge on establishing a vascular network in the graft before using it for therapy. For the first strategy, the constructed biomaterial composition, structure, and biomolecules (such as growth factors) should allow cells to attach and migrate to promote angiogenesis. Stem cells and endothelial cells are examples of cells used in such strategies.[42–44]

Pre-vascularization strategies can be further divided into *in vitro* and *in vivo* pre-vascularization approaches. In the *in vitro* approach, the cells that are supposed to form blood vessels, such as endothelial cells, are seeded into the scaffold *in vitro* before transplantation. Another *in vitro* approach that does not rely on scaffolds relies on the formation of cell clusters, called spheroids, and cell sheets that can be used to induce angiogenesis. Cell sheets were able to promote neovascularization and cardiomyocyte proliferation to treat myocardial infarction in a rat model.[45] In the *in vivo* approach, the scaffold is first transplanted into a highly vascularized tissue, and after the establishment of blood vessels, the scaffold is moved to the target site. However, this technique is time consuming and requires several surgical procedures.[46] The 3D bioprinting technique can also be used to manufacture pre-vascularized scaffolds since it allows precise control of the spatial allocation of cells within the construct.[47,48]

When designing scaffolds for BTE, two important terms are often discussed: osteoconductivity and osteoinductivity. A scaffold with osteoinductive properties can induce the differentiation of stem cells into an osteogenic lineage. An osteoconductive scaffold allows bone tissue to grow on its surface or inside its pores, meaning that it supports osteoblasts' function.[49] An example of an osteoinductive and osteoconductive scaffold is calcium phosphate ceramic, which has been studied intensely for BTE. Yet there are also other biomaterials and strategies to improve these properties such as the use of bone morphogenetic proteins.[50–57]

Hence, biomaterials with ROS scavenging properties are sought after for generating scaffolds with antioxidative properties.[58] Such a material is lignin, which will be discussed in this chapter.

It goes without saying that infections are a serious complication in any medical procedure. For scaffolds that are applied through surgical procedures, it is also crucial to consider the antimicrobial properties of the materials used. Scaffolds can acquire antimicrobial properties by using materials with innate antimicrobial activities, or by designing scaffolds with controlled release of antimicrobial drugs.[59] Lignin as a biomaterial is known for its antimicrobial properties, which will be discussed in the next section.

8.2 Lignin as a Biomaterial

This part of the chapter focuses on three aspects: (1) lignin, (2) lignin bioactivity, and (3) lignin-based materials developed and studied as additives and/or components for TE applications. It importantly focuses on the structure–property relationships: *e.g.*, the influence of the biomass resource and isolation process on the chemical structure of lignin and therefore its properties, such as antioxidant and antimicrobial activities, and its applications for TE. Finally, lignin-derived materials for TE applications will be presented considering most recently published studies.

8.2.1 Lignin Bioactivity

Bioactive substances extracted from plants have been widely used as antioxidant, anticancer, and antiviral agents in TE and biomedicine due to their enhanced chemical stability, good processability, and bioactivity.[60] Lignin, being a natural bioactive compound and the second most abundant bio-based polymer, has been of great interest in the last few decades owing to its sustainability and green and biodegradable nature in addition to having a variety of other biological activities, such as antioxidant, antimicrobial, antitumor, antifungal,[61] and antiviral activities in medical treatment. Additionally, it provides health benefits, enhances enzymatic hydrolysis, enhances UV screening, and improves drought resistance in plants. All these properties open the door to using lignin as an innovative biologically active material.[62–68]

Recently, lignin has been studied for wound healing dressings by developing lignin-containing hydrogels with chitosan/poly(vinyl chloride) (PVC),[69] alginate,[70] silver nanoparticles,[71] and poly(lactic acid) (PLA).[72] More benefits were observed when using lignin or its derivatives as additives for decreasing cholesterol in human liver cells and possibly avoiding coronary heart disease.[73] Lignin also shows potential for diabetes treatment,[74] TE,[75] UV shielding,[76] and drug delivery.[77]

8.2.1.1 Correlation Between Lignin Structure and Bioactivity

Lignin is associated with several functional groups including phenolic and aliphatic hydroxyl groups, carbon–carbon double bonds, methoxy groups, aromatic units, *etc.*[78] Lignin's properties strongly depend on the content of these functional groups.[79] Sulfite and kraft lignins, both by-products of the pulp and paper industry, are more commonly used lignins to synthesize pharmaceutical products.[80] Figure 8.2 illustrates some lignin's structural characteristics and its corresponding bioactivity.[73,79–85]

The mainly studied bioactivities of lignin are its antioxidant and antimicrobial activities. This is due to the importance of antioxidant and antimicrobial agents in our daily lives. The applications of such materials range

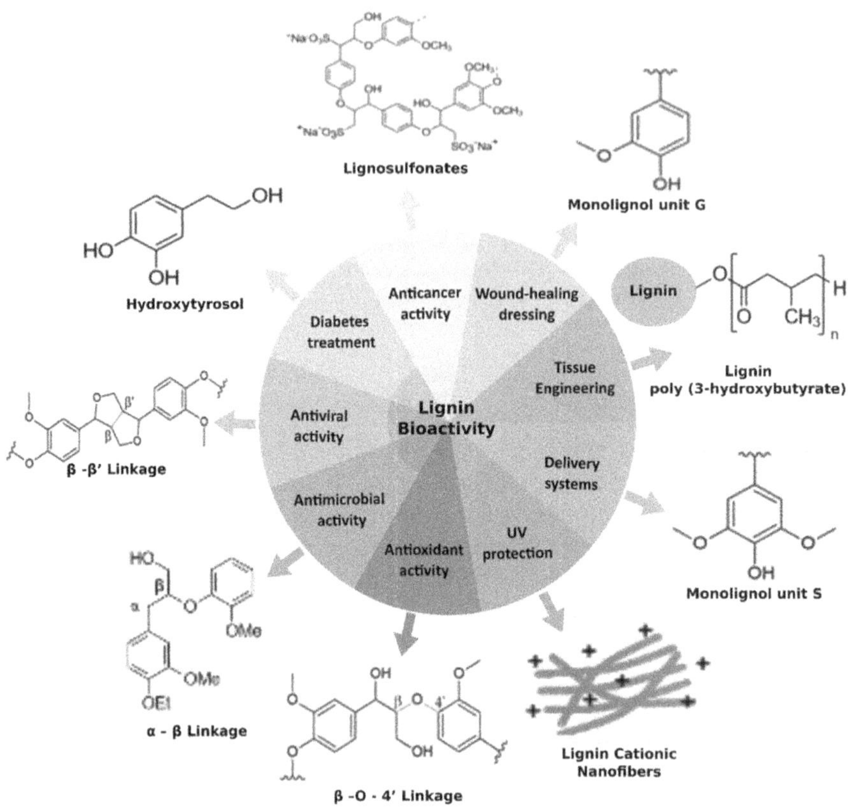

Figure 8.2 Structural fragments assigned to lignin bioactivity. The mainly studied bioactivities of lignin are its antioxidant and antimicrobial activities. This arises from the importance of antioxidants and antimicrobial agents in our daily lives. The applications of such materials range from food chemistry to biomaterials for several applications including biomedicines and pharmaceuticals.

from food chemistry to biomaterials for several applications including biomedicines and pharmaceuticals.

Based on the studies reported so far, lignin's antioxidant activity seems to be linked with the presence of free phenolic hydroxyl groups in its structure.[66] Auto-oxidation of polymers is reduced by accepting a proton from the phenolic hydroxyl group, resulting in blocking the oxidation reaction. This is supported by the presence of the π-electronic system supplied by the aromatic ring that enhances the deprotonation of the phenolic hydroxyl group more than in aliphatic ones.[86] The polarity of the solvent used for lignin extraction plays a significant role in the antioxidant activity where the yield of phenolic antioxidants increases as the polarity of the solvent increases.[87] Lignins which are rich in monolignol units S and G (see Figure 8.2) and a low carbonyl group content usually show a higher antioxidant[88] ability. Likewise, lignins with lower molecular weight (weight average M_w, number average M_n)

and polydispersity (PD) values have a higher scavenging ability for ROS.[89] Other factors affecting the antioxidant capacity of lignin are the pulping process, the source of biomass, the presence of low molecular weight fragments of lignin, *etc.* Lignins with high purity and fewer fragments, which can be produced during the pulping process, exhibit higher antioxidant activities than those with low molecular weight fragments.[62,90]

Furthermore, lignin has been found to have good antimicrobial properties due to the large number of polyphenolic compounds that make up the structure of lignin. These polyphenols can destroy the cell membranes, causing death of both Gram-positive and -negative bacteria. Even without fully and exactly understanding the mechanism for cell dissolution, it is believed that the phenolic compounds in lignin can act as ionophores known to increase the ion permeability of the cell membrane, leading to cell death.[65] Usually, the antibacterial activity is related to the origin of the lignin. It is particularly caused by the presence of phenolic species and different oxygen-containing functional groups (methoxyl and epoxy groups) in its structure.[91]

The antimicrobial mechanisms by which lignin inhibits bacterial growth may include the production of localized heat and different ROS after light irradiation. The adhesion of lignin to the bacterial membrane could involve the sugar content of the lignin (*i.e.*, sugars can be attracted to the peptidoglycan layer present in the bacterial cell wall). Besides, the nanocomposites attached appear to increase the activity against *Staphylococcus aureus*, while they remain harmless to *Escherichia coli*. This shows good selectivity towards *S. aureus*.[64] As suggested in previous studies, lignins containing many low molecular weight fragments showed higher antimicrobial activities than those with fewer low molecular weight fragments.[63,92] Both, a high carbohydrate content including xylan and glucan and lignin of low molecular weight, affected the antimicrobial activity of lignin against *S. aureus* and *Listeria monocytogenes* (Gram-positive bacteria),[62] which agrees with the previous study.[64]

8.2.1.2 Benefits of Lignin-derived Materials Over Common Materials

Biomaterials are utilized for the assessment, treatment, improvement, repair, or substitution of tissues or organs of the body. Conventional materials are naturally inert, and these materials lack interaction with the nearby tissues and are indeed toxic to humans.[93] Traditional metallic biomaterials, for example, including titanium alloys, stainless steels, and cobalt–chromium alloys, have been widely used clinically for load bearing hard tissue reconstruction and regeneration. They exhibit superior mechanical properties, durability, and machinability. They are, however, considered to be non-degradable, and a second removal surgery is necessary due to the long-term clinical complications that might occur.[94]

Biomaterials contribute in a wide variety of healthcare areas and have aided great enhancements in different biomedical fields, such as medical implants, TE, immunotherapies, and drug delivery.[95] Biomaterials are classified based on their chemical structure (metals, ceramics, polymers, and composites) and on their interaction with the biological environment (inert, bioactive, bioresorbable).[96]

Natural polymers, both protein- and polysaccharide-based, display enormous potential in TE. In contrast, traditional materials previously designed no longer meet the needs of modern medicine, where biomaterials are designed to make extensive biological and physical interactions with tissues and cells.[97–99]

Lignin's aromaticity, various reactive functional groups, excellent ultraviolet and oxidation resistance, high thermal stability, non-toxicity, biodegradability, renewability, and low costs made it a remarkable feedstock for a broad variety of applications.[100] Hence, lignin-based nanomaterials, hydrogels, and films, and some other lignin-based materials have also been rapidly developed for biomedical applications.[101] However, there were other challenges regarding the use of lignin-derived biomaterials. Table 8.2 lists some of the pros and cons of using lignin-derived biomaterials and lignin's effects on these developed materials.

These challenges should be addressed thoroughly and the development of new lignin-based materials, as future alternatives to conventional materials, is needed to widen the scope of lignin usage and to transfer it from the research state to an applicable state as an enhancement of our life.

8.2.2 Lignin-derived Materials for TE Applications

In terms of a sustainable shift in thinking, it is more important than ever to use biomaterials or biopolymers to produce regenerative tissue for medical purposes.[102] These compounds play a supporting role in designing ECMs and stabilizing 3D cell constructs. A wide range of synthetic and natural polymers are used as basic scaffolds for TE.[103] In addition to synthetic polymers (such as poly(ethylene glycol) (PEG), poly(glycolic acid) (PGA), poly(lactic-*co*-glycolic acid) (PLGA), PCL, and poly(vinyl alcohol) (PVA)), various ceramics (*e.g.*, alumina, hydroxyapatite, or tetracalcium phosphate) are also used.[104,105] Some biocompatible proteins like gelatin, elastin, and even keratin have already been used due to their elastic and bioactive properties. Biopolymers such as hyaluronic acid, alginates, and chitosan are widely used as hydrogels due to their low toxicity, high adsorption capacity, and helpful swelling properties.[106] However, with almost all the biopolymers mentioned, it is extremely difficult to achieve a balance between good mechanical stability and good antioxidant and antimicrobial properties. This is exactly the starting point for the use of lignin as an aromatic scaffold-building biopolymer.[107] Besides the characteristics listed earlier, it is also frequently used in the design of biomedical applications, since it is

Table 8.2 Pros and cons of lignin-derived biomaterials.

Lignin-derived biomaterial	Pros	Cons	Ref.
Hydrogels	1. Biodegradability 2. Inherent biocompatibility 3. Non-toxicity 4. Sustainability promotion 5. Affordability 6. Comparable performance as commonly used materials 7. Excellent mechanical strength and related properties 8. Porous 3D network structure	9. Complex structure of randomly cross-linked building blocks 10. Rigidity due to aromatic components 11. Phase separation	220–222
Nanoparticles	12. Antioxidant activity 13. Antimicrobial activity 14. Optical properties 15. Porous structure 16. No cytotoxicity 17. Biocompatibility 18. Encapsulation efficiency 19. Good UV resistance 20. Thermal stability 21. Improvement in mechanical properties	22. Heterogeneity 23. Complex structure 24. Expensive to prepare (due to low yields)	100, 101, 223 and 224
Scaffolds	25. Antioxidant and antimicrobial activity 26. No cytotoxicity 27. Low stiffness 28. Good cell adhesion 29. Enhanced porosity 30. Economically feasible 31. Enhanced mechanical properties	32. Complex structure 33. High polydispersity	16, 40, 46, 75, 101 and 225

extremely favourable in terms of high thermal stability, useful stiffness, and inherent antibacterial activity.[108] In addition to mimicking the ECM by various lignins alone, the use of lignin as a composite material is also very popular due to its high content of phenolic hydroxyl groups.[109]

8.2.2.1 Current Tissue Targets Using Lignin-derived Materials

Lignin has been proven in many manufacturing technologies and materials to prevent pollution and reduce the consumption of synthetic scaffolds (such as bone, skin, nerve, and muscle), and has also had very positive effects on various TE applications.[75] In the following section, the use of various lignins as templates to produce different types of tissues and composite materials is reported. Focus is given to the most recently reported studies of lignin exploited for TE applications (see Table 8.3).

8.2.2.2 Criteria to be Addressed Using Lignin in TE

There are a variety of tissue types that differ in their morphology and function. A suitable lignin, to address a performance spectrum, must cover the specific criteria of the tissue type. This chapter discusses four basic tissue types that are currently discussed in relevant literature (see Table 8.3): cartilage and bone, skin, nerve, and cardiovascular tissue. The prerequisites for new tissue formation are a structural scaffold, cells (mostly MSCs), and specifically secreted proteins that serve as growth factors.[110] The starting point for lignin is clearly the structural scaffold to provide its desired input to the tissue-specific environment. General criteria to be addressed were discussed in detail previously and are listed as follows:

- Mechanical stability while bridging defects
- Applicable handling (even in a sterile environment)
- Biocompatibility and ingrowth into the surrounding tissue
- Degradation into equilibrium with replacement by autologous tissue
- Cost effectiveness.

In general, the goal is to match the synthetic lignin composite mechanically/morphologically to the original tissue matrix, so that the cells produce the ECM and thus provide increased structural stability to the tissue.[111] The lignin composite scaffold is then gradually degraded. After complete resorption, only the newly created functional cell–matrix tissue remains. In addition to the general requirements for successful TE, the structural scaffold must be adapted with specific criteria for the tissue to be treated; Figure 8.3 provides an overview of these criteria.

A promising approach to the general suitability of lignin is provided by Menima-Medzogo *et al.*[241] For this purpose, the authors investigated the cytotoxic effect of organosolv lignin on various human primary cell types commonly used in TE. From the obtained data, no cytotoxic behaviour was observed. Furthermore, parameters such as stiffness, cell adhesion, and viscosity were evaluated when organosolv lignin was added, and its suitability as a carrier material in TE was confirmed. Figure 8.4 shows the results of this test series once with agarose as a carrier material and once with agarose and added lignin.

8.2.2.2.1 Cartilage and Bone Engineering. In principle, the production of lignin scaffolds enables making them as similar as possible to the original bone matrix. In addition to the most important requirement, biocompatibility (immunological defense reaction), the extracellular lignin composite should ideally have comparable mechanical properties to the

Table 8.3 Most recently reported studies on lignin-based materials investigated for tissue engineering applications.

Composition	Lignin type	Desired input of lignin	Application potential	Ref.
Electrospun PCL–lignin nanofibers	Alkali lignin	Adjustment of the mechanical and antioxidant properties *via* the lignin ratio	Bone tissue engineering	114
Lignin/PCL nanofibers	Kraft lignin	Hydroxyl groups of the lignin induce similar growth or nucleation of the hydroxyapatite to that in natural bone and tooth formations, good biocompatibility	Bone tissue engineering	154
Lignin/PCL films	Lignin*	Increased antibacterial properties and increased compatibility with tissue cells	Skin tissue engineering	226
Lignin–chitosan–poly(vinyl alcohol) composite hydrogels	Lignosulfonate	Increase of tensile stress and protein adsorption capacity – favours skin stimulation	Skin tissue engineering	69
Lignin-*co*-gelatin cryogels	Organosolv lignin	Improved the mechanical properties, shape memory, and injectability of cyrogels and created an effective inhibition zone against colonization by *E. coli* and *S. aureus*. It also resulted in an increase in antioxidant properties	Skin tissue engineering	227
Lignin embedded in poly(urethane)/poly(dimethylsiloxane) nanogels	Organosolv lignin	Good biocompatibility and non-cytotoxic, suitable antioxidant effect	Skin tissue engineering	116
Lignin into electrospun poly(vinyl alcohol) and poly(glycerol sebacate) fibers	Kraft lignin	Decrease of fiber diameter, increase of elastic modulus, improvement of cell proliferation, and positive effect on neuronal cell differentiation	Nerve tissue engineering	121
Electrospun PCL fibers embedded in lignin	Kraft lignin	Decrease in glial fibrillary acidic protein (Gfap) expression with concomitant reduction in Tubb III and Map2. Lignin provided better cell differentiation. The number and length of neurites increased and the elastic modulus also increased	Nerve tissue engineering	120

Table 8.3 (*Continued*)

Composition	Lignin type	Desired input of lignin	Application potential	Ref.
Lignin–carbohydrate complex/cellulose nanofibril composites	Organosolv lignin	The biocompatibility of human coronary artery endothelial cells increased with growing lignocellulose content	Heart tissue engineering	123
Lignin cross-linked chitosan hydrogels	Lignosulfonate	After laccase-induced cross-linking with chitosan, the molecular weight of the water-insoluble hydrogel increased extremely	Tissue engineering	228
Lignin/poly(acrylamide) hydrogels	Alkali lignin	With increasing lignin nanoparticle content, mechanical properties such as fracture stress and critical strain could be improved	Tissue engineering	229
Lignin/poly(3-hydroxybutyrate) copolymer	Alkali lignin	*In vitro* and *in vivo* improvement of bioactivity, and antioxidant and mechanical properties (tensile strength, stiffness, elongation)	Tissue engineering	230
Lignin bimetallic nanocomplexes (silver- and gold-based) embedded in poly(acrylic acid)-based hydrogels	Kraft lignin	Serves as a filling center and complexing agent due to the abundant hydroxyl groups on the surface, antimicrobial properties	Tissue engineering	231
Hydrogels with lignin-based carbon dots	Organosolv lignin	Increase of fluorescence lifetime and adjustment of mechanical properties (modulus of elasticity)	Tissue engineering	232

*, not specified.

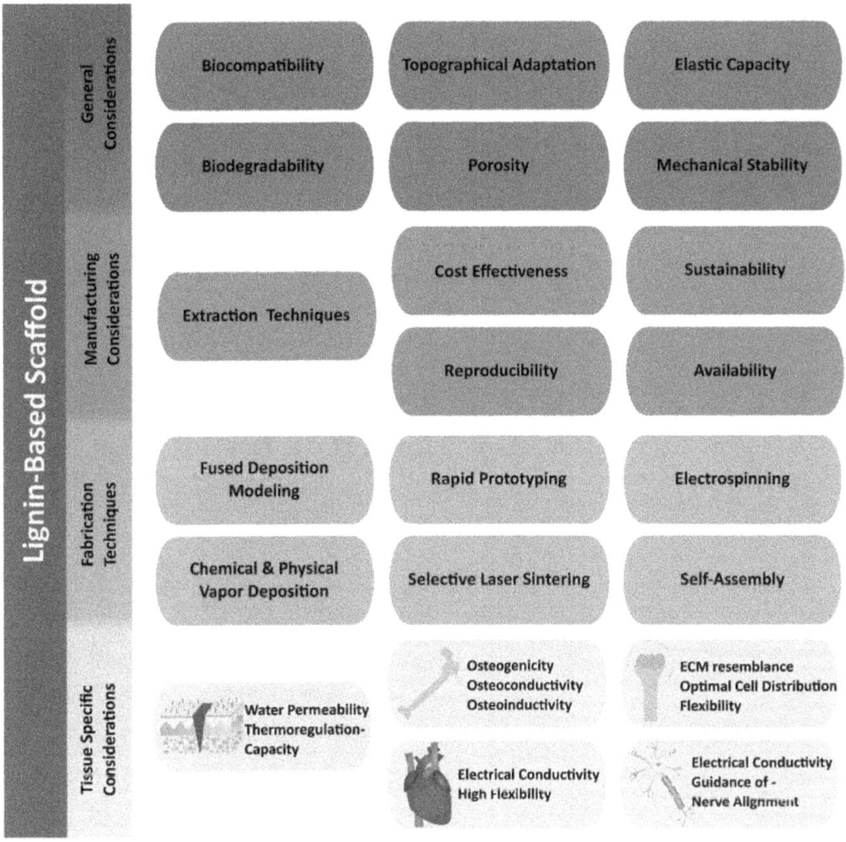

Figure 8.3 Scaffold design criteria to be addressed for tissue engineering applications.[206–209]

original bone material since bones are permanently loaded *in vivo*. In addition to the mechanical ability of the lignin scaffold, it is also important that it has morphologically typical bone substance characteristics so that the osteoconductivity is sufficient to provide an adequate interface for cells.[112] The pore size of the scaffold is central to the ingrowth of osteoblasts and their even distribution across the graft, as well as to cell nutrition and scaffold resorption. Wang *et al.* investigated the interaction between osteoblasts and supporting scaffolds in an *in vivo* application, validating lignin/PCL nanofibers.[113] For this purpose, they studied nanofibers coated with hydroxyapatite cultured with mouse preosteoblasts. After several days of cultivation, cell colonization and proliferation on the scaffold increased significantly compared to uncoated scaffolds. In addition, the lignin/PCL mineralized nanofiber films promoted the efficient adhesion and proliferation of osteoblasts by directing the expansion of filopodia. Liang *et al.* was able to further profit from the biocompatibility and antioxidant properties.[114] The study showed that nanofibers composed

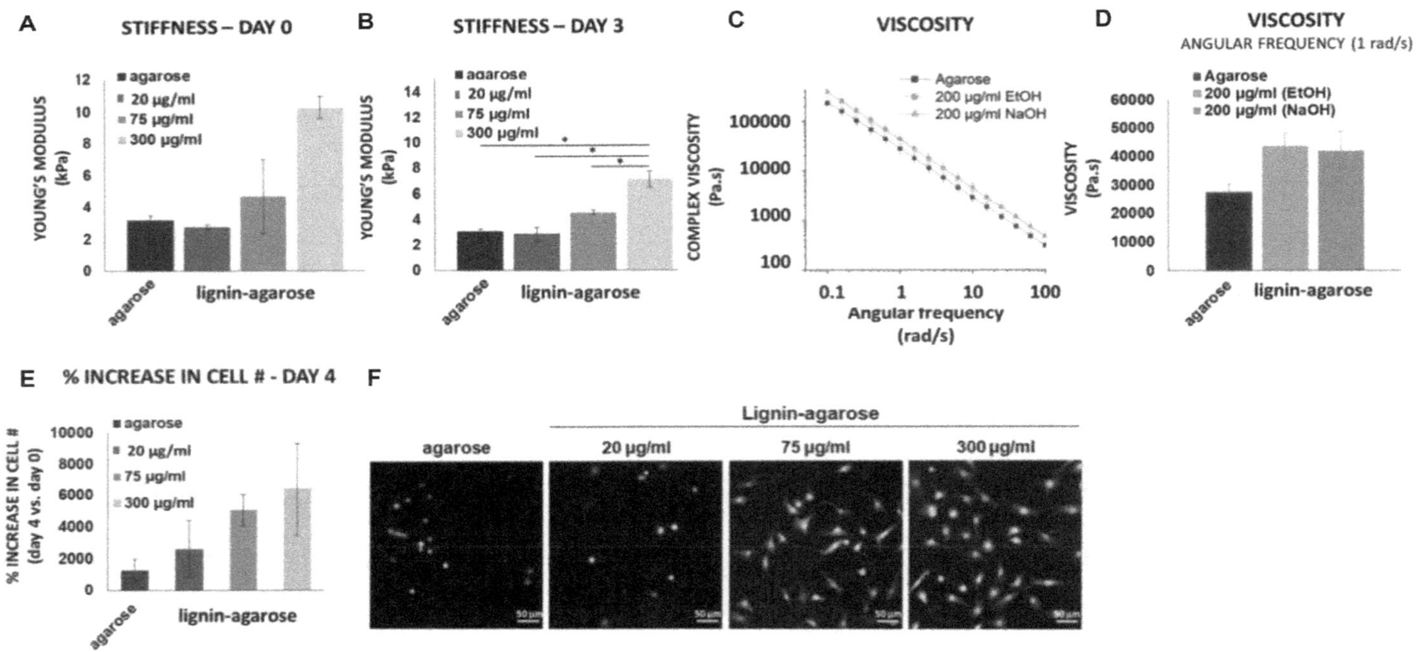

Figure 8.4 Results presented by Menima-Medzogo *et al.* of agarose and lignin–agarose hydrogels. Parts A and B show the increase in Young's modulus after increasing the lignin concentration. Graphs C and D represent the viscosity change of the agarose hydrogels *vs.* lignin–agarose hydrogels dissolved in sodium hydroxide solution and ethanol, respectively. Graph E shows the percentage increase in chondrocytes within 4 days and (F) fluorescence microscopy images of stained cells. Reproduced from ref. 242, https://doi.org/10.3390/biology11050696, under the terms of the CC BY 4.0 license https://creativecommons.org/licenses/by/4.0/.

of PCL and PCL-grafted lignin (PCL-*g*-lignin) copolymer can be used preventively for cartilage degradation and osteoarthritis. Results of the study confirmed that oxidative stress arising from osteoarthritis could be alleviated by varying lignin. The intrinsic antioxidant activity of the composite resulted in low cytotoxicity and an anti-inflammatory effect, which was validated using H_2O_2-stimulated human chondrocytes and an osteoarthritis rabbit model. Moreover, PCL/lignin nanofibers can effectively suppress cartilage degradation, as demonstrated by both macroscopic and histological *in vivo* studies.

8.2.2.2.2 Skin Regeneration and Wound Healing. The human skin essentially consists of three different layers. The epidermis and dermis in particular play important roles in the production of artificial skin. In humans, keratinocytes are in the epidermis. Keratins give the skin its resistance and firmness, and the keratinocytes attach themselves to the gel-like basic structure of collagen and form new skin structures. If one now tries to use lignin as a carrier material in the form of a composite, hydrogel, *etc.*, attention must be paid to mechanical strength, UV barrier properties, antioxidant and antimicrobial properties, and a very specific water permeability. Wound dressings that retain moisture could effectively enhance cell proliferation and migration, thus promoting wound healing. Lignin has often been able to demonstrate its potential in STE.[107]

Zhang *et al.* succeeded in improving the mechanical properties of an existing PVA/chitosan system by adding lignosulfonate.[69] The tensile stress could be increased by up to 8 MPa by the addition of lignin. The 3D structure of lignin and the large number of hydrogen and ionic bonds contribute to the high tensile strength and large elongation of the hydrogels. On top of this, the Zhang group demonstrated the accelerated wound healing ability of the lignin hydrogel (see Figure 8.5).

The lignin enhanced the re-epithelialization in this composite through its antioxidant activity. The arrangement of the connective tissue and the regeneration of the epithelium were morphologically more regular than in the hydrogel without lignin. In a very similar vein, Ravishankar *et al.* presented another promising ionotropic hydrogel system.[115] The xerogels made of chitosan/alkaline lignin gel exhibit a low cytotoxicity of 99% and are thus considered to be biocompatible and non-toxic. In addition, good cell adhesion on the surface of the hydrogel could be validated by fluorescence microscopy. In addition, the authors performed a so-called highly application-oriented scratch wound healing test with mouse fibroblast cell lines (NIH 3T3). They were able to show that the cells were already completely distributed on the trauma after 24 h, which suggests a rapid extracellular cell mobilization. Xu *et al.* used an organosolv lignin to prepare a composite of artificial polymers, forming a nanogel with strong antioxidant capacity.[116] Its application potential on skin was established by its suitable cytotoxicity and good biocompatibility, which was tested on human hepatocyte cells. Very promising are the test series on burn wounds of mice,

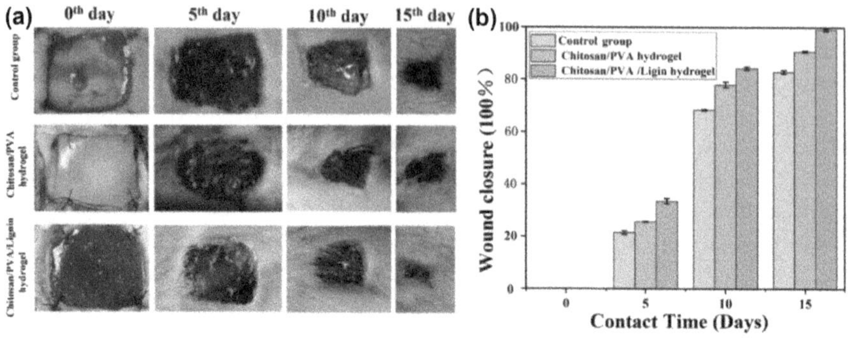

Figure 8.5 (a) Photographs show wound healing progression from a control group, and groups treated with a chitosan/PVA hydrogel with and without lignin modification from 0 to 15 days. (b) Diagram showing the relative wound healing rates of the control group and the groups treated with chitosan/PVA with and without lignin addition. Reproduced from ref. 243 with permission from Elsevier, Copyright 2019.

as this represents an altered trauma and is not only mechanical in nature. To this end, the lignin nanogel was compared to other commonly used compounds such as Vitamin C and Trolox (both commercial antioxidants) during wound healing to see how effectively ROS could be scavenged. Results confirmed comparable antioxidant capacity for the lignin-based nanogel. In addition, the wound healing process was increased by higher cell proliferation activity (Ki67 expression).

8.2.2.2.3 Nerve Regeneration. Cells of the nervous system are highly specialized in transmitting electrical impulses in the body. There are two major cell types in the nervous system: nerve cells and glial cells. Nerve cells tend to have a large cell body and long extensions. These are called dendrites and axons, respectively. Glial cells are the supporting cells of the nervous tissue and thus outnumber nerve cells. Among other functionalities, they serve as a kind of nerve sheath to ensure a high degree of insulation of the information transmitter.[117] The progress in research suggests that materials with low immunogenicity, high biodegradability, excellent biocompatibility, and beneficial microenvironments are of great importance for regeneration.[118] In the development of tissue-engineered scaffolds that simulate the ECM, it must be possible to provide an appropriate mix of morphology, chemical support, and electrical simulation for the neuron.[119] Electrospinning has great potential for producing fibrous scaffolds with controllable composition and structure that exhibit various biomimetic properties at the molecular and nanometer levels. Worth mentioning here is the approach of Amini *et al.* who added lignin to electrospun PCL fibers.[120] The addition of lignin sustainably improved the number and length of neurites in different cell types (PC12 and MSCs), and cell differentiation also directly benefited. The expression of neuronal markers for differentiated cells was upregulated by increasing the amount of lignin. The *in vivo* study also showed that P15L promotes peripheral nerve regeneration and can be used as a material to promote axonal budding, which is very important for nerve regeneration and reconstruction. Saudi *et al.* also prepared electrospun PVC/poly(glycerol sebacate) (PGS) fibers with an addition of lignin.[121] The fibers were also able to gain from the positive properties of lignin, as they exhibited a flat morphology and uniform diameter after addition. The mechanical properties (*e.g.*, increasing modulus of elasticity), degradation measurements, and water absorption behaviour of the scaffolds indicated that the inclusion of lignin was suitable for the development of neural tissue. The mRNA expression level for Gfap, Tubb III, and Map2 as much as immunocytochemistry showed the positive impact of lignin on neuronal cell development[79] (see Figure 8.6).

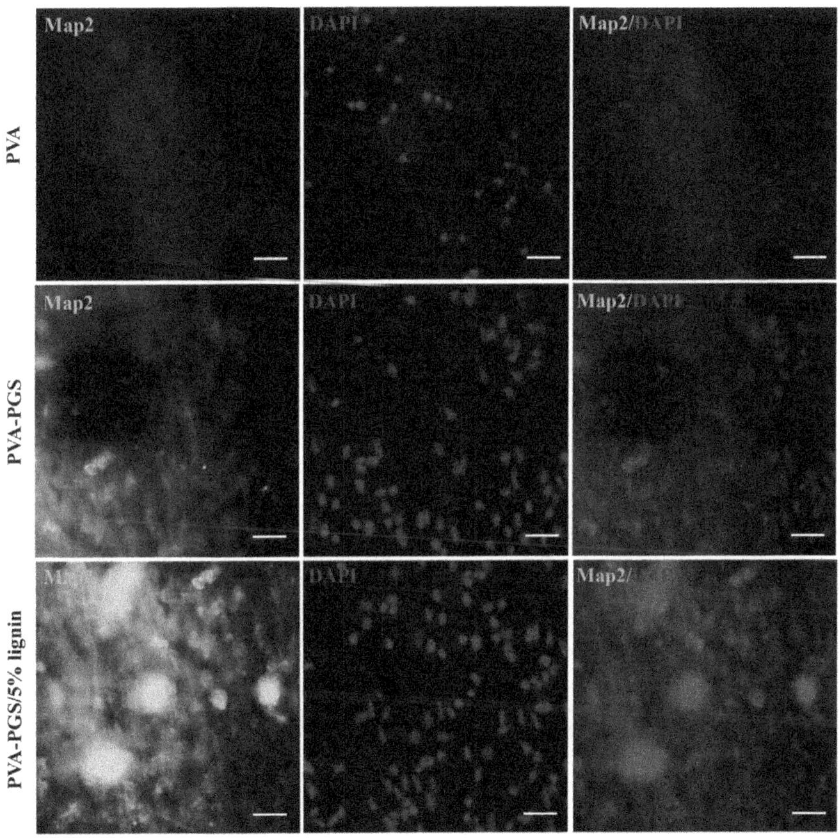

Figure 8.6 Scans of differentiated PC12 cells from an immunocytochemical photograph showing expression of the Map2 marker on PVA, PVA–PGS, and PVA–PGS/5% lignin electrospun fibers. The scale bar is 50 μm. Reproduced from ref. 244, with permission from Elsevier, Copyright 2019.

8.2.2.2.4 Muscle TE and CTE. Muscle tissues are mainly used for mechanical movement and contraction. Two types of musculatures are distinguished according to morphological aspects: striated musculature and smooth musculature. The striated musculature is in turn divided into skeletal musculature and cardiac musculature. Smooth muscle is found embedded in the interior of hollow organs and to a lesser extent in other structures. Muscle cells have the unique property of being contractile, or capable of contraction. This contraction results from the interaction of actin and myosin filaments. Muscle tissue plays an important role in the human body. It accounts for more than 50% of body mass and controls force generation, body locomotion, and internal organ function.[122] Injury to muscle tissue leads to movement disorders and organ dysfunction. This requires a method to create a lignin scaffold that mimics the ECM to provide muscle tissue or a regenerative environment, thereby simulating the original myocardium. Mimicking the nanofibrous structure and conductivity of the ECM is to mimic the electrical distribution of the original myocardium and should be used for cardiac scaffolds and cardiomyocyte-based bio-actuators to be tailored.[104] Zhao *et al.* prepared a potential heart tissue material from a lignin and cellulose nanofibrils.[123] This tubular porous support is an excellent cultivation site for human coronary artery endothelial cells (HCAECs) which was optically recorded by scanning electron microscopy (SEM). The addition of lignin increased the cell proliferation rate and metabolic activity. However, biocompatibility directly benefited from lignin addition. In parallel, these tubular structures were very stable at different pH values. In this context, Figure 8.7 shows the growth of HCAECs on the porous support of lignin and cellulose fibrils.

The properties of lignin show great promise for TE applications as a building block for biomaterials/scaffolds. Even though there remain challenges that need to be addressed, several forms of lignin-based materials have been developed. And considering that a TE construct doesn't depend only on the choice of biomaterials, it is necessary to consider the other components also.

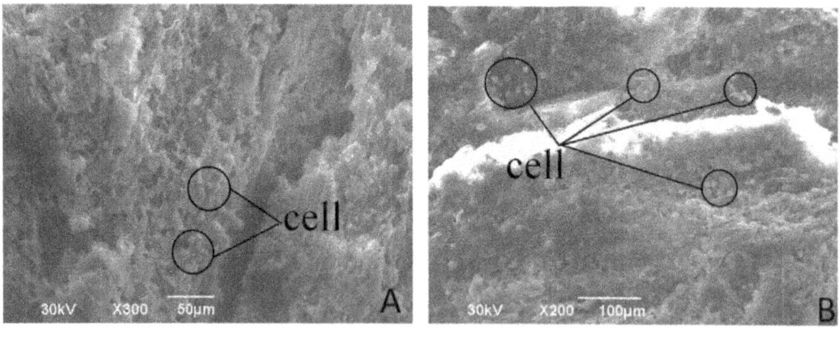

Figure 8.7 SEM images of the growth of HCAECs on a porous lignin/cellulose fibril composite. Parts A and B show composites with differently pre-treated lignins. Reproduced from ref. 245, with permission from Elsevier, Copyright 2019.

8.3 Cells for TE

As mentioned previously in this chapter, when designing TE constructs there are crucial points that need to be taken into consideration to produce a suitable graft that can recapitulate the native tissue and allow regeneration. In TE approaches that incorporate cells into the scaffold there are several choices for cells which can be considered. These cells can basically be divided into immature (stem or progenitor) cells and mature tissue-specific cells. Using cells in TE has several benefits, such as improving tissue regrowth, regeneration, and angiogenesis. Additionally, some cells, such as MSCs, elicit beneficial effects through their paracrine activity, which is due to their natural function within the body for maintenance and repair. These features are thought to have a major therapeutic effect also in the TE. In this section the cellular population of target tissues will be introduced briefly and the choices for cells that can be used for tissue engineering for the respective target tissues will be discussed.

8.3.1 Current Tissue Targets for Lignin-derived Biomaterials

Understanding the cellular composition of the target tissue provides an insight into which cell types are suitable to be incorporated in a TE construct. Hard tissue such as bone might seem lifeless, but it contains a dynamic cellular population and mineral composition, which is renewed within the body depending on regular turnover and strain. Much of the weight of bone tissue is composed of inorganic minerals such as hydroxyapatite, and the remainder of the acellular weight is water and matrix proteins. The outer layer of bone tissue is the periosteum, which is a bilayer with an outer fibrous collagenous part and a highly cellularized inner part, which contains osteoprogenitor cells.[124] The periosteum surrounds the osseous tissue which, depending on its function, can be divided into cortical (compact) and cancellous (trabecular/spongy) bone structures.[125] The endosteum surrounds the osseous tissue from the inside and contains progenitor cells such as MSCs. In the middle is the bone marrow: a spongy tissue that is responsible for the formation of blood cells.[126] On a cellular level, cells in bone tissue (excluding bone marrow) can be divided into bone forming cells, bone resorbing cells, and regulatory cells. Bone forming cells are called osteoblasts, which originate from MSCs in the bone.[127] Osteoblasts form bone by releasing minerals and ECM proteins. When the osteoblast has reached the end of its functional life span, it will either undergo apoptosis or become entrapped in the newly formed bone as osteocytes, which have a regulatory function in controlling bone mass.[127,128] Bone resorbing cells are called osteoclasts; these originate from hematopoietic stem cells, specifically from monocyte–macrophage progenitors. Osteoclasts resorb bone by releasing enzymes that degrade the organic matrix and acids that dissolve the inorganic matrix.[127,129] The balance between the activity of bone forming and resorbing cells is crucial for maintaining healthy bone, and any imbalance results in diseases such as osteoporosis and osteopetrosis.[130,131]

While cartilage is usually found attached to bone in the body, its cellular population, chemical composition, and physical properties are different. While the majority of bone is made of inorganic matrix, the majority of cartilaginous tissue is made of collagenous ECM that contains proteins such as glycosaminoglycans, proteoglycans, and primarily collagen.[132] Cartilage is a connective tissue that can be described as tough, semi-transparent, elastic, and hypocellular tissue. It can be classified into three types: elastic cartilage, hyaline cartilage, and fibrocartilage, which differ in relative amounts of collagen and proteoglycan.[133] The cellular population makes up about 5% of cartilage tissue, which is mainly chondrocytes grouping up within lacunae. Chondrocytes are responsible for the secretion of matrix proteins and matrix modelling proteins. They originate from MSCs. During embryonic skeletal development, MSCs condense and differentiate into chondrocytes.[134] Cartilage tissue contains no nerves or blood vessels and as such is not easily restored. In this avascular environment, chondrocytes gain their nutrients through diffusion from the surface which limits their regeneration ability in case of cartilage damage.[135]

In comparison to cartilage, cardiovascular tissues are complex. The heart in itself is an organ that is composed of four differently structured chambers. In general, the bulk of the adult heart wall is composed of the myocardium, which is one of the muscle types found in the body. The myocardium of the heart is surrounded by an outer membrane called the epicardium and an inner membrane called the endocardium. A large-scale single-cell and single-nucleus transcriptome analysis of the adult human heart showed a detailed percentile cellular composition. The cells include cardiomyocytes (arterial and ventricular), fibroblasts, endothelial cells, and mural cells (pericytes and vascular smooth muscle cells), among others.[136] The result also showed that in the atrial tissue 30.1% of the cells were cardiomyocytes and 24.3% were fibroblasts, while the ventricle contained 49.2% cardiomyocytes, with the next higher population being mural cells (making up to 21.2%). These data underline the complexity and variation in these tissues. While it is known that cardiomyocytes constitute most of the volume of the human heart, other cellular populations are essential for cardiac function as well.[137] An example is the fibroblasts found in cardiac tissue that are involved in maintaining the myocardial structure. The cross talk between cardiomyocytes and fibroblasts is vital for myocyte growth and ECM turnover. Fibroblasts are also involved in pathological conditions affecting cardiac tissue.[138,139] Endothelial cells make up the endothelium that lines the inner walls of the heart and blood vessels. They are important for angiogenesis and cardiovascular homeostasis.[140] Mural cells are important for maintaining vascular integrity and controlling blood vessel constriction and dilation.[141]

The skin is sometimes referred to as "the largest organ in the body"; even if that is only true for the surface, it certainly has a complex physicochemical tissue gradient necessary to perform several functions. The skin is a sensory and secretory organ. It is the body's first line of defense against microbes and helps in regulating body temperature among other functions.[32,142] The skin is composed of three main layers: the epidermis, the dermis, and the hypodermis, which contain epithelial and connective tissues. The epidermis is composed of cells such as keratinocytes, progenitor cells, and melanocytes.[143] The thickest layer is the dermis; this connective tissue is composed of an ECM that contains fibroblasts and vascular endothelial cells in addition to hair follicles, glands, nerve endings, and blood vessels.[144] Fibroblasts are the main cell type in the dermis and they are responsible for the production of ECM proteins such as collagen and elastin.[144] The hypodermis layer connects the skin to the inner organs and tissues (such as muscles) and it is rich in fat tissue and sensory nerves.[32]

Sensory nerves are an example of nerves that belong to the PNS. Peripheral nerve tissues can be divided into nerves and ganglion tissues. Neurological processes in the PNS depend on complex interactions between neural cells. Neurons that mainly make up these tissues have a unique structure. A neuron has three main parts: dendrites, an axon, and a cell body (or soma). The ganglion tissue contains the collection of neural cell bodies,[145] while neuronal axons make up the major part of the nerves. Neurons in the PNS can be either myelinated or unmyelinated and are accompanied by neuroglial cells (satellite cells and Schwann cells). Neurons and nerve fibers are surrounded by three layers of connective tissues: the outer layer, the epineurium which surrounds the nerve fibers, and the blood vessels. The perineurium surrounds a bundle of nerve fibers called nerve fascicles, and the endoneurium surrounds the nerve axon.[146,147] Neuroglial cells found in peripheral nerves are not involved in signal transmission directly but are vital for providing the neurons with structural support and maintaining the microenvironment surrounding the neurons. In the CNS, the neuroglial cells are oligodendrocytes, astrocytes, and microglia, while the PNS has satellite cells, macrophages, and Schwann cells. Schwann cells are of importance as they produce the myelin sheath that surrounds the neural axons.[146,147] They also support the nerves and have a role in nerve repair.[148,149]

The cellular composition of the tissue gives clues on the cell types that could be used as an *in vitro* model for creating and testing TE constructs as well as choices suitable for seeding in these constructs for further transplantation. The major choice of cells is either using mature tissue-specific cells or stem/progenitor cells. In the next sections, the cell types used as models to test lignin-based biomaterials will be introduced and an insight into other possible suitable cell sources that could be investigated in the future is provided.

8.3.2 Mature Tissue-specific Cells

TE constructs are usually tested *in vitro* and *in vivo* using various cellular and animal models that more or less represent the target tissue to determine their efficacy. For example, scaffolds designed for BTE are commonly tested with either osteoblastic cell lines or primary osteoblasts isolated from bone tissues.[150,151]

As discussed previously, the choice of cells depends on the target tissue and the ability of the cells. Examples of typical cell choices (allogenic or autogenic) for the respective tissue types are shown in Table 8.4.

So far lignin-based biomaterials have been tested for their biocompatibility and efficacy for TE only with a few cellular and animal models. For cartilage tissue regeneration applications, Liang and colleagues designed a scaffold with an antioxidant copolymer PCL-grafted lignin (PCL-*g*-lignin) for treatment of osteoarthritis. They investigated its biocompatibility *in vitro* using a human chondrocyte cell line and *in vivo* using a rabbit model. The addition of lignin to PCL showed an improvement of cell viability, underlining its biocompatibility. Additionally, *in vitro* analysis of its antioxidant activity showed that a PCL–lignin scaffold containing ~20% lignin was able to effectively protect chondrocytes from oxidative stress caused by H_2O_2. *In vivo* investigations confirmed that a PCL-lignin membrane applied to the tissue effectively protected the tissue from degradation and cell death caused by oxidative stress.[152] Kouhia and colleagues showed the suitability of an injectable hydrogel containing lignin for CTE that was able to enhance the proliferation and differentiation of chondrocytes *in vitro*.[153] For BTE applications, a lignin/PCL scaffold was fabricated with incorporated hydroxyapatite. The scaffold loaded with cells from a mouse preoteoblastic cell line was biocompatible and exhibited enhanced adhesion and proliferation compared to the controls.[154]

For neural tissue, Wang and colleagues fabricated lignin–PCL nanofibers that exhibited no cytotoxicity towards bone marrow MSCs. The fibers were able to support the growth of Schwann cells with myelin expression. The authors also demonstrated that the nanofibers could support the growth and proliferation of mouse primary neurons.[155] In 2019, Saudi and colleagues designed a scaffold with lignin and electrospun PVA and PGS fibers. The *in vitro* biocompatibility tests with rat pheochromocytoma cells (PC12) showed that the scaffold had no cytotoxic effects and improved the

Table 8.4 Cell types used as cellular models and possible choices for tissue engineering.

Tissue(s)	Cell types	Ref.
Bone	Osteoblasts, endothelial cells, co-culture of osteoclasts & osteoblasts, stem cells	233 and 234
Cartilage	Chondrocytes, preosteoblasts, stem cells	235 and 236
Cardiovascular tissues	Cardiomyocytes, vascular smooth muscle cells, endothelial cells, stem cells	28 and 237
Skin	Fibroblasts, keratinocytes, melanocytes, endothelial cells, stem cells	238 and 239
Neural tissues	Schwann cells, macrophages, stem cells	240 and 241

proliferation and differentiation of the cells.[121] In 2020, the authors fabricated aligned PCL fibers with various percentages of lignin nanoparticles and investigated them *in vitro* and *in vivo*. *In vitro* analysis with the PC12 cell line and human adipose MSCs revealed that the scaffolds promoted cell differentiation and growth with an increasing lignin percentage in the scaffold material. An *in vivo* analysis was done in that study in a 10 mm long nerve defect, which confirmed that 15% lignin showed most tissue regeneration.[120]

For skin tissues, chitin nanofibrils were combined with nanolignin to produce a compound which was tested *in vitro* for biocompatibility, and anti-inflammatory and antimicrobial activity. The effect was investigated with the human keratinocyte cell line (HaCaT cells) and human MSCs, which revealed a decrease in proinflammatory cytokines in HaCaT cells and an increase in expression of an antimicrobial peptide. The compound also didn't have any adverse effects on the viability or function of the MSCs.[156] Regarding CVTE, Zhao and colleagues fabricated nanocellulose and lignin–carbohydrate complex composite biological carriers which were tested *in vitro* with human coronary artery endothelial cells. The cells were able to adhere and grow on the carriers and those carriers that contained lignin outperformed the control in improving cell proliferation and metabolic activities such as albumin content and glucose consumption.[123]

While generally autologous and allogenic cells can be used in TE graft constructs, they both have several drawbacks such as eliciting an immune response, limited availability, or limited proliferation capacity. Therefore cell-based TE approaches have recently leancd heavily on methods utilizing stem cells, which have higher regenerative potential as differentiated tissue cells. The next section light will be shed on some of the major therapeutic benefits of stem cells.

8.3.3 Stem Cells

Stem cells are naïve unspecialized cells that have the ability to produce specialized cell types of the body, such osteoblasts and cardiomyocytes. Also stem cells have the self-renewal ability.[157–159] They can be found in embryonic, fetal, and adult tissues. Stem cells can also be reprogrammed from mature differentiated cells, which are called iPSCs. Although stem cells exist in many types and tissues, not all types of stem cells are suitable for successful and safe TE applications. ESCs are pluripotent stem cells, meaning that they have very high proliferative and differentiation potential and can generate cells from all three germinal layers: ectoderm, endoderm, and mesoderm. However, they are not a commonly used option for TE because of ethical restraints, the risk of immune rejection, and teratoma formation. Some adult stem cells (ASCs) are multipotent. They do not face the same ethical limitations as ESCs, albeit having a lower potency for various differentiation lineages in comparison to ESCs. ASCs can be found in most of the tissues of the body, including, but not limited to, bone marrow, adipose tissue, dental tissue, and blood.[160–164] ASCs include hematopoietic stem cells, intestinal stem cells, neural stem cells, and MSCs. MSCs are an attractive choice for TE. They can be isolated from various tissue sources providing interesting options for various tissue replacement strategies. They can be differentiated into several of the cell types mentioned throughout this chapter, including osteoblasts, chondrocytes, endothelial cells, and cardiomyocytes.[52,157,161,165–168] MSCs can modulate the immune system, which reduces the risk of rejection.[169–171] They also promote tissue regeneration and angiogenesis.[172] This among many other advantages has made MSCs an alluring choice for TE.[20,173,174] Many stem cell therapy clinical trials were already conducted with MSCs.[175,176]

However, translation of stem cell therapies from research into clinical settings requires a stable and consistent reproducible supply of MSCs. For this, MSCs have some drawbacks such as donor heterogeneity. They are also affected by aging and epigenetic changes, and they cannot be grown indefinitely *in vitro*.[177–179] This points to the need for a better standardized stem cell source. One alternative is using iPSCs. iPSCs were discovered in 2006 by Yamanaka and colleagues, which is considered to be a breakthrough in stem cell research.[180] As their name entails, they are pluripotent stem cells generated in the laboratory by introducing several transcription factors that reprogram mature specialized cells, such as fibroblasts, back to a stem cell-like type of cell.[180,181] This technique provided the scientific community with pluripotent stem cells behaving similarly to embryonic stem cells but without the need to sacrifice embryos. iPSCs themselves might not be the best option to use directly in TE; since they are pluripotent, they can lead to teratoma formations similar to ESCs. However, iPSCs can be used to produce neural cells, osteoblasts, cardiomyocytes, and chondrocytes, which then provide a safer option.[182–185] They can also be differentiated into induced mesenchymal stem cells (iMSCs) that are more homogenous and pertain a rejuvenation signature from when they were reprogramed. However, iPSCs are still at the initial stage of investigation for their therapeutic potential. But *in vitro* and *in vivo* animal studies conducted so far seem to have a favourable outcome.[186–189]

The therapeutic potential of stem cells is not limited to their ability to produce specialized tissue cells. Their paracrine activities which are means for cell communications are thought to have a major role in their therapeutic abilities. Stem cells secrete biomolecules into their surrounding environment that can stimulate themselves and other cells. These biomolecules are known collectively as the secretome and can be present in the surroundings of the stem cells in a soluble state and packed inside extracellular vesicles.[190] As such, the secretome of stem cells is gaining interest as an acellular therapy option that provides the benefits of stem cells without the complications that arise with using the cells themselves like immune rejection or the formation of unwanted cells or tissues (*e.g.*, teratoma formation). Figure 8.8 provides an outlook on the therapeutic potential of stem cells. Some examples of the therapeutic abilities of stem cells and their products

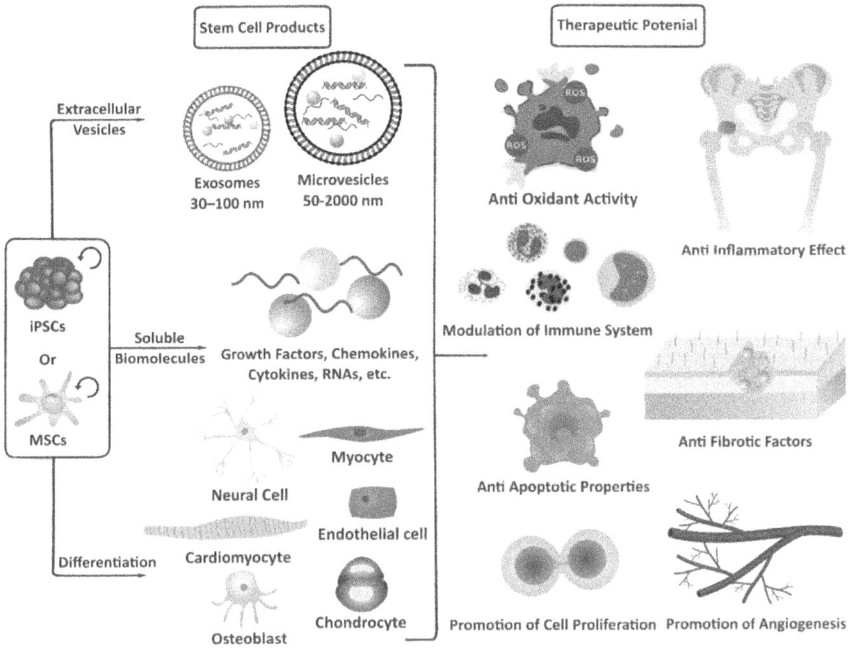

Figure 8.8 Therapeutic potential of stem cells for tissue engineering applications. Stem cells such as mesenchymal stem cells or induced pluripotent stem cells have the ability to reproduce tissue by generating specialized cells of the target tissue. Additionally stem cells produce a secretome that contains biomolecules (*e.g.*, growth factors, cytokines, nucleic acids) that can have therapeutic effects. These biomolecules are secreted in the environment of the stem cells either in a soluble state or packed in extracellular vesicles. Extracellular vesicles are divided into exosomes and microvesicles depending on their size. Stem cell products have various applications for tissue regeneration and engineering.

include, but are not limited to, improving tissue regeneration and cell proliferation, promoting angiogenesis, immune modulation, and anti-fibrotic, anti-inflammatory, antioxidant, and antiapoptotic abilities.[191–194] The therapeutic potential of stem cells is immense. And if provided with suitable delivery systems, and chemical and physical cues, they can be a powerful tool for TE and regeneration.

8.4 Conclusion and Future Prospects

TE holds promise for an effective and personalized treatment. Considering the shortcomings of current conventional treatments for tissue regeneration and repair, it is not surprising that it has gained focus in the scientific community. The interplay of its major components namely scaffolds, cells, and biomolecules enables generation of various tissue regeneration or repair strategies. Given that such a TE product is meant for medical applications, a scaffold should meet certain criteria related to its safety. Other criteria are related to its structure and function which are also vital, since the aim is to fabricate it in a manner which is suitable for the targeted tissue to ensure successful regeneration or repair. Since quality assurance is absolutely essential for the approval of any biomaterial used in TE, this aspect will become a key criterion particularly for lignin-based products. Currently, any commercial application of lignin-derived materials including those for tissue regeneration is limited due to difficulties in 3D lignin structure elucidation. This in turn prevents final quality control and assurance. Problems in structural analysis are caused not only by the very complex 3D structure consisting of the lignin macromolecule but also by limitations in analytical techniques. Furthermore, common methods such as IR and NMR spectroscopy are restricted due to signal overlap, even for most recent two-dimensional methods. To solve this problem, novel chemometric data processing methods combined with machine-learning tools are used for structure elucidation for complex chemical compounds.[195,196] As recently reported by our group, multivariate methods will play a key role in lignin specification and future quality assurance.[197,198] Most analytical measurement techniques, including NMR spectroscopy, are rather expensive and therefore limited to specialized laboratories. This situation is just changing since affordable benchtop NMR devices have been developed. Very recently, our group could show that benchtop 1H NMR data correlate with the molecular weight of lignin, a key parameter in lignin specification.[199,200] Moreover, we could confirm that a calibration transfer of partial least squares (PLS) regressions (lignin molecular weight data) between high-field and low-field (benchtop) NMR devices is possible.[201] These results open new perspectives for lignin analysis and thus commercial lignin exploitation including lignin-based biomaterials for TE applications. Only when the lignin structure can be elucidated, the molecular weight can be determined, and the nature and number of functional groups can be precisely analysed will it be possible to derive correlations with the bioactivity behaviour. Then answers can be given as to how antimicrobial or antiviral properties can be specifically controlled and improved – and how lignin-derived materials can be tailored for future tissue regeneration.

Besides the choice of biomaterial to be used for scaffolds, the choice of cells is also of major importance in TE approaches. Understanding the tissue composition on a cellular basis gives an insight into suitable options. Mature, specialized cells and stem cells have both been used for testing the biocompatibility of newly fabricated lignin scaffolds. However, when considering their potential for therapeutic applications, stem cells and their cellular products have the upper hand compared to mature cells. Furthermore, the paracrine activity of stem cells exerted through their secretome is believed to have a major therapeutic influence. This is backed by the fact that most stem cells if introduced into the target tissue without scaffold disappear from the tissue shortly after administration, but their therapeutic outcome remains.[202,203] Also, safety concerns related to cellular therapy are a cause for concern, while the secretome released by stem cells provides a safer and cell-free alternative.[204,205] Taken together, the orchestrated interplay of the tissue matrix, tissue cells, and biomolecules must be understood in more detail for successful fabrication of a scaffold loaded with the respective cells and biomolecules for future successful TE.

Abbreviations

ASCs	Adult stem cells
BTE	Bone tissue engineering
CNS	Central nervous system
CTE	Cartilage tissue engineering
CVD	Cardiovascular disease
CVTE	Cardiovascular tissue engineering
3D	Three-dimensional
DPPH assay	2,2-Diphenyl-1-picryl-hydrazyl-hydrate assay
ECM	Extracellular matrix
EHD-jetting	Electrohydrodynamic jet 3D printing
ESC	Embryonic stem cell
iMSC	Induced mesenchymal stem cell
iPSC	Induced pluripotent stem cell
MPa	MegaPascal
MSC	Mesenchymal stem cell
NGC	Nerve guidance catheter/channel/conduit
NTE	Neural tissue engineering
PCL	Poly(caprolactone)
PEDOT:PSS	Poly(3,4-ethylene dioxythiophene):poly(4-styrenesulfonate)
PEG	Poly(ethylene glycol)
PGA	Poly(glycolic acid)
PGS	Poly(glycerol sebacate)
PLA	Poly(lactic acid)
PLGA	Poly(lactic-*co*-glycolic acid)
PNS	Peripheral nervous system
PPy	Poly(pyrrole)
PVA	Poly(vinyl alcohol)
PVC	Poly(vinyl chloride)
ROS	Reactive oxygen species
STE	Skin tissue engineering
TE	Tissue engineering

Acknowledgements

This work was supported by the Bundesministerium für Bildung und Forschung (BMBF) FHprofUnt to E. T., FKZ: 13FH012PB2; Bundesministerium für Bildung und Forschung (BMBF) IngenieurNachwuchs to E. T., FKZ: 03FH019IX5; Ministerium für Innovation, Wissenschaft und Forschung (MIWF) NRW FH Zeit für

Forschung to E. T., FKZ: 005-1703-0017; and Bundesministerium für Bildung und Forschung (BMBF) IngenieurNachwuchs, FKZ: 13FH569IX6. A PhD scholarship was given by the Graduate Institute (GI) of the Bonn-Rhein-Sieg University of Applied Sciences to S. H. S. and A. A.

References

1. H. Walles and T. Walles, in *Comprehensive Biomaterials*, ed. P. Ducheyne, Elsevier, Oxford, 2011, pp. 361–367.
2. E. Tobiasch, in *Forschungsspitzen und Spitzenforschung*, Springer, 2009, pp. 329–338.
3. P. F. Ottensmeyer, M. Witzler, M. Schulze and E. Tobiasch, *Int. J. Mol. Sci.*, 2018, **19**, 3601.
4. M. Schulze and E. Tobiasch, in *Tissue engineering III: cell-surface interactions for tissue culture*, Springer, 2011, pp. 153–194.
5. D. Schipper, P. Babczyk, F. Elsayed, S. E. Klein, M. Schulze and E. Tobiasch, in *Nanostructures for novel therapy*, Elsevier, 2017, pp. 567–589.
6. R. R. Betz, *Orthopedics*, 2002, **25**, s561–s570.
7. J. J. Marler, J. Upton, R. Langer and J. P. Vacanti, *Adv. Drug Delivery Rev.*, 1998, **33**, 165–182.
8. C. M. Revell and K. A. Athanasiou, *Tissue Eng., Part B*, 2009, **15**, 1–15.
9. D. Yu, X. Wang and L. Ye, *J. Cardiovasc. Dev. Dis.*, 2021, **8**, 153.
10. N. Liu, X. Ye, B. Yao, M. Zhao, P. Wu, G. Liu, D. Zhuang, H. Jiang, X. Chen, Y. He, S. Huang and P. Zhu, *Bioact. Mater.*, 2021, **6**, 1388–1401.
11. H. Amani, H. Kazerooni, H. Hassanpoor, A. Akbarzadeh and H. Pazoki-Toroudi, *Artif. Cells, Nanomed., Biotechnol.*, 2019, **47**, 3524–3539.
12. S. B. Habre, G. Bond, X. L. Jing, E. Kostopoulos, R. D. Wallace and P. Konofaos, *Ann. Plast. Surg.*, 2018, **80**, 252–261.
13. F. S. Wong, B. P. Chan and A. C. Lo, *Int. J. Mol. Sci.*, 2014, **15**, 10669–10723.
14. L. B. Dahlin and G. Lundborg, *Neurosurg. Clin. North Am.*, 2001, **12**, 341–352.
15. M. C. Dodla, M. Alvarado-Velez, V. J. Mukhatyar and R. V. Bellamkonda, in *Principles of Regenerative Medicine*, ed. A. Atala, R. Lanza, A. G. Mikos and R. Nerem, Academic Press, Boston, 3rd edn, 2019, pp. 1223–1236.
16. M. Alonzo, F. A. Primo, S. A. Kumar, J. A. Mudloff, E. Dominguez, G. Fregoso, N. Ortiz, W. M. Weiss and B. Joddar, *Curr. Opin. Biomed. Eng.*, 2021, **17**, 100248.
17. E. Roddy, M. R. DeBaun, A. Daoud-Gray, Y. P. Yang and M. J. Gardner, *Eur. J. Orthop. Surg. Traumatol. Orthop. Traumatol.*, 2018, **28**, 351–362.
18. G. D. Barabaschi, V. Manoharan, Q. Li and L. E. Bertassoni, *Adv. Exp. Med. Biol.*, 2015, **881**, 79–94.
19. T. Stampoultzis, P. Karami and D. P. Pioletti, *Curr. Res. Transl. Med.*, 2021, **69**, 103299.

20. C. A. Limbach, M. Lange, M. Schulze and E. Tobiasch, *Recent Pat. Regener. Med.*, 2012, **2**, 75–102.
21. C. A. Limbach, M. Lange, M. Schulze and E. Tobiasch, *Recent Pat. Regener. Med.*, 2012, **2**, 75–102.
22. S. Samavedi, L. K. Poindexter, M. Van Dyke and A. S. Goldstein, in *Regenerative medicine applications in organ transplantation*, Elsevier, 2014, pp. 81–99.
23. D. F. Williams, *Progress in biomedical engineering: definitions in biomaterials*, Elsevier, 1987.
24. S. Prasadh and R. C. W. Wong, *Oral Sci. Int.*, 2018, **15**, 48–55.
25. M. F. Griffin, B. C. Leung, Y. Premakumar, M. Szarko and P. E. Butler, *J. Otolaryngol.-Head Neck Surg.*, 2017, **46**, 33.
26. É. Lakatos, L. Magyar and I. Bojtár, *J. Med. Eng.*, 2014, **2014**, 470539.
27. J. A. Apablaza, M. F. Lezcano, A. Lopez Marquez, K. Godoy Sánchez, G. H. Oporto and F. J. Dias, *Polymers*, 2021, **13**, 2563.
28. R. Chaudhuri, M. Ramachandran, P. Moharil, M. Harumalani and A. K. Jaiswal, *Mater. Sci. Eng., C*, 2017, **79**, 950–957.
29. M. Roshandel and F. Dorkoosh, *Polym. Adv. Technol.*, 2021, **32**, 2290–2305.
30. Q. L. Loh and C. Choong, *Tissue Eng., Part B*, 2013, **19**, 485–502.
31. S. Vijayavenkataraman, S. Zhang, S. Thaharah, G. Sriram, W. F. Lu and J. Y. H. Fuh, *Polymers*, 2018, **10**, 753.
32. M. Rahmati, J. J. Blaker, S. P. Lyngstadaas, J. F. Mano and H. J. Haugen, *Mater. Today Adv.*, 2020, **5**, 100051.
33. S. Misra and S. Raghuwanshi, in *Fundamental Biomaterials: Metals*, ed. P. Balakrishnan and S. Thomas, Woodhead Publishing, 2018, pp. 355–370.
34. D. A. Ahmad Ruzaidi, M. M. Mahat, S. A. Shafiee, Z. Mohamed Sofian, A. S. Mohmad Sabere, R. Ramli, H. Osman, H. H. Hamzah, Z. Zainal Ariffin and K. K. Sadasivuni, *Polymers*, 2021, **13**, 3395.
35. A. Burnstine-Townley, Y. Eshel and N. Amdursky, *Adv. Funct. Mater.*, 2020, **30**, 1901369.
36. H. Esmaeili, A. Patino-Guerrero, M. Hasany, M. O. Ansari, A. Memic, A. Dolatshahi-Pirouz and M. Nikkhah, *Acta Biomater.*, 2022, **139**, 118–140.
37. K. Roshanbinfar, L. Vogt, B. Greber, S. Diecke, A. R. Boccaccini, T. Scheibel and F. B. Engel, *Adv. Funct. Mater.*, 2018, **28**, 1803951.
38. S. Vijayavenkataraman, S. Kannan, T. Cao, J. Y. Fuh, G. Sriram and W. F. Lu, *Fron. Bioeng. Biotechnol.*, 2019, 266.
39. E. C. Novosel, C. Kleinhans and P. J. Kluger, *Adv. Drug Delivery Rev.*, 2011, **63**, 300–311.
40. D. Hielscher, C. Kaebisch, B. J. V. Braun, K. Gray and E. Tobiasch, *Stem Cell Rev. Rep.*, 2018, **14**, 642–667.
41. P. Babczyk, C. Conzendorf, J. Klose, M. Schulze, K. Harre and E. Tobiasch, *J. Clin. Med.*, 2014, **3**, 39–87.
42. V. Mastrullo, W. Cathery, E. Velliou, P. Madeddu and P. Campagnolo, *Front. Bioeng. Biotechnol.*, 2020, 8.

43. M. D. Sarker, S. Naghieh, N. K. Sharma, L. Ning and X. Chen, *J. Healthcare Eng.*, 2019, **2019**, 1–20.
44. A. M. Ross, Z. Jiang, M. Bastmeyer and J. J. S. Lahann, *Small*, 2012, **8**, 336–355.
45. K. Dergilev, Z. Tsokolaeva, P. Makarevich, I. Beloglazova, E. Zubkova, M. Boldyreva, E. Ratner, D. Dyikanov, M. Menshikov, A. Ovchinnikov, F. Ageev and Y. Parfyonova, *BioMed Res. Int.*, 2018, **2018**, 3536854.
46. S. V. Lopes, M. N. Collins, R. L. Reis, J. M. Oliveira and J. Silva-Correia, *ACS Appl. Bio Mater.*, 2021, **4**, 2941–2956.
47. I. Matai, G. Kaur, A. Seyedsalehi, A. McClinton and C. T. Laurencin, *Biomaterials*, 2020, **226**, 119536.
48. F. Simunovic and G. Finkenzeller, *Cells*, 2021, **10**, 1749.
49. T. Albrektsson and C. Johansson, *Eur. Spine J.*, 2001, **10**, S96–S101.
50. T. Ghassemi, A. Shahroodi, M. H. Ebrahimzadeh, A. Mousavian, J. Movaffagh and A. Moradi, *Arch. Bone Jt Surg.*, 2018, **6**, 90–99.
51. M. Witzler, D. Büchner, S. H. Shoushrah, P. Babczyk, J. Baranova, S. Witzleben, E. Tobiasch and M. Schulze, *Biomolecules*, 2019, **9**, 840.
52. W. Götz, E. Tobiasch, S. Witzleben and M. Schulze, *Pharmaceutics*, 2019, **11**, 117.
53. V. Grotheer, M. Schulze and E. Tobiasch, *Trends in Bone Tissue Engineering: Proteins for Osteogenic Differentiation and the Respective Scaffolding*, iConcept Press, 2016.
54. N. Zippel, C. A. Limbach, N. Ratajski, C. Urban, C. Luparello, A. Pansky, M. U. Kassack and E. Tobiasch, *Stem Cells Dev.*, 2012, **21**, 884–900.
55. B. El Khaldi-Hansen, F. El-Sayed, D. Schipper, E. Tobiasch, S. Witzleben and M. Schulze, *Front. Stem Cell Regener. Med. Res.*, 2017, **4**, 3–58.
56. A. Leiendecker, S. Witzleben, M. Schulze and E. Tobiasch, *Curr. Stem Cell Res. Ther.*, 2017, **12**, 103–123.
57. M. Mittal, M. R. Siddiqui, K. Tran, S. P. Reddy and A. B. Malik, *Antioxid. Redox Signaling*, 2014, **20**, 1126–1167.
58. M. Shafiq, Y. Chen, R. Hashim, C. He, X. Mo and X. Zhou, *Front. Bioeng. Biotechnol.*, 2021, **9**, 821288.
59. A. R. Unnithan, R. S. Arathyram and C. S. Kim, in *Nanotechnology Applications for Tissue Engineering*, ed. S. Thomas, Y. Grohens and N. Ninan, William Andrew Publishing, Oxford, 2015, pp. 103–123.
60. S. Feroz, N. Muhammad, J. Ranayake and G. Dias, *Bioact. Mater.*, 2020, **5**, 496–509.
61. C. M. L. de Melo, I. J. Da Cruz Filho, G. F. de Sousa, G. A. de Souza Silva, D. K. D. do Nascimento Santos, R. S. Da Silva, B. R. de Sousa, R. G. de Lima Neto, M. do Carmo Alves de Lima and G. J. de Moraes Rocha, *Int. J. Biol. Macromol.*, 2020, **162**, 1725–1733.
62. A. Alzagameem, B. E. Khaldi-Hansen, D. Büchner, M. Larkins, B. Kamm, S. Witzleben and M. Schulze, *Molecules*, 2018, **23**, 2664.
63. A. Alzagameem, S. E. Klein, M. Bergs, X. T. Do, I. Korte, S. Dohlen, C. Hüwe, J. Kreyenschmidt, B. Kamm, M. Larkins and M. Schulze, *Polymers*, 2019, **11**, 670.

64. D. M. Rocca, J. P. Vanegas, K. Fournier, M. C. Becerra, J. C. Scaiano and A. E. Lanterna, *RSC Adv.*, 2018, **8**, 40454–40463.
65. R. M. Kalinoski and J. Shi, *Ind. Crops Prod.*, 2019, **128**, 323–330.
66. M. Guo, T. Jin, N. P. Nghiem, X. Fan, P. X. Qi, C. H. Jang, L. Shao and C. Wu, *Appl. Biochem. Biotechnol.*, 2018, **184**, 350–365.
67. Y. Yan, P. Wang, Y. Lu, Y. Bai, Y. Wei, G. Liu and H. Shi, *Plant J.*, 2021, **107**, 847–860.
68. L. Yao, H. Yang, X. Meng and A. J. Ragauskas, *Front. Energy Res.*, 2022, **9**, 804086.
69. Y. Zhang, M. Jiang, Y. Zhang, Q. Cao, X. Wang, Y. Han, G. Sun, Y. Li and J. Zhou, *Mater. Sci. Eng., C*, 2019, **104**, 110002.
70. D. Spasojević, D. Zmejkoski, J. Glamočlija, M. Nikolić, M. Soković, V. Milošević, I. Jarić, M. Stojanović, E. Marinković, T. Barisani-Asenbauer, R. Prodanović, M. Jovanović and K. Radotić, *Int. J. Antimicrob. Agents*, 2016, **48**, 732–735.
71. M. Li, X. Jiang, D. Wang, Z. Xu and M. Yang, *Colloids Surf., B*, 2019, **177**, 370–376.
72. J. Domínguez-Robles, N. K. Martin, M. L. Fong, S. A. Stewart, N. J. Irwin, M. I. Rial-Hermida, R. F. Donnelly and E. Larrañeta, *Pharmaceutics*, 2019, **11**, 165.
73. I. Spiridon, *Environ. Chem. Lett.*, 2020, **18**, 771–785.
74. B. Sarrouh, R. O. A. de Souza, R. H. da Silva Florindo, R. C. Z. Lofrano and A. M. de Oliveira, *Waste Biomass Valorization*, 2021, **12**, 787–794.
75. M. Witzler, A. Alzagameem, M. Bergs, B. E. Khaldi-Hansen, S. E. Klein, D. Hielscher, B. Kamm, J. Kreyenschmidt, E. Tobiasch and M. Schulze, *Molecules*, 2018, **23**, 1885.
76. M. J. Mehta and A. Kumar, *Chem. - Eur. J.*, 2019, **25**, 1269–1274.
77. L. Cheng, B. Deng, W. Luo, S. Nie, X. Liu, Y. Yin, S. Liu, Z. Wu, P. Zhan, L. Zhang and J. Chen, *J. Agric. Food Chem.*, 2020, **68**, 5249–5258.
78. M. Bergs, X. T. Do, J. Rumpf, P. Kusch, Y. Monakhova, C. Konow, G. Völkering, R. Pude and M. Schulze, *RSC Adv.*, 2020, **10**, 10740–10751.
79. R. Kumar, A. Butreddy, N. Kommineni, P. G. Reddy, N. Bunekar, C. Sarkar, S. Dutt, V. K. Mishra, K. R. Aadil, Y. K. Mishra, D. Oupicky and A. Kaushik, *Int. J. Nanomed.*, 2021, **16**, 2419–2441.
80. J. Gil-Chávez, P. Gurikov, X. Hu, R. Meyer, W. Reynolds and I. Smirnova, *Biomass Convers. Biorefin.*, 2021, **11**, 2387–2403.
81. O. Yu and K. H. Kim, *Appl. Sci.*, 2020, **10**, 4626.
82. R. Li, R. Ouda, C. Kimura, R. Narita, H. Nishimura, T. Fujita and T. Watanabe, *ACS Sustainable Chem. Eng.*, 2021, **9**, 9248–9256.
83. M. R. V. Bertolo, L. B. Brenelli de Paiva, V. M. Nascimento, C. A. Gandin, M. O. Neto, C. E. Driemeier and S. C. Rabelo, *Ind. Crops Prod.*, 2019, **140**, 111591.
84. B. Jiang, Y. Zhang, L. Gu, W. Wu, H. Zhao and Y. Jin, *Int. J. Biol. Macromol.*, 2018, **116**, 513–519.
85. J. A. Sirviö, M. Y. Ismail, K. Zhang, M. V. Tejesvi and A. Ämmälä, *J. Mater. Chem. A*, 2020, **8**, 7935–7946.

86. F. Shu, B. Jiang, Y. Yuan, M. Li, W. Wu, Y. Jin and H. Xiao, *Biomacromolecules*, 2021, **22**, 4905–4918.
87. I. M. Savic, I. M. Savic Gajic, M. G. Milovanovic, S. Zerajic and D. G. Gajic, *Antioxidants*, 2022, **11**, 297.
88. H. Sadeghifar and D. S. Argyropoulos, *ACS Sustainable Chem. Eng.*, 2015, **3**, 349–356.
89. J. L. Espinoza-Acosta, P. I. Torres-Chávez, B. Ramírez-Wong, C. M. López-Saiz and B. J. B. Montaño-Leyva, *BioResources*, 2016, **11**, 5452–5481.
90. L. Zheng, G. Lu, W. Pei, W. Yan, Y. Li, L. Zhang, C. Huang and Q. Jiang, *Int. J. Biol. Macromol.*, 2021, **190**, 291–300.
91. R. Kaur, S. K. Uppal and P. Sharma, *Sugar Tech*, 2017, **19**, 675–680.
92. T. V. Lourençon, G. G. d. Lima, C. S. P. Ribeiro, F. A. Hansel, G. M. Maciel, K. Da Silva, S. M. B. Winnischofer, G. I. B. d. Muniz and W. L. E. Magalhães, *Int. J. Biol. Macromol.*, 2021, **166**, 1535–1542.
93. H. Qu, H. Fu, Z. Han and Y. Sun, *RSC Adv.*, 2019, **9**, 26252–26262.
94. Y. Su, I. Cockerill, Y. Wang, Y.-X. Qin, L. Chang, Y. Zheng and D. Zhu, *RSC Adv.*, 2019, **37**, 428–441.
95. E. Mariani, G. Lisignoli, R. M. Borzì and L. Pulsatelli, *Int. J. Mol. Sci.*, 2019, **20**, 636.
96. E. Marin, F. Boschetto and G. Pezzotti, *J. Biomed. Mater. Res., Part A*, 2020, **108**, 1617–1633.
97. E. Troy, M. A. Tilbury, A. M. Power and J. G. Wall, *Polymers*, 2021, **13**, 3321.
98. M. Brovold, J. I. Almeida, I. Pla-Palacín, P. Sainz-Arnal, N. Sánchez-Romero, J. J. Rivas, H. Almeida, P. R. Dachary, T. Serrano-Aulló, S. Soker and P. M. Baptista, *Adv. Exp. Med. Biol.*, 2018, **1077**, 421–449.
99. P. S. Bakshi, D. Selvakumar, K. Kadirvelu and N. S. Kumar, *Int. J. Biol. Macromol.*, 2020, **150**, 1072–1083.
100. Q. Tang, Y. Qian, D. Yang, X. Qiu, Y. Qin and M. Zhou, *Polymers*, 2020, **12**, 2471.
101. R. Liu, L. Dai, C. Xu, K. Wang, C. Zheng and C. Si, *ChemSusChem*, 2020, **13**, 4266–4283.
102. H. Jahangirian, E. G. Lemraski, R. Rafiee-Moghaddam and T. J. Webster, *Int. J. Nanomed.*, 2018, **13**, 5953–5969.
103. P. Szczepańczyk, M. Szlachta, N. Złocista-Szewczyk, J. Chłopek and K. Pielichowska, *Polymers*, 2021, **13**, 946.
104. S. Iravani and R. S. Varma, *Green Chem.*, 2019, **21**, 4839–4867.
105. S. Kargozar, R. K. Singh, H.-W. Kim and F. Baino, *Acta Biomater.*, 2020, **115**, 1–28.
106. M. Jouyandeh, H. Vahabi, N. Rabiee, M. Rabiee, M. Bagherzadeh and M. R. Saeb, *Emerg. Mater.*, 2022, **5**, 603–620.
107. L. Dai, R. Liu and C.-L. Si, in *Lignin-Based Materials for Biomedical Applications*, Elsevier, 2021, pp. 327–370.
108. S. Sugiarto, Y. Leow, C. L. Tan, G. Wang and D. Kai, *Bioact. Mater.*, 2022, **8**, 71–94.
109. R. Burger, S. Lindner, J. Rumpf, X. T. Do, B. W. K. Diehl, M. Rehahn, Y. B. Monakhova and M. Schulze, *J. Pharm. Biomed. Anal.*, 2022, **212**, 114649.

110. C. Colnot, *Tissue Eng., Part B*, 2011, **17**, 449–457.
111. X. Wang, C. Shi, X. Hou, S. Song, C. Li, W. Cao, W. Chen and L. Li, *J. Biomater. Appl.*, 2022, **36**, 1484–1502.
112. P. M. Sivakumar, A. A. Yetisgin, S. B. Sahin, E. Demir and S. Cetinel, *Carbohydr. Polym.*, 2022, **283**, 119142.
113. D. Wang, J. Jang, K. Kim, J. Kim and C. B. Park, *Biomacromolecules*, 2019, **20**, 2684–2693.
114. R. Liang, J. Zhao, B. Li, P. Cai, X. J. Loh, C. Xu, P. Chen, D. Kai and L. Zheng, *Biomaterials*, 2020, **230**, 119601.
115. K. Ravishankar, M. Venkatesan, R. P. Desingh, A. Mahalingam, B. Sadhasivam, R. Subramaniyam and R. Dhamodharan, *Mater. Sci. Eng., C*, 2019, **102**, 447–457.
116. J. Xu, J. J. Xu, Q. Lin, L. Jiang, D. Zhang, Z. Li, B. Ma, C. Zhang, L. Li, D. Kai, H.-D. Yu and X. J. Loh, *ACS Appl. Bio Mater.*, 2021, **4**, 3–13.
117. L. R. Doblado, C. Martínez-Ramos and M. M. Pradas, *Front. Nanotechnol.*, 2021, **3**, 643507.
118. X. Gu, *Engineering*, 2021, 7, 1700–1703.
119. S. Amini, H. Salehi, M. Setayeshmehr and M. Ghorbani, *Polym. Adv. Technol.*, 2021, **32**, 2267–2289.
120. S. Amini, A. Saudi, N. Amirpour, M. Jahromi, S. S. Najafabadi, M. Kazemi, M. Rafienia and H. Salehi, *Int. J. Biol. Macromol.*, 2020, **159**, 154–173.
121. A. Saudi, S. Amini, N. Amirpour, M. Kazemi, A. Z. Kharazi, H. Salehi and M. Rafienia, *Mater. Sci. Eng., C*, 2019, **104**, 110005.
122. R. Dong, P. X. Ma and B. Guo, *Biomaterials*, 2020, **229**, 119584.
123. H.-K. Zhao, X.-Y. Wei, Y.-M. Xie and Q.-H. Feng, *Int. J. Biol. Macromol.*, 2019, **137**, 1161–1168.
124. J. R. Dwek, *Skeletal Radiol.*, 2010, **39**, 319–323.
125. T. M. Keaveny and W. C. Hayes, *Bone*, 1993, 7, 285–344.
126. B. Q. Le, V. Nurcombe, S. M. Cool, C. A. Van Blitterswijk, J. De Boer and V. L. S. LaPointe, *Materials*, 2018, **11**, 14.
127. T. Bellido, L. I. Plotkin and A. Bruzzaniti, in *Basic and Applied Bone Biology*, ed. D. B. Burr and M. R. Allen, Academic Press, 2nd edn, 2019, pp. 37–55.
128. G. J. Atkins and D. M. Findlay, *Osteoporosis Int.*, 2012, **23**, 2067–2079.
129. B. El Khaldi-Hansen, M. Witzler, M. Schulze, P. Ottensmeyer, J. Baranova and T. Edda, *Frontiers in Stem Cell and Regenerative Medicine Research: Volume 10*, 2020, pp. 1–38.
130. A. L. Boskey, in *Handbook of Biomineralization*, 2007, pp. 59–79.
131. C. H. Tonk, S. H. Shoushrah, P. Babczyk, B. El Khaldi-Hansen, M. Schulze, M. Herten and E. Tobiasch, *Int. J. Mol. Sci.*, 2022, **23**, 1393.
132. C. W. Archer and P. Francis-West, *Int. J. Biochem. Cell Biol.*, 2003, **35**, 401–404.
133. J. Parvizi and G. K. J. H. Y. O. Kim, *High Yield Orthop.*, 2010, 80–81.
134. M. B. Goldring, *Ther. Adv. Musculoskeletal Dis.*, 2012, **4**, 269–285.
135. H. Akkiraju and A. Nohe, *J. Dev. Biol.*, 2015, **3**, 177–192.

136. M. Litviňuková, C. Talavera-López, H. Maatz, D. Reichart, C. L. Worth, E. L. Lindberg, M. Kanda, K. Polanski, M. Heinig, M. Lee, E. R. Nadelmann, K. Roberts, L. Tuck, E. S. Fasouli, D. M. DeLaughter, B. McDonough, H. Wakimoto, J. M. Gorham, S. Samari, K. T. Mahbubani, K. Saeb-Parsy, G. Patone, J. J. Boyle, H. Zhang, H. Zhang, A. Viveiros, G. Y. Oudit, O. A. Bayraktar, J. G. Seidman, C. E. Seidman, M. Noseda, N. Hubner and S. A. Teichmann, *Nature*, 2020, **588**, 466–472.
137. H. Vliegen, A. Van der Laarse, C. Cornelisse and F. Eulderink, *Eur. Heart J.*, 1991, **12**, 488–494.
138. P. Camelliti, T. K. Borg and P. Kohl, *Cardiovasc. Res.*, 2005, **65**, 40–51.
139. C. Hall, K. Gehmlich, C. Denning and D. Pavlovic, *J. Am. Heart Assoc.*, 2021, **10**, e019338.
140. D. L. Brutsaert, *Physiol. Rev.*, 2003, **83**, 59–115.
141. A. Armulik, G. Genové and C. Betsholtz, *Dev. Cell*, 2011, **21**, 193–215.
142. M. Venus, J. Waterman and I. J. S. McNab, *Surgery*, 2010, **28**, 469–472.
143. C. Harvey, *Orthop. Nurs.*, 2005, **24**, 143–157.
144. K. Vig, A. Chaudhari, S. Tripathi, S. Dixit, R. Sahu, S. Pillai, V. A. Dennis and S. R. Singh, *Int. J. Mol. Sci.*, 2017, **18**, 789.
145. G. A. Mihailoff and D. E. Haines, in *Fundamental Neuroscience for Basic and Clinical Applications*, ed. D. E. Haines and G. A. Mihailoff, Elsevier, 5th edn, 2018, pp. 15–33.
146. D. E. Haines and G. A. Mihailoff, *Fundamental neuroscience for basic and clinical applications E-book*, Elsevier Health Sciences, 2017.
147. P. R. Hof, G. Kidd, J. DeFelipe, J. de Vellis, M. A. G. Sosa, G. A. Elder and B. D. Trapp, in *From Molecules to Networks*, Elsevier, 2014, pp. 3–21.
148. J. A. Gomez-Sanchez, K. S. Pilch, M. van der Lans, S. V. Fazal, C. Benito, L. J. Wagstaff, R. Mirsky and K. R. Jessen, *J. Neurosci.*, 2017, **37**, 9086–9099.
149. T. Gordon, *Int. J. Mol. Sci.*, 2020, **21**, 8652.
150. A. Przekora, *Mater. Sci. Eng., C*, 2019, **97**, 1036–1051.
151. E. Mazzoni, C. Mazziotta, M. R. Iaquinta, C. Lanzillotti, F. Fortini, A. D'Agostino, L. Trevisiol, R. Nocini, G. Barbanti-Brodano, A. Mescola, A. Alessandrini, M. Tognon and F. Martini, *Front. Dev. Biol.*, 2021, **8**, 610570.
152. R. Liang, J. Zhao, B. Li, P. Cai, X. J. Loh, C. Xu, P. Chen, D. Kai and L. J. B. Zheng, *Biomaterials*, 2020, **230**, 119601.
153. M. Kouhi, J. Varshosaz, B. Hashemibeni and A. Sarmadi, *Mater. Sci. Eng., C*, 2020, **115**, 111114.
154. D. Wang, J. Jang, K. Kim, J. Kim and C. B. Park, *Biomacromolecules*, 2019, **20**, 2684–2693.
155. J. Wang, L. Tian, B. Luo, S. Ramakrishna, D. Kai, X. J. Loh, I. H. Yang, G. R. Deen, X. J. C. Mo and S. B. Biointerfaces, *Colloids Surf., B*, 2018, **169**, 356–365.
156. S. Danti, L. Trombi, A. Fusco, B. Azimi, A. Lazzeri, P. Morganti, M.-B. Coltelli and G. Donnarumma, *Int. J. Mol. Sci.*, 2019, **20**, 2669.

157. E. Tobiasch, in *Stem Cell Engineering*, Springer, 2011, pp. 61–77.
158. Y. Zhang, D. Khan, J. Delling and E. Tobiasch, *Sci. World J.*, 2012, **2012**, 793823.
159. C. H. Tonk, M. Witzler, M. Schulze and E. Tobiasch, in *Essential Current Concepts in Stem Cell Biology*, Springer, 2020, pp. 21–39.
160. S. H. Shoushrah, J. L. Transfeld, C. H. Tonk, D. Büchner, S. Witzleben, M. A. Sieber, M. Schulze and E. Tobiasch, *Int. J. Mol. Sci.*, 2021, **22**, 6387.
161. B. Brand-Saberi, *Essential Current Concepts in Stem Cell Biology*, Springer, 2020.
162. E.-M. Haddouti, M. Skroch, N. Zippel, C. Müller, B. Birova, A. Pansky, C. Kleinfeld, M. Winter and E. Tobiasch, *Materialwiss. Werkstofftech.*, 2009, **40**, 732–737.
163. P. Babczyk, M. Winter, C. Kleinfeld, A. Pansky, C. Oligschleger and E. Tobiasch, *Appl. Sci.*, 2022, **12**, 2031.
164. D. Khan, C. Kleinfeld, M. Winter and E. Tobiasch, *Tissue Regeneration – From Basic Biology to Clinical Application*, 2012, pp. 325–335.
165. A. Pfützner, D. Schipper, A. Pansky, C. Kleinfeld, B. Roitzheim and E. Tobiasch, *Int. J. Stem Cells*, 2017, **10**, 154–159.
166. Y. Zhang, P. Babczyk, A. Pansky, M. U. Kassack and E. Tobiasch, *Int. J. Mol. Sci.*, 2020, **21**, 6210.
167. C. Kaebisch, D. Schipper, P. Babczyk and E. Tobiasch, *Comput. Struct. Biotechnol. J.*, 2014, **13**, 75–84.
168. A. Longo, M. Librizzi, F. Naselli, F. Caradonna, E. Tobiasch and C. Luparello, *Biochimie*, 2013, **95**, 1888–1896.
169. K. Jayaramayya, I. Mahalaxmi, M. D. Subramaniam, N. Raj, A. A. Dayem, K. M. Lim, S. J. Kim, J. Y. An, Y. Lee, Y. Choi, A. Raj, S.-G. Cho and B. Vellingiri, *BMB Rep.*, 2020, **53**, 400–412.
170. D. Ma, K. Xu, G. Zhang, Y. Liu, J. Gao, M. Tian, C. Wei, J. Li and L. Zhang, *Int. Immunopharmacol.*, 2019, **74**, 105687.
171. A. R. R. Weiss and M. H. Dahlke, *Front. Immunol.*, 2019, **10**, 1191.
172. H. Tao, Z. Han, Z. C. Han and Z. Li, *Stem Cells Int.*, 2016, **2016**, 1314709.
173. N. Zippel, M. Schulze and E. Tobiasch, *Recent Pat. Biotechnol.*, 2010, **4**, 1–22.
174. J. Baranova, D. Büchner, W. Götz, M. Schulze and E. Tobiasch, *Int. J. Mol. Sci.*, 2020, **21**, 4031.
175. U. Galderisi, G. Peluso and G. Di Bernardo, *Stem Cell Rev. Rep.*, 2022, **18**, 23–36.
176. I. Razeghian-Jahromi, A. G. Matta, R. Canitrot, M. J. Zibaeenezhad, M. Razmkhah, A. Safari, V. Nader and J. Roncalli, *Stem Cell Res. Ther.*, 2021, **12**, 361.
177. C. M. Dunn, S. Kameishi, D. W. Grainger and T. Okano, *Acta Biomater.*, 2021, **133**, 114–125.
178. S. H. Lee, *Osteoporosis Sarcopenia*, 2018, **4**, 150.
179. F. M. Pérez-Campo and J. A. Riancho, *Curr. Genomics*, 2015, **16**, 368–383.
180. K. Takahashi, K. Tanabe, M. Ohnuki, M. Narita, T. Ichisaka, K. Tomoda and S. Yamanaka, *Cell*, 2007, **131**, 861–872.

181. D. Wu, B. Hamilton, C. Martin, Y. Gao, M. Ye and S. Yao, *J. Visualized Exp.*, 2009, **34**, 1553.
182. L. Han, J. Mich-Basso and B. Kühn, in *Cardiac Regeneration*, Springer, 2021, pp. 125–139.
183. R. Castro-Vinuelas, C. Sanjurjo-Rodriguez, M. Pineiro-Ramil, S. Rodriguez-Fernandez, I. Fuentes-Boquete, F. Blanco-Garcia and S. D. Prado, *Osteoarthritis Cartilage*, 2020, **28**, S34.
184. M. Bordoni, F. Rey, V. Fantini, O. Pansarasa, A. M. Di Giulio, S. Carelli and C. Cereda, *Int. J. Mol. Sci.*, 2018, **19**, 3972.
185. G. Bilousova, D. H. Jun, K. B. King, S. De Langhe, W. S. Chick, E. C. Torchia, K. S. Chow, D. J. Klemm, D. R. Roop and S. M. Majka, *Stem Cells*, 2011, **29**, 206–216.
186. X. Liu, W. Chen, C. Zhang, W. Thein-Han, K. Hu, M. A. Reynolds, C. Bao, P. Wang, L. Zhao and H. H. K. Xu, *Tissue Eng., Part A*, 2017, **23**, 546–555.
187. L.-S. Spitzhorn, M. Megges, W. Wruck, M. S. Rahman, J. Otte, Ö. Degistirici, R. Meisel, R. V. Sorg, R. O. C. Oreffo and J. Adjaye, *Stem Cell Res. Ther.*, 2019, **10**, 100.
188. W. Wruck, N. Graffmann, L.-S. Spitzhorn and J. Adjaye, *Front. Cell Dev. Biol.*, 2021, **9**, 717772.
189. Q. Lian, Y. Zhang, J. Zhang, H. K. Zhang, X. Wu, Y. Zhang, F. F. Lam, S. Kang, J. C. Xia, W. H. Lai, K. W. Au, Y. Y. Chow, C. W. Siu, C. N. Lee and H. F. Tse, *Circulation*, 2010, **121**, 1113–1123.
190. S. H. Ranganath, O. Levy, M. S. Inamdar and J. M. Karp, *Cell Stem Cell*, 2012, **10**, 244–258.
191. K.-H. Lee, W.-C. Tseng, C.-Y. Yang and D.-C. Tarng, *Int. J. Mol. Sci.*, 2019, **20**, 3529.
192. X. Bian, K. Ma, C. Zhang and X. Fu, *Stem Cell Res. Ther.*, 2019, **10**, 158.
193. Y. Guo, Y. Yu, S. Hu, Y. Chen and Z. Shen, *Cell Death Dis.*, 2020, **11**, 349.
194. M. Librizzi, E. Tobiasch and C. Luparello, *Cell Biochem. Funct.*, 2016, **34**, 7–15.
195. A. Alzagameem, J. Bergrath, J. Rumpf and M. Schulze, in *Micro and Nanolignin in Aqueous Dispersions and Polymers*, ed. D. Puglia, C. Santulli and F. Sarasini, Elsevier, 2022, pp. 131–171.
196. J. Rumpf, X. T. Do, R. Burger, Y. Monakhova and M. Schulze, in *Lignin-Based Materials for Biomedical Applications*, ed. H. Santos and P. Figueiredo, Elsevier, 2021, pp. 105–158.
197. J. Rumpf, X. T. Do, R. Burger, Y. B. Monakhova and M. Schulze, *Biomacromolecules*, 2020, **21**, 1929–1942.
198. M. Bergs, X. T. Do, J. Rumpf, P. Kusch, Y. Monakhova, C. Konow, G. Völkering, R. Pude and M. J. R. A. Schulze, *RSC Adv.*, 2020, **10**, 10740–10751.
199. R. Burger, S. Lindner, J. Rumpf, X. T. Do, B. W. K. Diehl, M. Rehahn, Y. B. Monakhova and M. Schulze, *J. Pharm. Biomed. Anal.*, 2022, **212**, 114649.
200. R. Burger, J. Rumpf, X. T. Do, Y. B. Monakhova, B. W. K. Diehl, M. Rehahn and M. Schulze, *ACS Omega*, 2021, **6**, 29516–29524.

201. S. Lindner, R. Burger, D. N. Rutledge, X. T. Do, J. Rumpf, B. W. K. Diehl, M. Schulze and Y. B. Monakhova, *Anal. Chem.*, 2022, **94**, 3997–4004.
202. E. Eggenhofer, V. Benseler, A. Kroemer, F. Popp, E. Geissler, H. Schlitt, C. Baan, M. Dahlke and M. Hoogduijn, *Front. Immunol.*, 2012, **3**, 297.
203. R. H. Lee, A. A. Pulin, M. J. Seo, D. J. Kota, J. Ylostalo, B. L. Larson, L. Semprun-Prieto, P. Delafontaine and D. J. Prockop, *Cell Stem Cell*, 2009, **5**, 54–63.
204. P. R. Baraniak and T. C. McDevitt, *Regener. Med.*, 2010, **5**, 121–143.
205. G. Maguire, *Commun. Integr. Biol.*, 2013, **6**, e26631.
206. T. Biswal, *Mater. Today: Proc.*, 2021, **41**, 397–402.
207. X. Zhang, L. Li, J. Ouyang, L. Zhang, J. Xue, H. Zhang and W. Tao, *Nano Today*, 2021, **39**, 101196.
208. M. Qu, X. Jiang, X. Zhou, C. Wang, Q. Wu, L. Ren, J. Zhu, S. Zhu, P. Tebon, W. Sun and A. Khademhosseini, *Adv. Healthcare Mater.*, 2020, **9**, e1901714.
209. H. Zhao, M. Liu, Y. Zhang, J. Yin and R. Pei, *Nanoscale*, 2020, **12**, 14976–14995.
210. I. 10993-1, International Organization for Standardization Geneva, Switzerland, 2018.
211. F. Zhang and M. W. King, *Adv. Healthcare Mater.*, 2020, **9**, 1901358.
212. T.-R. Kim, M.-S. Kim, T. S. Goh, J. S. Lee, Y. H. Kim, S.-Y. Yoon and C.-S. Lee, *Appl. Sci.*, 2019, **9**, 1965.
213. J. Wieding, A. Jonitz and R. J. M. Bader, *Materials*, 2012, **5**, 1336–1347.
214. A. Shahini, M. Yazdimamaghani, K. Walker, M. Eastman, H. Hatami-Marbini, B. Smith, J. Ricci, S. Madihally, D. Vashaee and L. Tayebi, *Int. J. Nanomed.*, 2014, **9**, 167–181.
215. S. V. Lopes, M. N. Collins, R. L. Reis, J. M. Oliveira and J. Silva-Correia, *ACS Appl. Bio Mater.*, 2021, **4**, 2941–2956.
216. R. Grimmig, P. Babczyk, P. Gillemot, K.-P. Schmitz, M. Schulze and E. Tobiasch, *Appl. Sci.*, 2019, **9**, 4414.
217. A. Barradas, H. Yuan, C. A. van Blitterswijk and P. Habibovic, *Eur. Cell. Mater.*, 2011, **21**, 407–429.
218. S. Antenucci, L. Panzella, H. Farina, M. A. Ortenzi, S. Martinotti, E. Ranzato, B. Burlando, M. d'Ischia, A. Napolitano and L. Verotta, *RSC Adv.*, 2015, **6**, 2993–3002.
219. S. Saravanan, S. Nethala, S. Pattnaik, A. Tripathi, A. Moorthi and N. Selvamurugan, *Int. J. Biol. Macromol.*, 2011, **49**, 188–193.
220. D. Rico-García, L. Ruiz-Rubio, L. Pérez-Alvarez, S. L. Hernández-Olmos, G. L. Guerrero-Ramírez and J. L. Vilas-Vilela, *Polymers*, 2020, **12**, 81.
221. Y. Meng, J. Lu, Y. Cheng, Q. Li and H. Wang, *Int. J. Biol. Macromol.*, 2019, **135**, 1006–1019.
222. A. Zerpa, L. Pakzad and P. Fatehi, *ACS Omega*, 2018, **3**, 8233–8242.
223. D. Kai, M. J. Tan, P. L. Chee, Y. K. Chua, Y. L. Yap and X. J. Loh, *Green Chem.*, 2016, **18**, 1175–1200.
224. A. Grossman and W. Vermerris, *Curr. Opin. Biotechnol.*, 2019, **56**, 112–120.

225. P. Terzioğlu, F. N. Parın and Y. Sıcak, in *Lignin*, ed. S. Sharma and A. Kumar, Springer International Publishing, Cham, 2020, pp. 253–273.
226. A. Ogan, B. Yüce-Dursun, D. Abdullah, A. Beyler-Çiğil, M. V. Kahraman, P. Çağlayan, M. Birbir, Ö. Mutlu and N. Gülsoy, *Polym. Adv. Technol.*, 2020, **31**, 2222–2228.
227. T. Abudula, T. Colombani, T. Alade, S. A. Bencherif and A. Memić, *Biomacromolecules*, 2021, **22**, 4110–4121.
228. B. Beer, M. J. Bartolome, L. Berndorfer, G. Bochmann, G. M. Guebitz and G. S. Nyanhongo, *Int. J. Biol. Macromol.*, 2020, **161**, 1440–1446.
229. Y. Chen, K. Zheng, L. Niu, Y. Zhang, Y. Liu, C. Wang and F. Chu, *Int. J. Biol. Macromol.*, 2019, **128**, 414–420.
230. D. Kai, K. Zhang, S. S. Liow and X. J. Loh, *ACS Appl. Bio Mater.*, 2019, **2**, 127–134.
231. S. Chandna, N. S. Thakur, R. Kaur and J. Bhaumik, *Biomacromolecules*, 2020, **21**, 3216–3230.
232. L. Sun, Z. Mo, Q. Li, D. Zheng, X. Qiu and X. Pan, *Int. J. Biol. Macromol.*, 2021, **175**, 516–525.
233. M. N. Collins, G. Ren, K. Young, S. Pina, R. L. Reis and J. M. Oliveira, *Adv. Funct. Mater.*, 2021, **31**, 2010609.
234. M. Witzler, P. F. Ottensmeyer, M. Gericke, T. Heinze, E. Tobiasch and M. Schulze, *Int. J. Mol. Sci.*, 2019, **20**, 3565.
235. W. Wei and H. Dai, *Bioact. Mater.*, 2021, **6**, 4830–4855.
236. X. Zhao, D. A. Hu, D. Wu, F. He, H. Wang, L. Huang, D. Shi, Q. Liu, N. Ni, M. Pakvasa, Y. Zhang, K. Fu, K. H. Qin, A. J. Li, O. Hagag, E. J. Wang, M. Sabharwal, W. Wagstaff, R. R. Reid, M. J. Lee, J. M. Wolf, M. El Dafrawy, K. Hynes, J. Strelzow, S. H. Ho, T.-C. He and A. Athiviraham, *Front. Bioeng. Biotechnol.*, 2021, **9**, 603444.
237. P. Montero, M. Flandes-Iparraguirre, S. Musquiz, M. Pérez Araluce, D. Plano, C. Sanmartín, G. Orive, J. J. Gavira, F. Prosper and M. M. Mazo, *Front. Bioeng. Biotechnol.*, 2020, **8**, 955.
238. A. B. Mohd Hilmi and A. S. Halim, *World J. Stem Cells*, 2015, **7**, 428–436.
239. H. Bi and Y. Jin, *Burns Trauma*, 2013, **1**, 63–72.
240. P. Sensharma, G. Madhumathi, R. D. Jayant and A. K. Jaiswal, *Mater. Sci. Eng., C*, 2017, **77**, 1302–1315.
241. Y. Li, Z. Ma, Y. Ren, D. Lu, T. Li, W. Li, J. Wang, H. Ma and J. Zhao, *Front. Neurol.*, 2021, **12**, 768267.
242. J. A. Menima-Medzogo, K. Walz, J. C. Lauer, G. Sivasankarapillai, F. R. Gleuwitz, B. Rolauffs, M.-P. Laborie and M. L. Hart, *Biology*, 2022, **11**, 696.
243. Y. Zhang, M. Jiang, Y. Zhang, Q. Cao, X. Wang, Y. Han, G. Sun, Y. Li and J. Zhou, *Mater. Sci. Eng., C*, 2019, **104**, 110002.
244. A. Saudi, S. Amini, N. Amirpour, M. Kazemi, A. Zargar Kharazi, H. Salehi and M. Rafienia, *Mater. Sci. Eng., C*, 2019, **104**, 110005.
245. H.-K. Zhao, X.-Y. Wei, Y.-M. Xie and Q.-H. Feng, *Int. J. Biol. Macromol.*, 2019, **137**, 1161–1168.

CHAPTER 9

Lignin and Its Composites for Wound Dressing

HORACIO BACH

Department of Medicine, Division of Infectious Diseases, University of British Columbia, 410-2660 Oak Street, Vancouver, BC, V6H3Z6, Canada
Email: hbach@mail.ubc.ca

9.1 Wound Healing

Upon skin injury, wound healing will commence, regenerating normal skin. Skin healing involves complex sequential and overlapping processes where different systems work together. The healing process can be divided into the following mechanisms: (1) coagulation, (2) inflammation, (3) proliferation, and (4) tissue remodeling.[1] The healing process also includes a variety of cell populations that are orchestrated to complete the process. For example, the replication and migration of skin fibroblasts and keratinocytes are crucial to reducing the wound area. However, when skin injuries involve extensive surfaces, wound healing might not finalize with a complete repair of the tissues. This damage might occur to the dermis and epidermis (second-degree burn) or destruction of the skin (third-degree burn and chronic diabetic foot ulcers).

Wound healing represents a burden to health authorities, with an upper cost estimated close to USD100 billion in 2014.[2] Hence, proper wound healing is important to reduce healthcare costs and prevent chronic wound progression.

Biomaterials Science Series No. 16
Lignin-based Materials: Health Care and Medical Applications
Edited by Kuruvilla Joseph, Runcy Wilson, Gejo George and Saritha Appukuttan

Published by the Royal Society of Chemistry, www.rsc.org

When developing a wound dressing, a series of characteristics are desirable to improve wound healing. For example, an ideal wound dressing should maintain the moisture of the injured area to facilitate fast healing and scarring. Other factors to consider are protecting against infections, eliminating wound exudates, and reducing pain, especially when extensive burns are treated. Moreover, wound dressing may contain specific agents to help wound healing, such as growth factors or nitric oxide (NO).[3]

Traditionally, wound dressings, such as hydrocolloids, are the most commonly used because they can address the desired functions of moisturizing, controlling infections, and absorbing exudates.[4]

With the advent of nanotechnology, new developments in wound dressing have been pursued. For instance, the fact that specific nanoparticles are effective for infection control opened a new venue for improving wound dressing. In addition, it is desired that the nanoparticles used are biocompatible with injured tissue. In this regard, the survival, migration, and proliferation of keratinocytes (90% of the total epidermis cells) should not be affected by their exposure to nanoparticle-containing wound dressings. Furthermore, nanocomposites such as nanofibers are attractive because their three-dimensional structure mimics skin topography, as shown by cell proliferation stimulation, increased wound repair,[5] and absorption of wound exudates.[6]

9.2 Desired Wound Care Properties

Many types of wound dressings have been developed to address wound care, such as foams, nanofibers, and hydrogels.[7] Hydrogels are one of the most studied systems because of their unique properties that include flexibility (to adapt to the topography of the wound), elasticity, biocompatibility, and high water content.[8] On the other hand, the use of hydrogels can be limited as a result of their insufficient mechanical strength and weak antimicrobial activity. In this regard, antibacterial activity is crucial for the adequate property of hydrogel-based wound dressing. However, antimicrobial agents, such as silver and antibiotics, can fulfill this deficiency. Also, some positively charged biopolymers, such as chitosan, can affect the stability of the negatively charged bacterial cell wall.

Besides the need for antimicrobial activity, mechanical properties are also crucial for wound dressings.[9] In other words, these dressings should be able to adapt to the expansion and contraction that happen during wound healing involving blood vessels, nerves, skin, and tissue changes.[10] A lack of response to these changes can lead to micro-cracks and deformities with a concomitant loss of properties. These problems can be exacerbated as a result of wound dehydration, as the presence of moisture is important for the healing process.[11] The primary activity of other traditional dry wound dressings (*e.g.*, gauze and bandages) is that they can absorb wound exudates but can provide a moist environment.

9.3 Lignin

Lignin is a heterogeneous and natural polysaccharide obtained from wood and tree bark with promising application in biomedicine. This polymer is an attractive and compatible biomaterial due to its high strength with antioxidant and anti-inflammatory properties. It can be converted into nanofibers by different methods, including electrospinning and mechanical and chemical treatment.[12,13]

9.3.1 Structure, Reactive Groups, and Modifications

Lignin is a phenolic biopolymer with a complex amorphous three-dimensional structure that originates from linkages between heterologous molecules of phenylpropanol. These molecules are *p*-hydroxyl-phenyl, syringyl, and guaiacyl propanols (Figure 9.1).

These units form an intricate network as a result of carbon–carbon (β-5, 5-5, β-1, and β-β) and carbon–oxygen (β-O-4, α-O-4, and 4-O-5) bonds.[14] The chemical groups present in the lignin molecule, such as hydroxyl, carboxyl, and methoxyl groups, can be used to functionalize the molecule with ample combinations to design new lignin-based composites. In addition, depending on their isolation procedure, different types of lignin are commercialized. The most common versions are alkali lignin, kraft lignin, organosolv lignin, and lignosulfonates[15] (Figure 9.2).

The alkaline process isolates lignin from wood (its primary source). The alkali used in the process breaks down the non-phenolic β-aryl-ether bonds in the wood, releasing lignin fragments. The alkali lignin lacks sulfur content, limiting its use due to its low reactivity.[16] The alkali lignin's phenylglycerol and styryl ether contents were higher than those of the kraft lignin (discussed later).[17]

The kraft process is an industrial process by which wood chips are treated with a hot mixture of NaOH and Na_2S. This process releases lignin, hemicellulose, and cellulose. This method targets the aromatic rings rather than the aliphatic groups, and it introduces thiol groups that can be used for

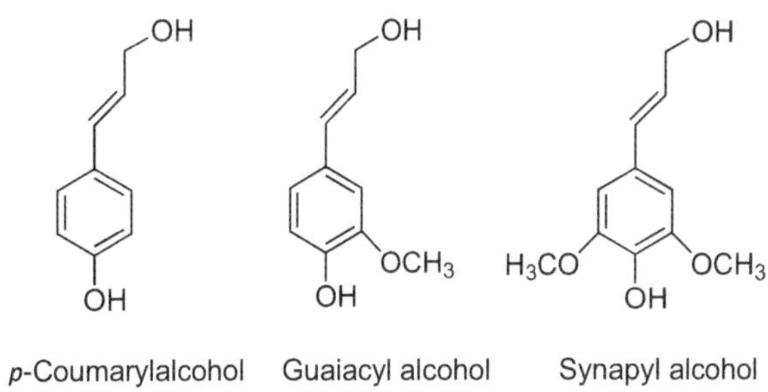

Figure 9.1 Lignin monomers.

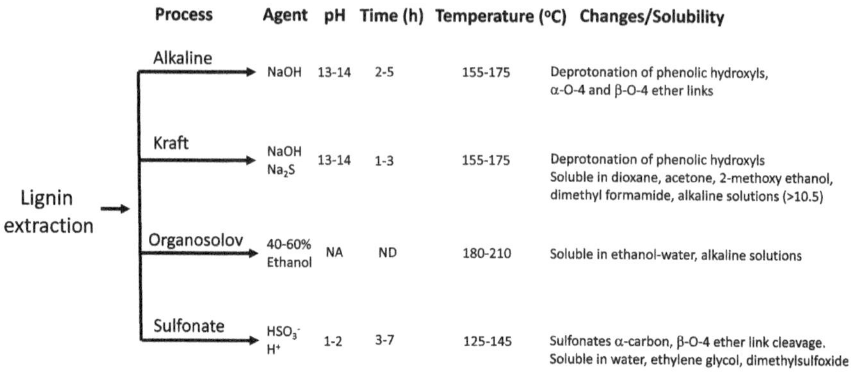

Figure 9.2 Extraction and characterization of lignin using different technologies.

further functionalization of the lignin. The lignin fragments obtained by this process range from 2000 to 3000 Da.[18]

Organosolv is obtained by dissolving wood in organic solvents, such as acetone, methanol, ethanol, and butanol. This type of lignin produces more reactive soluble lignin with higher purity.[19] In addition, the treated lignin has a higher proportion of oxidized and phenolic hydroxyl groups suitable for further functionalization.[19]

Another group of lignins is lignosulfonates, which are obtained by a sulfonation process. HSO_3^- breaks down the ether bonds, generating 1000–140 000 Da fragments. The sulfonation process targets only the side chain of the lignin but not the aromatic rings.[18]

The most reactive functional group in lignin is the hydroxyl. It can, for example, react with diisocyanates, generating polyurethanes.[20] Also, aliphatic hydroxyl (R–OH) and phenolic (Ph–OH) groups are the most abundant in the lignin molecule. Then, modification methods have been developed to alter the lignin structure. These methods include hydroxypropylation,[21] esterification,[22] and acetylation.[23]

Another modification of lignin is lignosulfonates. These water-soluble anionic polymer byproducts are obtained after treating wood chips with a mixture of sulfites.[24] It has been shown that lignosulfonates, in combination with collagen, change the mechanical properties and decrease the cytotoxic and immunological responses during tissue repair.[25] Further modification of lignosulfonates by methacrylation and thiolene reaction provides a tunable cross-linking with enhanced stiffness. In addition, the results of this reaction suggest that aliphatic hydroxyl groups incorporate more thiol groups than phenolic hydroxyl groups.[26,27]

Enzymatic modifications have shown that it is possible to increase the phenolic content in lignin with an increase in its reactivity. This modification can be performed using laccases, which are copper-based oxidases able to oxidize phenolic substrates.[28] However, due to the enzymatic activity,[29] the phenolic content was duplicated.[30]

9.3.2 Lignin Nanocomposites

9.3.2.1 Hydrogels

Since lignin lacks gelation properties, the synthesis of lignin-containing hydrogels should be accompanied by the addition of a compound that will increase the formation of molecular associations by hydrogen bonding or van der Waals forces of attraction between the molecules. These forces cause interactions between hemiacetal oxygen, methyl, and hydroxyl and sugar residues of the polysaccharide. Therefore, different materials have been used for the fabrication of hydrogels (Table 9.1).

9.3.2.1.1 Single Molecules – Curcumin. Curcumin is a component of turmeric, a spice amply used worldwide. This compound has shown several activities, including those in treatments for skin-related disorders.[31,32] Moreover, some studies performed *in vitro* and *in vivo* showed an enhancement of wound repair and reduced oxidative stress.[33–35] Furthermore, the mechanisms by which curcumin affects tissue repair upon wound damage have been reported, including upregulation of the endothelial-extracellular matrix and the transforming growth factor-beta (TGFB) expression in wounds.[36] Finally, curcumin accelerates proliferative healing, enhancing *de novo* vascularization, epithelialization, increased collagen deposition, and tissue development.[37]

9.3.2.1.2 Polysaccharides. A few polysaccharides have been used to produce hydrogels in combination with lignin. These biopolymers include the following:

Agarose is a linear polysaccharide obtained from different species of red seaweeds. The mechanical properties of hydrogel-based agarose combined with lignin were improved by adding silk fibroin, an insoluble protein obtained from the larvae of *Bombyx mori*,[38] and $ZnCr_2O_4$.[39]

Bacterial cellulose (BC) is a natural polymer mainly synthesized by the Gram-negative bacterium *Acetobacter xylinum*.[40] This polysaccharide has shown excellent mechanical properties with potential applications in biotechnology.[41] In addition, other bacterial strains have been reported to produce cellulose, such as *Sarcina ventriculi*,[42] *Agrobacterium tumefaciens*,[43] and *Rhizobium trifolii*.[44] Furthermore, BC polymer complies with most of the desired characteristics of a hydrogel for wound care. For instance, it retains a high water content (up to 99%), it has high wet strength with high chemical purity, and its structure is resistant to sterilization. In addition, it is a highly biocompatible material with proven applications. Moreover, other studies reported its effect on the rapid regeneration of tissues and promoting angiogenesis in injured areas.[45,46]

One of the drawbacks of BC is the lack of antimicrobial activities. However, studies have reported that adding antimicrobial agents such

Table 9.1 Materials used in the fabrication of hydrogels.

Polymer	Structure	Unit
Natural		
Agarose	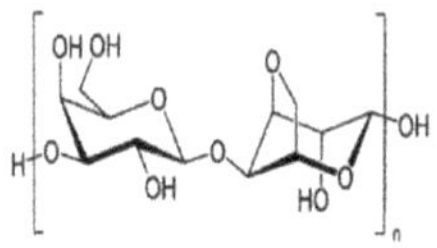	Agarose (a disaccharide of D-galactose and 3,6-anhydro-L-galactopyranose)
Bacterial cellulose	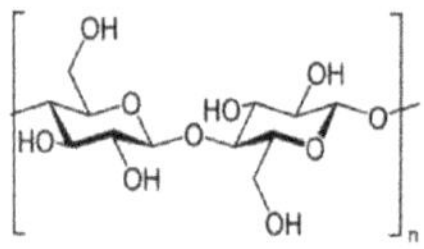	β1→4 D-glucose
Carrageenan	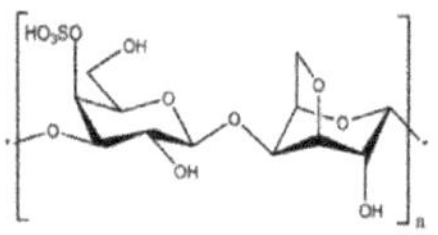	3-β-D-Galactopyranose (G-unit) and 4-α-D-galactopyranose (D-units) or 4-3,6-anhydro-α-D-galactopyranose (DA-units)
Chitosan	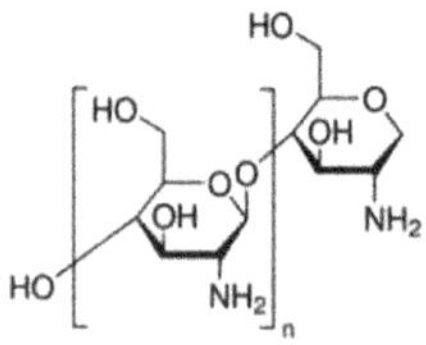	β-(1→4)-D-Glucosamine (deacetylated unit) and *N*-acetyl-D-glucosamine (acetylated unit)
Dextran	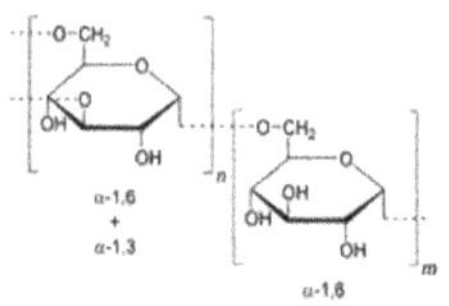	Branched poly-α-D-glucoside with predominantly C-1→C-6 with branches from α-C-1→C3 linkages
Synthetic		
PEG	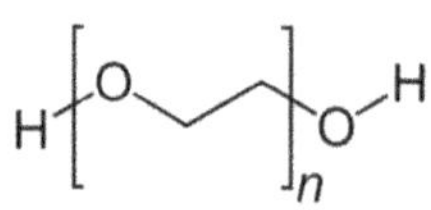	Polyethylene glycol
Polyurethane	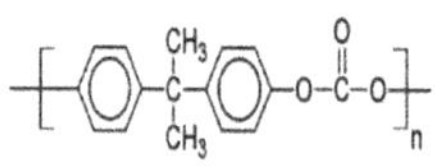	
Poly(lactic acid)	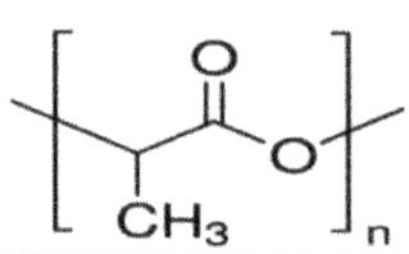	Lactic and lactide units

as benzalkonium chloride and metal nanoparticles (*e.g.*, Ag, Cu, and ZnO) is feasible.[47–50]

Carrageenan is a high-molecular-weight linear sulfated polysaccharide extracted from red seaweeds, forming flexible molecules with helical structures. There are three types of carrageenan: kappa, iota, and lambda, with variable numbers and positions of the sulfate groups. The type of carrageenan dictates the strength of gels; for instance, higher levels of ester sulfate groups lower the strength of gels.

When preparing carrageenan-based hydrogels, adding a divalent cation like Ca^{2+} generates a bridge between the helices of carrageenan through electrostatic bonds.[51] These bonds are important to impart stabilization and gelling conditions to the hydrogel.

Chitosan is a linear polysaccharide obtained by the deacetylation of chitin (using NaOH), a polysaccharide that constitutes the skeleton of crustaceans and the cell wall of fungi. The addition of different percentages of lignin to chitosan together with polyvinyl alcohol (PVA) affected the porosity of the hydrogel with a concomitant reduction in the porous size as the concentration increased.[52] The formation of two different types of bonding can explain this change: hydrogen and ionic bonding generated between the PVA and the lignin and the amino group from the chitosan and the sulfonic group of lignin, respectively. Another approach to producing hydrogels with this polysaccharide is the cross-linking of alkali lignin with chitosan.[53] In this approach, an electrostatic complex is formed with hydrogel characteristics upon using at least 10% w/v alkali lignin, which increases the shear strength and the viscosity of the hydrogel.[53]

Dextran is a branched glucan that was discovered in wine. It is a versatile biopolymer that can be used in hydrogel synthesis. In this respect, it was combined in a colloidal solution with lignin and glycol and then soaked in cotton fibers. Finally, the reactions between the cotton fibers and the colloidal solution improved the mechanical properties of the hydrogel in terms of its tensile strength and total elongation.[54]

9.3.2.1.3 Synthetic Polymers. Several studies reported the use of synthetic polymers for the synthesis of hydrogels combined with lignin. These synthetic polysaccharides include polyethylene glycol (PEG), polyurethane, and poly(lactic acid) (PLA).

9.3.2.1.3.1 PEG and Gelatin. The composition of this type of hydrogel requires modifications of the raw materials used for its synthesis. These modifications will increase the functionality of individual components with improvements to the final nanocomposite. One example is the use of thiolated lignosulfonate combined with diacrylate polyethylene glycol and fish gelatin.[55] Interestingly, the use of UV light as the cross-linker did not achieve the expected results as only the surface of the nanocomposite was cross-linked. Instead, the synthesis using a phosphine derivative cross-linking conferred superior antioxidant activity to the hydrogel for tissue repair.[55] Although incorporating the thiolated lignosulfonate with the fish gelatin decreased the elasticity of the hydrogel, the addition of PEG restored it to around 80% of the original value.[55] This final addition to this hydrogel enhanced the human dermal fibroblast adherence.

9.3.2.1.3.2 PLA. This is a synthetic thermoplastic aliphatic polymer formed by the condensation of two monomers: lactic acid and lactide. PLA is a biodegradable polymer with increased use as a biomaterial.[56]

Lignin and PLA are immiscible,[57] and modifications of the polymers are necessary to improve their compatibility. For example, different types of commercially available lignins were used to enhance their compatibility. In the case of alkali lignin, esterification of the hydroxyl groups by acetylation improved its compatibility and thermal stability.[58] Similar esterification was achieved with raw lignin modified by butyration,[59] whereas kraft lignin was treated with butyric anhydride, and an increase in its mechanical properties was measured.[60]

9.3.2.2 Nanofibers

Nanofibers are structures ranging between 100 and 1000 nanometers. This type of nanocomposite is characterized by pore sizes in the nanometer range as well as a large surface area-to-volume ratio. Nanofibers are produced from various raw materials, including polymers, sol–gels, and ceramics.

Electrospinning is a technique to produce nanofibers.[61,62] Furthermore, these nanofibers can be modified by anchoring functional elements, such as nanoparticles.[63] In this regard, adding nanoparticles to nanofibers is an attractive method to fulfill the antibacterial activity of wound dressing. Recently, the use of lignin as a raw material for the synthesis of nanofibers has shown to be adequate to stabilize silver, gold, and silica nanoparticles.[64]

Lignin nanofibers have the potential to absorb water and be transformed into a viscous gel.[65,66] Thus, these nanofibers can be used in the preparation of wound dressings. The addition of wound healing agents onto the nanofibers has been shown to have an effect on wound healing. For example, the gaseous agent NO functions as a signaling molecule in organisms.[67] The NO synthase enzyme synthesizes NO from L-arginine, oxygen, and NADPH. NO can modulate multiple processes, including inflammation, re-epithelialization, blood vessel formation, and increased collagen deposition.

Regarding wound healing, it has been demonstrated that the *in situ* release of NO positively affected the wound healing process.[68] Thus, adding one of the components needed to synthesize NO on the nanofiber could contribute to NO production in the wounded area. As mentioned earlier, NO synthesis requires L-arginine, oxygen, and NADPH. L-Arginine is the molecule to be incorporated into the nanofibers as NADPH is a very expensive co-factor, and the oxygen is abundant on the injured skin. A previous study reported the electrostatic binding of L-arginine to lignin.[66] This binding is achieved because L-arginine is a basic amino acid positively charged at neutral pH (pI = 10.8).

On the other hand, crude non-reactive lignin should be transformed to increase its solubility. As mentioned earlier, lignins are modified with different reagents, rendering them a negatively charged compound. One of the methods that can support the charge of modified lignin is the zeta potential.

The zeta potential is defined as the difference in the dispersion medium's electric potential and the dispersant's first layer. Thus, upon electrostatic binding of the positively charged arginine to the negatively charged lignin would increase the zeta potential value.[66] It is crucial to maintain a specific amount of L-arginine attached to the nanofibers as concentrations over 200 $mg\,g^{-1}$ gel result in increased inflammation with impairment of wound healing.[69]

Another modification to lignin nanofibers that can increase their applicability to wound healing is their binding to nanoparticles with antibacterial activities. The lignin structure has polyphenols capable of binding and complexing with silver cations as a result of cation–hydroxyl and cation–π bonding.[70] Also, the presence of available hydroxyl groups on these phenol groups can facilitate the reduction of silver cations to nanoparticles (silver metal).[71] In summary, using lignin conjugated to silver nanoparticles is an attractive method as lignin can be used to reduce silver nitrate to form the nanoparticles.[71]

In a previous study, lignin:silver nanoparticles (AgNPs) were synthesized by sonochemistry, and antibacterial tests showed their superior antimicrobial activity. In this study AgNPs were synthesized using lignin to reduce power.[71] Thus, these nanoparticles can be used as part of nanofiber fabrication to provide an additional function for wound healing. Another study reported the synthesis of copper nanoparticles on cellulose nanofibers using a process where lignin induced deacetalization of cellulose nanofibers.[72]

9.4 Antimicrobial Activity

The synthesis of lignin-based nanocomposites with antibacterial activity is desired for wound healing. This is important because any mechanical injury to the skin would drag bacteria inside the wounds. These wounds would provide the bacteria with the necessary environment for division and proliferation. This event is more problematic when burns occur.

Nanoparticles with antibacterial activity have been incorporated into the lignin nanocomposite. The antibacterial activity of published nanocomposites is detailed in Table 9.2.

The antibacterial activity is reported using different dimensions, making it difficult to compare them. The best performance is observed for carrageenan, PLA, and polyurethane nanocomposites. This claim is based on a $\geq$5 log reduction of the colony-forming units. Antibacterial activity expressed in the zone of inhibition (in mm) is not an accurate measurement as this test is based on the diffusibility of the active components of the nanocomposite. In the case of the nanoparticles, this diffusibility is very low because nanoparticles are in suspension or attached to the core of the nanocomposite. Similar issues are encountered when the activity is expressed in percentage reduction of the optical density. This is not a quantitative method, and percentages are only a reduction of up to two-fold.

Table 9.2 Antibacterial activity of nanocomposites.[a]

Nanocomposite	BC	BS	EC	LM	PA	SA	Units	Ref.
Natural								
Alginate	230	—	700	2	20	450	$mg\,mL^{-1}$	73
Carrageenan	—	—	7	—	—	6	log reduction CFU per mL	74
Curcumin	—	99	79	—	85	95	Reduction percentage	75
Dextran/glycol	—	—	50	—	—	50	CFU reduction	54
Synthetic								
Polyacrylonitrile	—	—	19	—	—	21	mm	76
Poly(lactic acid)	—	—	—	—	10	10	log reduction CFU per mL	77
Polyurethane	—	—	—	—	5.5	5	log reduction CFU per mL	25

[a]BC, Bacillus cereus; BS, Bacillus subtilis; EC, Escherichia coli; EF, Enterococcus faecalis; LM, Listeria monocytogenes; PA, Pseudomonas aeruginosa; SA, Staphylococcus aureus; CFU, colony-forming units.

One of the resistant systems that bacteria develop is the synthesis of biofilms. The production of biofilms allows bacterial communities to resist adverse environmental conditions and be protected from antibiotics or other antiseptic methods. However, it is difficult to eradicate biofilms because the protection mainly provided by exopolysaccharides hermitizes the bacterial community, and higher doses of antibiotics are required.[73] A study evaluating the prevention of *Pseudomonas aeruginosa* biofilm formation found a significant inhibition of biofilm synthesis when a hydrogel containing lignin, agarose, fibroin, and $ZnCr_2O_4$ was assessed. The study reported that a reduction of 80% in the biofilm synthesis was measured.[39]

9.5 Antioxidant Activity

Upon skin injury, there is a mobilization of immune cells that arrive at the injury site to elicit an inflammatory response as a defense process. These immune cells produce reactive oxygen species (ROS), powerful short-life radicals that target microorganisms.[74] The regulation of ROS production is crucial to avoid self-damage to the tissues surrounding the injury.[75,76] However, an increased level of ROS production has been measured in chronic wounds, leading to oxidative stress with subsequent damage to the extracellular matrix surrounding the area of the injury and prolonging the wound repair.[77,78] Thus, reducing the ROS level is important to avoid further damage (Figure 9.3).

The composition of the lignin-based nanocomposite can provide an answer to minimize the diffusion of ROS. For example, incorporating polyurethane foams with AgNPs reduced the ROS by 85% compared to controls.[30] This study indicated that the scavenging activity of the polyurethane foam was responsible for their absorption with a subsequent reduction in oxidative stress. Moreover, although lignin, a known antioxidant polysaccharide, is suggested to significantly contribute to the antioxidant activity, this assumption was proven invalid because the lignin used in the nanocomposite synthesis did not show an increase in the antioxidant

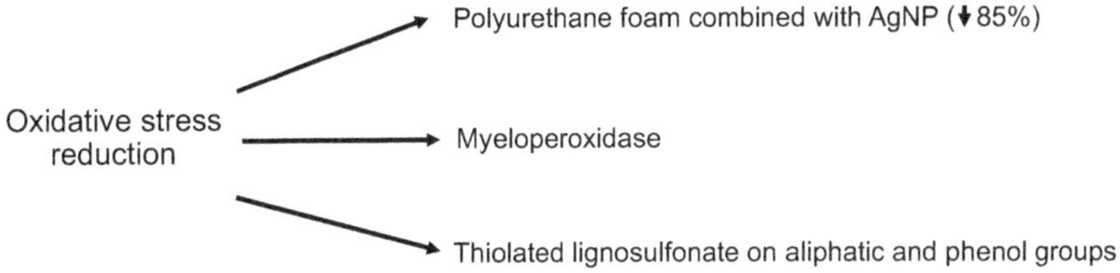

Figure 9.3 Reduction of oxidative stress in a treated wound using lignin-based nanocomposites.

activity.[79] The authors of this study concluded that the number of phenolic groups on the lignin molecule was decreased partly because of the partial oxidation in the synthesis of AgNPs and partly because of a reaction with isocyanate used in the synthesis of polyurethane.

Another way to reduce oxidative stress is the reduction of the enzyme myeloperoxidase. This enzyme is mainly synthesized in neutrophils, the first immune cells arriving at the injured site. Myeloperoxidase produces hypochlorous acid from H_2O_2, a molecule with potent antimicrobial activities.[80] When dysregulation of hypochlorous acid production occurs,[81] the synthesis of toxic radicals ensues with detrimental effects on chronic wound healing.[82] Thus, a reduced level of myeloperoxidase could accelerate the healing process. For example, a study showed that the myeloperoxidase inhibition was more significant than 80% when AgNPs were added to the polyurethane.[30] This study suggested that the enzyme was retained inside the nanocomposite with a concomitant activity reduction.

The antioxidant capacity of the nanocomposite could add another value to the formulation. This capacity is reflected by neutralizing or quenching toxic radicals produced during wound healing, especially in chronic wounds.

The compound 2,2-diphenyl-1-picrylhydrazyl (DPPH) is a radical used as a trap for other radicals and used for the antioxidant capacity measurement.[83] In the case of the thiolated lignosulfonate, there is a significantly higher DPPH inhibition compared to the control.[55] Furthermore, the thiolated lignosulfonate showed a high radical scavenger activity, probably because of the addition of the thiol groups. Thus, the thiolation on aliphatic hydroxyl groups and the phenolic groups conferred superior antioxidant activity compared to the control.[55]

9.6 Wound Healing

Wound healing comprises steps involving multiple growth factors and different types of cells. For the physiological details of this process, the readers are invited to read specific studies.[84,85]

Several genes are upregulated or modulated during skin regeneration to close the damage and avoid excessive scarring properly. Genes involved in the fibrotic response include collagen type I alpha 1 (*COL1A1*), actin alpha 2 (*ACTA*), tumor growth factor-b1 (*TGFB1*), and hypoxia-induced factor 1 subunit alpha (*HIF1A*).[86]

Analysis of the expression of these genes using thiolated lignin combined with fish gelatin (for cell adherence) and PEG (scaffold) showed that the fibrotic response was significantly reduced.[55] For example, the expression of *COL1A1* when fibroblasts were exposed to the nanocomposite was similar to that when fibroblasts were exposed to the lignosulfonate (control). In the case of *ACTA2*, an increase in its expression signifies an excess of myofibroblasts, cells responsible for fibrocontractive diseases, and an excess of granulation.[87] However, a significant attenuation in the expression of this gene was measured when the cells were grown in the presence of the nanocomposite.[55] The presence of an elevated number of myofibroblasts is not desired in chronic wounds because of the complications observed in the extracellular matrix surrounding the injury with permanent deformation of the tissues. The expression of *TGFB1* was also not significantly different from the controls, decreasing the likelihood of deformation of the healing tissues and scar contracture.[88,89] Finally, the expression of *HIF1A* was also not significant, indicating that the function of this gene in apoptosis is minimal.[90]

9.7 Cytotoxicity

Cytotoxicity assessment is a method used to predict the toxic effects of nanocomposites. Biomaterials are considered to be safe and biocompatible when the cell viability value is >70%.[91] The most common assay for the cytotoxicity assessment is based on the activity of mitochondria, giving an evaluation of the cell's metabolic activity. This is a colorimetric assay based on the conversion of 3-(4,5-dimethylthiazol-2-yl)-2,5-diphenyl-2*H*-tetrazolium bromide (MTT, yellow) into formazan (purple color) as a result of the activity of the mitochondrial NAD(P)H-dependent cellular oxidoreductase enzyme.

Other cytotoxic tests used in cytotoxicity assessment are the Alamar blue assay and calcein staining. The first test involves a non-toxic, membrane-permeable compound with weak fluorescence termed resazurin.[92] This compound is a phenoxazine dye sensible to redox changes and its color changes from blue to purple (resorufin) upon mitochondrial activity, similar to MTT.[93] The second test is based on the cleavage of the non-fluorescent calcein, a permeable dye, into green-fluorescent calcein upon the activity of intracellular esterases.[94]

Cytotoxic results are reported in several studies, and this is an important experiment when evaluating nanocomposites before animal model studies.

Cytotoxicity reports vary according to the cell line used for the test. Therefore, it is preferable to test the cells exposed to the nanoformulation instead of cell lines from other origins to mimic as close as possible the toxic effects. In addition, when dealing with chronic or open wounds, the presence of toxicants in the formulation might increase an inflammatory response with detrimental consequences to the tissue under repair. The compatibility of lignin-based nanoformulations is detailed in Table 9.3.

Table 9.3 *Ex vivo* compatibility of lignin-based formulations.[a]

Nanocomposite	Test	Cell type	Cell name	Origin	Viability (%)	Ref.
Natural						
Agarose	MTT	F	Hu02	Human	>87	39
Alginate	CAL	C	Hep2C	Human	>90	73
Carrageenan	MTT	F	L929	Human	>90	74
Chitosan	MTT	MS	C3H10T1/2	Mouse	>99	53
	MTT	O	MC3T3	Mouse	100	52
	MTT	F	NIH-3T3	Mouse	>80	100
Curcumin	MTT	K	HaCat	Human	91	75
Dextran	AB	F	NIH-3T3	Mouse	93	54
Synthetic						
PLA	MTT	Ki	Vero	Monkey	100	77
Polyurethane	AB	F	BJ5ta	Human	>80	25
	AB	K	HaCat	Human	>80	25

[a]AB, Alamar blue; C, cervix; CAL, calcein; F, fibroblast; K, keratinocyte; Ki, kidney; MS, mesenchymal stem cells; O, osteoblast.

The reported biocompatibilities of lignin-containing nanoformulations showed no significant cytotoxicity using different types of cells with measured compatibilities >80% (Table 9.3). So then, it looks like all of them were compatible with their cells and suitable to be used in the nanoformulations.

9.8 Lignin-containing Nanocomposites in Animal Models of Wound Healing

The final step is to test the efficacy of formulations in animal models. In the case of lignin-based formulations for wound care, some assessments should be considered.

9.8.1 Protein Synthesis

The induction of protein synthesis in the wound area commences shortly after injury. The main extracellular protein produced is collagen, which increases the wound strength, providing a tissue matrix upon upregulation in wound healing.[95] Its deposition is important for the stability and integrity of the matrix and is found mainly in connective tissues, such as skin, ligaments, tendons, and cartilages.

Collagen is produced as a helix forming fibrils and comprises an extensive family,[96] and the main type involved in wound healing is type I.

Results from an *in vivo* study using a rat model showed that curcumin-loaded lignin nanoparticles showed a higher deposition of collagen when compared to control, suggesting an upregulation of the production of this protein during the wound healing process.[97] As discussed earlier, another desired event during wound healing is a reduction in myeloperoxidase level. When suppressing or quenching the activity of this enzyme, the healing process gave a better outcome. In the case of the curcumin-loaded nanoparticles, a significant reduction in the level of myeloperoxidase was measured 12 days after the wounding event.[97] The authors concluded that this reduction is probably related to the anti-inflammatory activity of the curcumin, which might reduce the infiltration of neutrophils into the wound area.[98]

The amino acid arginine is another molecule that intensifies collagen deposition in wounded skin. This amino acid was supplemented in a lignin nanofiber formulation, which upregulated collagen deposition in an animal model of the wound 9 days after surgery.[66] The effect of the high deposition of collagen upon the addition of arginine to the wound site can be attributed to its metabolism upon entry into the mitochondrion. Thus, arginine is transformed into ornithine by the enzymatic activity of the arginase,[99] which increases the level of proline synthesis by the enzymatic activity of the ornithine aminotransferase.[99] Proline is the main amino acid in the collagen molecule, and once released into the cytosol, it can be used to produce collagen.[100] In addition, proline can also be made due to the metabolic transformation of glutamine into glutamate by the enzymatic activity of glutaminases.[100] In summary, arginine supplementation in the formulation increases the high demand for proline for collagen production as a result of wound healing.[101]

Another group of proteins abundant in wound tissues is the group of proteolytic enzymes called metalloproteinases.[102] These enzymes' role is to remodel the damaged matrix by eliminating damaged tissues and breaching the capillary basement membrane, permitting angiogenesis and cell migration. The synthesis of metalloproteinases is tightly regulated because an uncontrolled increase might destroy the matrix, leading to impaired healing, especially in chronic wounds.[103,104]

The expression of various metalloproteinases was upregulated under inflammatory and wound conditions (*e.g.*, treatment with interleukin-1 beta (IL-1β)). However, curcumin-loaded lignin nanoparticles suppressed the expression of the wound-relevant metalloproteinase 9.[97] This finding aligned with the reported curcumin-mediated anti-inflammatory activity directly linked to the metalloproteinase 9 downregulation.[105]

9.8.2 Wound Closure

The time required for a complete closure of a wound is a parameter used to measure the efficacy of a formulation. Wounds generated by burning in a murine model were treated with the lignin–agarose nanobiocomposite supplemented with silk fibroin and $ZnCr_2O_4$, described earlier.[39] The wounds were healed on the 5th day compared to the control, which registered a 91% healing. It is important to stress that the control showed no sign of infection, probably because of $ZnCr_2O_4$, which has antimicrobial activity.[39]

An increase in wound closure time was also measured in a rat model of a wound using a hydrogel composed of carrageenan, lignin, AgNPs, and $MgCl_2$

or $CaCl_2$.[106] The wound surface was almost completely healed (97%) 14 days post-treatment when $MgCl_2$ was present in the hydrogel. Interestingly, the replacement of $MgCl_2$ with $CaCl_2$ recorded a closure of 91%.[106] Overall, the fast healing can be attributed to eliminating harmful microorganisms by the AgNPs included in the formulation. Moreover, the presence of Mg^{2+} in the formulation appeared to increase the binding of fibroblasts to the extracellular matrix.[107–109]

A similar pattern in the healing speed was measured when a gel composed of lignin, arginine, and nanofibers was used in a rat model of a wound.[66] However, in this study, 93% of wound closure was observed on the 9th day post-surgery compared to only 70% recorded with the control (no arginine). In addition, the fast closure of the wound can result from the prevented infection as a result of the increased production of NO (by the presence of arginine), which is released *in situ* with potent antimicrobial activity.[110]

When excisional wounds were generated in a murine model, a wound closure of 87% was recorded on the 10th day post-surgery after the treatment with a nanocomposite containing lignin, chitosan, and PVA.[52] This increase in the closure time can be explained by the moist wound environment and the antioxidant activity of lignin.[52]

9.9 Conclusions and Future Research

The use of lignin as a component of nanoformulations has been proven to be effective. Moreover, the use of lignin fulfills a few of the requirements of wound dressing. This is the result of antioxidant activity, and lignin can be endowed with antimicrobial activities by supplementing it with antimicrobial agents such as AgNPs.

The new formulations synthesized should show different activities: for example, antioxidant activity, antimicrobial activity, optimal mechanical and rheological characteristics, and collagen deposition with a concomitant decrease of the closure time of the wound. Although these characteristics have been discussed in this chapter, the number of available publications is still low and more research is necessary in this specific field.

Another issue encountered in the research is the lack of reports of results using accepted microbiological tests. Most of the reported antibacterial activities were performed and reported using methodologies that do not follow microbiology standards. This issue is exacerbated when comparing the microbial activity of individual studies with that of the rest of the available literature.

Concerning animal models, it is preferable that the studies follow a flow chart to test specific parameters important in the healing process. For example, not all of the studies available measure all the parameters mentioned earlier, leaving discrepancies across the different studies again. In conclusion, the use of lignin accompanied by other materials is a topic that deserves more research regarding wound treatment.

References

1. H. N. Wilkinson and M. J. Hardman, *Open Biol.*, 2020, **10**, 200223.
2. S. R. Nussbaum, M. J. Carter, C. E. Fife, J. DaVanzo, R. Haught, M. Nusgart and D. Cartwright, *Value Health*, 2018, **21**, 27.
3. L. G. Ovington, *Clin. Dermatol.*, 2007, **25**, 33.
4. J. S. Boateng, K. H. Matthews, H. N. E. Stevens and G. M. Eccleston, *J. Pharm. Sci.*, 2008, **97**, 2892.
5. J. S. Choi, K. W. Leong and H. S. Yoo, *Biomaterials*, 2008, **29**, 587.
6. H. Li, M. Wang, G. R. Williams, J. Wu, X. Sun, Y. Lv and L.-M. Zhu, *RSC Adv.*, 2016, **6**, 50267.
7. K. C. Broussard and J. G. Powers, *Am. J. Clin. Dermatol.*, 2013, **14**, 449.
8. Y. S. Zhang and A. Khademhosseini, *Science*, 2017, **356**, eaaf3627.
9. K. Bankoti, A. P. Rameshbabu, S. Datta, P. P. Maity, P. Goswami, P. Datta, S. K. Ghosh, A. Mitra and S. Dhara, *Mater. Sci. Eng., C*, 2017, **81**, 133.
10. Z. Wang, L. Yuan, F. Jiang, Y. Zhang, Z. Wang and C. Tang, *ACS Macro Lett.*, 2016, **5**, 220.
11. G. D. Winter, *Nature*, 1962, **193**, 293.
12. D. Kai, S. Jiang, Z. W. Low and X. J. Loh, *J. Mater. Chem. B*, 2015, **3**, 6194.
13. H. Yousefi, M. Faezipour, T. Nishino, A. Shakeri and G. Ebrahimi, *Polym. J.*, 2011, **43**, 559.
14. F. G. Calvo-Flores and J. A. Dobado, *ChemSusChem*, 2010, **3**, 1227.
15. D. Kun and B. Pukánszky, *Eur. Polym. J.*, 2017, **93**, 618.
16. A. Tribot, G. Amer, M. Abdou Alio, H. de Baynast, C. Delattre, A. Pons, J.-D. Mathias, J.-M. Callois, C. Vial, P. Michaud and C.-G. Dussap, *Eur. Polym. J.*, 2019, **112**, 228.
17. C. Zhao, J. Huang, L. Yang, F. Yue and F. Lu, *Ind. Eng. Chem. Res.*, 2019, **58**, 5707.
18. T. Aro and P. Fatehi, *ChemSusChem*, 2017, **10**, 1861.
19. M. H. Hussin, A. A. Rahim, M. N. Mohamad Ibrahim and N. Brosse, *Ind. Crops Prod.*, 2013, **49**, 23.
20. A. Gandini, M. N. Belgacem, Z.-X. Guo and S. Montanari, in *Chemical Modification, Properties, and Usage of Lignin*, ed. T. Q. Hu, Springer US, Boston, MA, 2002, p. 57.
21. C. A. Cateto, M. F. Barreiro, A. E. Rodrigues and M. N. Belgacem, *Ind. Eng. Chem. Res.*, 2009, **48**, 2583.
22. X. Zhang, Y. Kim, I. Elsayed, M. Taylor, T. L. Eberhardt, E. B. Hassan and R. Shmulsky, *Ind. Crops Prod.*, 2019, **141**, 111797.
23. H. Jeong, J. Park, S. Kim, J. Lee, N. Ahn and H. Roh, *Fibers Polym.*, 2013, **14**, 1082.
24. J. D. Gargulak, S. E. Lebo and T. J. McNally, in *Kirk-Othmer Encyclopedia of Chemical Technology*, John Wiley & Sons, Ltd, 2015, pp. 1–26.
25. J. A. Belgodere, S. A. Zamin, R. M. Kalinoski, C. E. Astete, J. C. Penrod, K. M. Hamel, B. C. Lynn, J. S. Rudra, J. Shi and J. P. Jung, *ACS Appl. Bio. Mater.*, 2019, **2**, 3562.

26. C. Jin, X. Zhang, J. Xin, G. Liu, J. Chen, G. Wu, T. Liu, J. Zhang and Z. Kong, *Ind. Eng. Chem. Res.*, 2018, **57**, 7872.
27. J. Vandooren, P. E. Van den Steen and G. Opdenakker, *Crit. Rev. Biochem. Mol. Biol.*, 2013, **48**, 222.
28. E. I. Solomon, U. M. Sundaram and T. E. Machonkin, *Chem. Rev.*, 1996, **96**, 2563.
29. R. Hilgers, J.-P. Vincken, H. Gruppen and M. A. Kabel, *ACS Sustainable Chem. Eng.*, 2018, **6**, 2037.
30. A. G. Morena, I. Stefanov, K. Ivanova, S. Pérez-Rafael, M. Sánchez-Soto and T. Tzanov, *Ind. Eng. Chem. Res.*, 2020, **59**, 4504.
31. H. Gunes, D. Gulen, R. Mutlu, A. Gumus, T. Tas and A. E. Topkaya, *Toxicol. Ind. Health*, 2016, **32**, 246.
32. T. A. Nguyen and A. J. Friedman, *J. Drugs Dermatol.*, 2013, **12**, 1131.
33. R. Whetten and R. Sederoff, *Plant Cell*, 1995, **7**, 1001.
34. G. S. Sidhu, A. K. Singh, D. Thaloor, K. K. Banaudha, G. K. Patnaik, R. C. Srimal and R. K. Maheshwari, *Wound Repair Regen.*, 1998, **6**, 167.
35. M. Kulac, C. Aktas, F. Tulubas, R. Uygur, M. Kanter, M. Erboga, M. Ceber, B. Topcu and O. A. Ozen, *J. Mol. Histol.*, 2013, **44**, 83.
36. R. L. Thangapazham, S. Sharad and R. K. Maheshwari, *BioFactors*, 2013, **39**, 141.
37. M. Panchatcharam, S. Miriyala, V. S. Gayathri and L. Suguna, *Mol. Cell. Biochem.*, 2006, **290**, 87.
38. O. Hakimi, D. P. Knight, F. Vollrath and P. Vadgama, *Composites, Part B*, 2007, **38**, 324.
39. R. Eivazzadeh-Keihan, H. A. M. Aliabadi, F. Radinekiyan, M. Sobhani, F. Khalili, A. Maleki, H. Madanchi, M. Mahdavi and A. Esmail Shalan, *RSC Adv.*, 2021, **11**, 17914.
40. S. Kongruang, *Appl. Biochem. Biotechnol.*, 2008, **148**, 245.
41. S. Yamanaka, K. Watanabe, N. Kitamura, M. Iguchi, S. Mitsuhashi, Y. Nishi and M. Uryu, *J. Mater. Sci.*, 1989, **24**, 3141.
42. E. Canale-Parola and R. S. Wolfe, *Biochim. Biophys. Acta, Gen. Subj.*, 1964, **82**, 403.
43. A. G. Matthysse, K. V. Holmes and R. H. Gurlitz, *J. Bacteriol.*, 1981, **145**, 583.
44. C. Napoli, F. Dazzo and D. Hubbell, *Appl. Microbiol.*, 1975, **30**, 123.
45. L. Fu, P. Zhou, S. Zhang and G. Yang, *Mater. Sci. Eng., C*, 2013, **33**, 2995.
46. P. Muangman, S. Opasanon, S. Suwanchot and O. Thangthed, *J. Am. Coll. Certif. Wound Spec.*, 2011, 16.
47. B. Wei, G. Yang and F. Hong, *Carbohydr. Polym.*, 2011, **84**, 533.
48. J. Wu, Y. Zheng, W. Song, J. Luan, X. Wen, Z. Wu, X. Chen, Q. Wang and S. Guo, *Carbohydr. Polym.*, 2014, **102**, 762.
49. J. P. Ruparelia, A. K. Chatterjee, S. P. Duttagupta and S. Mukherji, *Acta Biomater.*, 2008, **4**, 707.
50. K. Ghule, A. V. Ghule, B.-J. Chen and Y.-C. Ling, *Green Chem.*, 2006, **8**, 1034.

51. V. L. Campo, D. F. Kawano, D. B. da Silva and I. Carvalho, *Carbohydr. Polym.*, 2009, 77, 167.
52. Y. Zhang, M. Jiang, Y. Zhang, Q. Cao, X. Wang, Y. Han, G. Sun, Y. Li and J. Zhou, *Mater. Sci. Eng., C*, 2019, **104**, 110002.
53. K. Ravishankar, M. Venkatesan, R. P. Desingh, A. Mahalingam, B. Sadhasivam, R. Subramaniyam and R. Dhamodharan, *Mater. Sci. Eng., C*, 2019, **102**, 447.
54. M. Dai, J. Zhang, N. Liu and X.-H. Zhang, *J. Polym. Environ.*, 2022, **30**, 206.
55. J. A. Belgodere, D. Son, B. Jeon, J. Choe, A. C. Guidry, A. X. Bao, S. A. Zamin, U. M. Parikh, S. Balaji, M. Kim and J. P. Jung, *ACS Biomater. Sci. Eng.*, 2021, 7, 2212.
56. P. Saini, M. Arora and M. N. V. R. Kumar, *Adv. Drug Delivery Rev.*, 2016, **107**, 47.
57. M. A. Rahman, D. De Santis, G. Spagnoli, G. Ramorino, M. Penco, V. T. Phuong and A. Lazzeri, *J. Appl. Polym. Sci.*, 2013, **129**, 202.
58. O. Gordobil, I. Egüés, R. Llano-Ponte and J. Labidi, *Polym. Degrad. Stab.*, 2014, **108**, 330.
59. M. Thunga, K. Chen, D. Grewell and M. R. Kessler, *Carbon*, 2014, **68**, 159.
60. C. Vila, V. Santos, B. Saake and J. C. Parajó, *BioResources*, 2016, **11**, 5322.
61. D. Li and Y. Xia, *Adv. Mater.*, 2004, **16**, 1151.
62. A. Greiner and J. H. Wendorff, *Angew. Chem., Int. Ed.*, 2007, **46**, 5670.
63. D.-N. Phan, N. Dorjjugder, M. Q. Khan, Y. Saito, G. Taguchi, H. Lee, Y. Mukai and I.-S. Kim, *Cellulose*, 2019, **26**, 6629.
64. S. Y. Yeo, H. J. Lee and S. H. Jeong, *J. Mater. Sci.*, 2003, **38**, 2143.
65. T. Yuan, J. Zeng, B. Wang, Z. Cheng and K. Chen, *Carbohydr. Polym.*, 2021, **254**, 117441.
66. F. Reesi, M. Minaiyan and A. Taheri, *Drug Delivery Transl. Res.*, 2018, **8**, 111.
67. L. J. Ignarro, G. Cirino, A. Casini and C. Napoli, *J. Cardiovasc. Pharmacol.*, 1999, **34**, 879.
68. K. Blecher, L. R. Martinez, C. Tuckman-Vernon, P. Nacharaju, D. Schairer, J. Chouake, J. M. Friedman, A. Alfieri, C. Guha, J. D. Nosanchuk and A. J. Friedman, *Nanomed. Nanotechnol. Biol. Med.*, 2012, **8**, 1364.
69. Y. Matsumoto and Y. Kuroyanagi, *J. Biomater. Sci., Polym. Ed.*, 2010, **21**, 715.
70. A. Lagutschenkov, R. K. Sinha, P. Maitre and O. Dopfer, *J. Phys. Chem. A*, 2010, **114**, 11053.
71. Y. Slavin, K. Ivanova, J. Hoyo, I. Perelshtein, G. Owen, A. Haegert, Y. Lin, S. LeBihan, A. Gedanken, U. Häfeli, T. Tzanov and H. Bach, *ACS Appl. Mater. Interfaces*, 2021, **13**, 22098.
72. M. K. Haider, A. Ullah, M. N. Sarwar, Y. Saito, L. Sun, S. Park and I. S. Kim, *Int. J. Biol. Macromol.*, 2021, **173**, 315.

73. H.-C. Flemming and J. Wingender, *Nat. Rev. Microbiol.*, 2010, **8**, 623.
74. Y. N. Slavin, J. Asnis, U. O. Häfeli and H. Bach, *J. Nanobiotechnol.*, 2017, **15**, 65.
75. M. Wlaschek and K. Scharffetter-Kochanek, *Wound Repair Regen.*, 2005, **13**, 452.
76. M. Schäfer and S. Werner, *Pharmacol. Res.*, 2008, **58**, 165.
77. M.-M. Azevedo, C. Lisboa, L. Cobrado, C. Pina-Vaz and A. Rodrigues, *Br. J. Nurs.*, 2020, **29**, S6.
78. R. G. Frykberg and J. Banks, *Adv. Wound Care*, 2015, **4**, 560.
79. B. Jiang, Y. Zhang, L. Gu, W. Wu, H. Zhao and Y. Jin, *Int. J. Biol. Macromol.*, 2018, **116**, 513.
80. J. Marcinkiewicz, B. Chain, B. Nowak, A. Grabowska, K. Bryniarski and J. Baran, *Inflammation Res.*, 2000, **49**, 280.
81. D. Odobasic, A. R. Kitching and S. R. Holdsworth, *J. Immunol. Res.*, 2016, **2016**, e2349817.
82. S. J. Klebanoff, *J. Leukocyte Biol.*, 2005, **77**, 598.
83. W. Brand-Williams, M. E. Cuvelier and C. Berset, *LWT – Food Sci. Technol.*, 1995, **28**, 25.
84. I. Pastar, O. Stojadinovic, N. C. Yin, H. Ramirez, A. G. Nusbaum, A. Sawaya, S. B. Patel, L. Khalid, R. R. Isseroff and M. Tomic-Canic, *Adv. Wound Care*, 2014, **3**, 445.
85. M. Rodrigues, N. Kosaric, C. A. Bonham and G. C. Gurtner, *Physiol. Rev.*, 2019, **99**, 665.
86. M. Jinnin, *J. Dermatol.*, 2010, **37**, 11.
87. K. Kis, X. Liu and J. S. Hagood, *Expert Rev. Mol. Med.*, 2011, **13**, e27.
88. A. E. Brissett and D. A. Sherris, *Facial Plast. Surg.*, 2001, **17**, 263.
89. M. J. Crowe, T. Doetschman and D. G. Greenhalgh, *J. Invest. Dermatol.*, 2000, **115**, 3.
90. K. Deschene, C. Céleste, D. Boerboom and C. L. Theoret, *J. Dermatol. Sci.*, 2012, **65**, 12.
91. ISO 10993-5, https://www.iso.org/cms/render/live/en/sites/isoorg/contents/data/standard/03/64/36406.html (accessed March 29, 2022).
92. J. O'Brien, I. Wilson, T. Orton and F. Pognan, *Eur. J. Biochem.*, 2000, **267**, 5421.
93. S. Anoopkumar-Dukie, J. B. Carey, T. Conere, E. O'sullivan, F. N. van Pelt and A. Allshire, *Br. J. Radiol.*, 2005, **78**, 945.
94. C. E. M. Krämer, A. Singh, S. Helfrich, A. Grünberger, W. Wiechert, K. Nöh and D. Kohlheyer, *PLoS One*, 2015, **10**, e0141768.
95. G. I. Broughton, J. E. Janis and C. E. Attinger, *Plast. Reconstr. Surg.*, 2006, **117**, 12S.
96. J.-Y. Exposito, U. Valcourt, C. Cluzel and C. Lethias, *Int. J. Mol. Sci.*, 2010, **11**, 407.
97. M. S. Alqahtani, A. Alqahtani, M. Kazi, M. Z. Ahmad, A. Alahmari, M. A. Alsenaidy and R. Syed, *J. Drug Delivery Sci. Technol.*, 2020, **60**, 102020.
98. R. Srivastava, *Agents Actions*, 1989, **28**, 298.

99. M. B. Witte and A. Barbul, *Wound Repair Regen.*, 2003, **11**, 419.
100. A. Barbul, *J. Nutr.*, 2008, **138**, 2021S.
101. P. Chithra, L. Suguna and G. Chandrakasan, *J. Clin. Biochem. Nutr.*, 1995, **18**, 111.
102. N. Philips, S. Auler, R. Hugo and S. Gonzalez, *Enzyme Res.*, 2011, **2011**, 427285.
103. N. J. Trengove, M. C. Stacey, S. Macauley, N. Bennett, J. Gibson, F. Burslem, G. Murphy and G. Schultz, *Wound Repair Regen.*, 1999, **7**, 442.
104. E. A. Rayment, Z. Upton and G. K. Shooter, *Br. J. Dermatol.*, 2008, **158**, 951.
105. Y.-H. Yen, C.-M. Pu, C.-W. Liu, Y.-C. Chen, Y.-C. Chen, C.-J. Liang, J.-H. Hsieh, H.-F. Huang and Y.-L. Chen, *Int. Wound J.*, 2018, **15**, 605.
106. L. Jaiswal, S. Shankar, J.-W. Rhim and D.-H. Hahm, *Int. J. Biol. Macromol.*, 2020, **159**, 859.
107. F. I. Wolf and A. Cittadini, *Front. Biosci. J. Virtual Libr.*, 1999, **4**, D607.
108. H. Rubin, *BioEssays*, 2005, **27**, 311.
109. S. Banai, L. Haggroth, S. E. Epstein and W. Casscells, *Circ. Res.*, 1990, **67**, 645.
110. M. A. DeGroote and F. C. Fang, in *Nitric Oxide and Infection*, ed. F. C. Fang, Springer US, Boston, MA, 2002, pp. 231–261.

CHAPTER 10

Lignin as a Biomaterial for Bioimaging

MONALISA MISHRA

Neural Developmental Biology Lab, Department of Life Science NIT Rourkela, Odisha, India
Email: mishramo@nitrkl.ac.in

10.1 Introduction

The name lignin is derived from "lignum", which is a Latin word having the meaning "wood". In nature it is the 2nd most available natural biopolymer after cellulose. It contains 30% biomass and is enriched with aromatic compounds.[1] It consists of three different types of monolignol like syringyl, guaiacyl, and *p*-hydroxyphenyl.[2] The monolignol units differ from each other based on their methoxy and aromatic groups. Lignin is formed by the condensation of three phenylpropanoids. This polymer is synthesised by using monolignol units to form a heterogeneous structure. The peroxidase-mediated dehydrogenation method is followed by taking C–C bonds and aryl–ether linkages.[3,4] The lignin types vary from one another based on altering monomeric units and linkages (Figure 10.1).

The presence of all these structural components provides different strengths and rigidity to the cell wall. Indirectly, lignin is a low-value by-product and a combustion fuel in biorefinery and chemical pulping processes.[5–7] More importantly, it has high carbon content, which provides hydrophobic properties and allows the development of carbon-based nanomaterials from lignin.[8–10] However, the strong carbon–carbon bond and complex aromatic structure of lignin do not allow easy synthesis of

Biomaterials Science Series No. 16
Lignin-based Materials: Health Care and Medical Applications
Edited by Kuruvilla Joseph, Runcy Wilson, Gejo George and Saritha Appukuttan

Published by the Royal Society of Chemistry, www.rsc.org

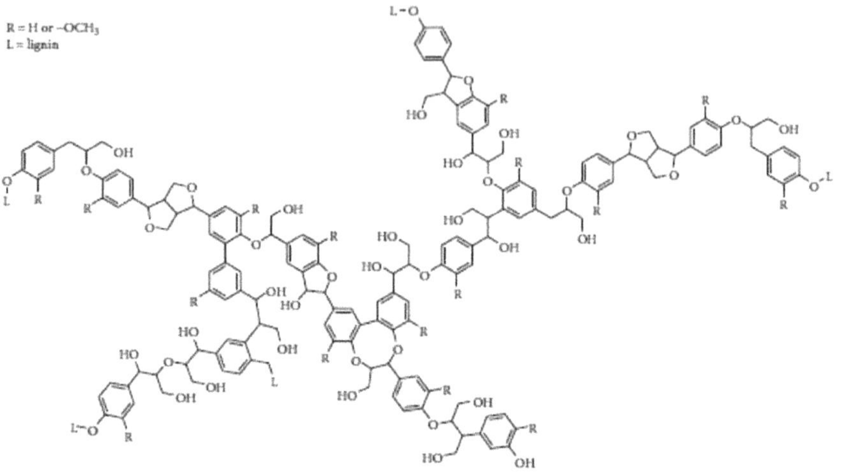

Figure 10.1 Typical structure of lignin. Reproduced from ref. 48, https://doi.org/10.1155/2017/8951658, under the terms of the CC BY 4.0 license https://creativecommons.org/licenses/by/4.0/.

carbon-based nanomaterials from lignin. Lignin-derived biomaterials are used (1) as sensors; (2) for drug release at targeted sites; (3) as memory materials; as (4) photodynamic therapy (PDT), (5) biomaging, and (6) drug delivery agents; (7) as UV absorbants; and (8) as antioxidants.[8–13] The applications of lignin in treatment of colon cancer cell lines have been extensively reviewed by several authors.[11–13] Many lignin-based biomaterials are less toxic to and biocompatible with cells, and show photoluminescent properties.[14–16] All these properties make lignin a suitable molecule for biomedical imaging. Biomedical imaging is an obligatory tool for detecting many life-threatening diseases including cancer.[14,16–18] This chapter summarises the application of lignin-based biomaterials for numerous imaging studies including photodynamic and photothermal therapy.

10.2 Localisation of Lignin

To check localisation of lignin in the plant cell, fluorescence techniques and confocal Raman microscopy were used.[19,20] Lignin is localised in the cell wall of a wild-type plant; as a result, it exhibits green fluorescence.[21] Confocal Raman microscopy allows the visualisation of lignin in a label-free manner[22] (Figure 10.2).

A higher amount of lignin is detected in the cell corner (CC) and compound middle lamella (CML).[23,24] The chemical images also reveal a high amount of lignin in the CML followed by that in the CC[24–26] (Figure 10.3).

The difference observed in the lignin is due to the unique lignin synthesis process during cell growth.[7,27–29] Lignin is synthesised within the cell membrane and then translocated to the cell wall[29–31] and the polymerisation is initiated at the end of cell growth.[32–34] Lignin polymerisation begins at the

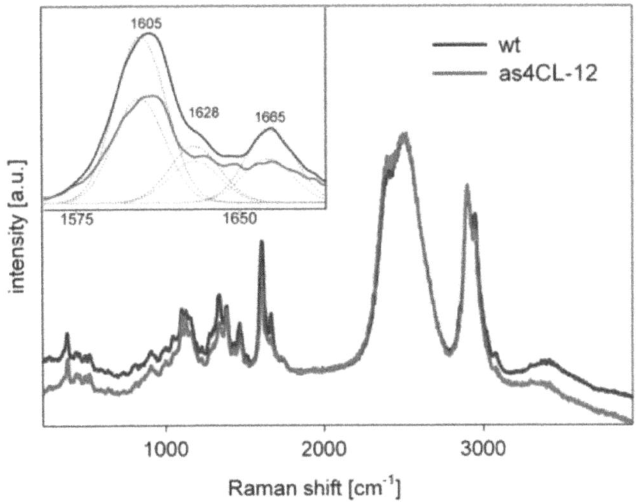

Figure 10.2 Raman spectra of *Populus trichocarpa* wild-type and transgenic 4CL-12. The inset peaks indicate the presence of lignin. Reproduced from ref. 19 with permission from Springer Nature, Copyright 2009.

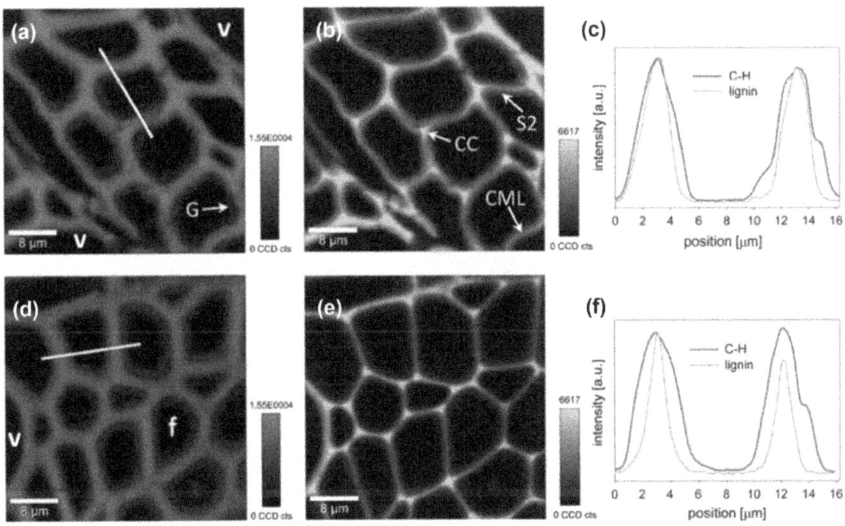

Figure 10.3 Chemical images based on C–H stretching bands. (a) Wild-type C–H stretching bands. (b) Lignin distribution in wild type. (c) Intensity graph of C–H stretching and lignin in wild type. (d) C–H stretching bands in transgenic 4CL-12. (e) Lignin distribution in 4CL-12. (f) Intensity graph of C–H stretching and lignin in transgenic 4CL-12. CC, cell corner; CML, compound middle lamella; S1 and S2, sublayers of the secondary cell wall; V, vessel; and G, gelatinous layer. Reproduced from ref. 19 with permission from Springer Nature, Copyright 2009.

CC and CML, which include the middle lamella and primary cell wall and extend to the secondary cell wall.[35–37] The secondary cell wall has three different layers, S1, S2, and S3, out of which S2 is the majority of the secondary cell wall.[38,39] Since lignin is associated with carbohydrates and distributed evenly within the cell wall, no exogenous labeling is needed for the localisation of lignin.[40–42] Raman microspectroscopy reveals the chemical composition of the lignin present in the cell wall.[43–45] Stimulated Raman scattering can detect the Raman vibration band at 1600 cm^{-1}.[46–48] The 1605 cm^{-1} band is due to the aryl ring stretching, the 1628 cm^{-1} band is due to coniferaldehyde, and the 1665 cm^{-1} band is due to coniferyl alcohol and coniferaldehyde.[19] Lignin is bonded with hemicellulose using ether, ester, and glycosidic bonds to form a lignin–carbohydrate complex.[49] Non-covalent interaction is present between lignin and hemicellulose but few interactions are present between lignin and cellulose.[50–52] The amount of lignin is negatively correlated with biodigestibility.[53] With the increase of carbohydrate content, the lifetime of the lignin also increases.[54] To check the molecular localisation of lignin, two mutant lines of *Populus trichocarpa*, 4CL-12 and 4CL-4, along with wild-type control were used.[55] The lignin content was found to be 20.5% in the control line. In the 4CL-12 and 4CL-4 lines, the lignin content was found to be 12.1 and 15.0, respectively. The cinnamate-4-hydroxylase (C4H) gene regulates the lignin content of the plant.[55]

10.3 Artificial Synthesis of Lignin

In the laboratory, lignin can be synthesised from *p*-coumaryl alcohol, coniferyl alcohol, and sinapyl alcohol monomers. This involves enzyme-guided dehydrogenation, radical coupling, and dimerisation which result in amorphous three-dimensional materials having ether and carbon–carbon bonds.[56–58] All these steps involved in the synthesis make it difficult to determine the exact molecular weight of lignin. However, lignin is always a mixture of polymers. Lignin monomers are called monolignols and consist of phenylpropanoic acid.[56,59] They are present in different forms: *p*-coumaryl alcohol, coniferyl alcohol, and sinapyl alcohol (lignin monomers from beyond the canonical monolignol biosynthetic pathway: another brick in the wall). Technically lignin can be divided into two types: (1) sulfur containing commercial lignin and (2) lignin without sulfur composition such as organosolv lignin and soda lignin.[60,61] Lignosulfates and kraft lignin are also produced in large quantities.[62,63] Lignin has a very high carbon content, thermal stability, biodegradability, and antioxidant and antimicrobial activity.[62,64–66] Lignin% is more in softwood than in hardwood.[67–69] Lignin is a difficult material to work on due to its molecular heterogeneity and complex macromolecular structure.[70–72] Besides this, many high end products are synthesised from lignin, which include vanillin, dimethyl sulfoxide, active carbon, and carbon fibers.[56,73,74] Lignin is synthesised by precipitation, solvent–antisolvent, sonication, emulsification, and aerosol methods.[75–77]

10.3.1 Synthesis of Lignin Nanoparticles

Lignin is a biodegradable, biocompatible material having phenolic and aliphatic hydroxyl groups. The presence of phenolic and hydroxyl groups allows us to synthesise lignin nanoparticles (LNPs).[78,79] The LNPs are further modified, so that they can interact with specific cells selectively.[80] LNPs were modified to treat cancer cells.[81] Carboxylated LNPs (CLNPs) were synthesised to treat cancer.[82] CLNPs were functionalised with NH2-PEG-PHIS.[83] The synthesised NPs were checked for cytotoxicity, surface charge, and stability. Benzoyl-loaded CLNPs can be used to treat many cancer cell lines.[82] The primary structure of lignin is formed by radical polymerisation of monolignol by peroxidase and lactases that lead to various inter-unit linkages.[84] The β-O-4 ether linkage is the most abundant and involves the binding of the β-carbon of one unit with the phenolic hydroxyl group of another unit.[85] The other linkages include α-O-4′ and 4-O-5′-diphenyl ether.[86] The proportion of the three monomer types influences inert unit linkage, which determines the branching of lignin.[70] The quality of lignin depends on the way it is processed.[87] Lignin and its derivatives are used to treat diabetes, obesity, and emphysema as coagulants. Lignin is also used for the production of plastic and composites.[88] Using lignin as a source, vanillin, dimethyl sulfoxide, and active carbon are synthesised.[56] Introduction of the amine group in lignin helps in the preparation of epoxy resins which are used for adhesives, composites, and coatings. Based on the extraction methods there are four different types of lignin: (1) soda lignin, (2) kraft lignin, (3) lignosulfonates, and (4) organosolv lignin.[89,90] Lignin is resistant to ultraviolet (UV) and pH.[91,92]

10.4 Applications of Lignin for Photoinduced Antimicrobial Activity

PDT involves a light-sensing chemical that can sense light and kill abnormal cells. This therapy is used to treat several diseases and microbial films. It is safer than the oral administration of medicines. LNPs are used for PDT. To achieve this, various functional groups are attached to LNPs. The common lignin modification techniques include acetylation and the incorporation of nanoparticles. The modification of different LNPs for PDT is described in the following sections.

10.4.1 Acetylation of Lignin

The presence of aliphatic and phenolic hydroxyl groups on the polymer surface allows multiple functionalisations.[93] Acetylation is a common surface modification technique that helps to modify the biopolymer into a plastic matrix for structural analysis.[94,95] The synthesis and physicochemical properties of acetylated lignin have been studied by many scientists.[96,97]

Lignin on acetylation resulted in radical production upon exposure to light. Light-induced yellowing of wood and wood-based materials is associated with the radical behaviour of acetylated lignins.[98,99] Acetylated lignin is a photosensitive material that can eradicate bacteria and fungi by generating reactive oxygen species (ROS).[18,98] Three different types of lignin were synthesised by different processes: (1) KL1, (2) KL-2, and (3) OL.[100] KL1 was synthesised by the addition of acidic solution, KL-2 by a LignoBoost process, and OL by a classical ethanol/water organosolv process. The acetylated lignins are known as AcKL-1, AcKL-2, and AcOL. In KL-2, 69.49% carbon was detected.[98] Acetylated lignin produces singlet oxygen at a 1.5 μM concentration, which makes lignin suitable for antimicrobial photodynamic therapy (ADT).[101] ROS production was observed by monitoring the luminescence of $^1O_2$.[102] With light irradiation $O_2^{\bullet-}$ is generated from lignin.[103] The superoxide radical generation is three times faster in acetylated lignin.[104] Blocking the antioxidant function by acetylation increases ROS with irradiation with light.[98]

10.4.2 Colloidal Lignin Coated with AuNPs and AgNPs

Based on the sugar content, lignin is classified into three different types: (1) alkali lignin (with low sulfonate content and no reducing sugar), (2) ZHL lignin (depolymerised and 27% sugar content), and (3) AL lignin (alkali extracted, 16% sugar content). Of the three, alkali lignin is soluble in water and ZHL and AL lignin are soluble in alkali solution.[105] Colloidal lignin coated with AuNPs or AgNPs[106] (AuNP@lignin or AgNP@lignin) was synthesised in a minute using thermal or photoinduced methods.[106,107] All three different nanocomposites had negative zeta potentials. The negative zeta potential nanoparticles were less cytotoxic and could adsorb the protein selectively. The synthesised nanoparticles had sizes of around 200 to 400 nm. Both the NPs had the potential to kill Gram-positive and Gram-negative bacteria. The LNPs could prevent the growth of *Staphylococcus aureus* 292S3 and *Escherichia coli* CF073.[106] The MNP@lignin nanocomposite and AuNP@lignin composite when exposed to white light at 78 W m^2 for six hours showed ROS production.[106] Lignin-stabilised noble metal nanoparticles are well known for their compatibility with a biological sample and light-induced antibacterial activity.[106] They further showed the bacteriostatic type of activity. AuNP@lignin was nontoxic to bacteria under dark conditions.[106] The MNP@lignin nanocomposite showed better activity in the presence of light, indicating its use in photodynamic antimicrobial chemotherapy (PACT).[106] The peptidoglycan layer of the bacterial membrane helped to attach to the lignin nanocomposite. All lignin nanocomposites were noncytotoxic.[106] 3T3 fibroblast cells were used to check the cell viability in the presence of MNP@lignin nanocomposites at a concentration of 0.1 mg ml^{-1}. MTT assay was conducted to detect the cell viability.[106]

10.4.3 Lignin Bimetallic Nanoconjugate Doped H-responsive Hydrogel

Hydrogels incorporating silver–LNPs were developed and checked for their activity against bacteria and in the healing of wounds.[108,109] Lignin-derived silver (Ag) and gold (Au) metallic as well as bimetallic nanocomplexes showed their activity against a wide range of microbes.[110–112] Lignin-based metallic nanoparticles are compatible with biological molecules and non-toxic in nature.[78,106,111] The surface functionalisation of lignin-based nanocomplexes is used to develop biodegradable antimicrobial hydrogels.[110,111] These hydrogels are used for PDT against microbial infections. PDT is a well-known technique to kill microbes by using the multiple actions of light, oxygen, and a photosensitiser. PDT is used for the treatment of acne and wounds and to kill bacteria.[113–115] Surface functionalised lignin-based nanocomplexes are used as biodegradable antimicrobial hydrogels to treat microbial infections. The photosensitiser present in the surface functionalised lignin-based nanocomposite generates ROS in response to light.[116–119] Rose Bengal (RB) is an anionic, hydrophobic xanthene dye. It is well known for its use in light sensing and PDT.[120,121] Under $<$532 nm light irradiation, RB produces ROS, which are required for the photodynamic inactivation of microbes.[113,122,123] RB is conjugated with lignin-based metallic and lignin-based bimetallic nanocomplexes.[111,124,125] This bioinspired photodynamic hydrogel is ideal for antibacterial, antifungal, and antiviral coatings.[111,126,127]

Lignin's role as a photosensitising agent also makes it useful for antioxidant activity.[128,129] Lignin-coated noble metallic nanoparticles were used for their antimicrobial activity against *S. aureus* and *E. coli*.[106] Lignin linked with a phenyl porphyrin resulted in a biopolymer with enhanced fluorescence and without any photodynamic approach.[130] LNPs were checked against *Chlamydomonas reinhardtii* and *Saccharomyces cerevisiae*.[131] LNPs were innocuous to caco-e cells,[132] a human colon carcinoma cell line. Acetylated lignin (AcLi) is an organosolv lignin and is a weak photosensitiser.[98] AcLi is a spherical nanoparticle capable of transporting active molecules.[98,133] 5,10,15,20-Tetrakis(4-hydroxyphenyl)-21*H*,23*H*-porphyrin (THPP) inside AcLi was synthesised and checked against *S. aureus* CIP76.25, *Staphylococcus epidermidis* CIP109562, *Enterococcus faecalis* CIP76.1170, *E. coli* CIP53.126, and *Pseudomonas aeruginosa* CIP76.11 under white LED light. THPP-loaded @AcLi was stable and could tolerate a wide range of pHs.[101]

10.4.4 Lignin-centred Carbon (LCQDs) and Graphene Quantum Dots (LGQDs) for Bioimaging

Lignin-based carbon quantum dots (LCQDs) are widely used for sensing metal ions, biosensors, and bioimaging.[134] LCQDs have advantages over other materials like less cytotoxicity and more biocompatibility.[134] Moreover, they are cost-effective, can be easily manufactured, can be synthesised from easily available non-food materials, and do not need complex reactions for their synthesis. Lignin varies from each other by having different oxygen groups, substituted aromatic moieties, and alkyl chains. Besides these dissimilarities, they share similar physicochemical properties such as π–π aggregation-induced molecular aggregation and aggregation-induced fluorescence emission. Several methods are used for the synthesis of LCQDs.[135–139]

The first LCQDs were synthesised by using a hydrothermal process using H_2O_2 as an oxidising agent.[134] The synthesised LCQDs were irregular in shape and exhibited poor luminescence.[140] The luminescent properties of these LCQDs were investigated in the HeLa cell line. Later microwave-assisted depolymerisation of sodium lignosulfonate was developed followed by carbonisation assisted by microwave irradiation.[141] The synthesised LCQDs were reduced with sodium borohydride ($NaBH_4$) to form r-LCQDs. The r-LCQDs can adsorb and release at neutral pH. These r-LCQDs can penetrate A549 and SW480 cell lines *via* membrane diffusion. However, $NaBH_4$ used for the synthesis of LCQDs is a harmful, expensive, and sensitive reducing agent. To overcome this issue, later ascorbic acid was used.

Green synthesis of CDs was developed by Niu *et al.*, and these CDs are referred to as LCDs.[15] The synthesised LCDs were biocompatible and could be easily taken up by HeLa cells. The synthesised NPs had an average size of around 1.4 nm. The LCDs were rich in an oxygen-containing group. The quantum yields of LCDs in water and the solid state were 1.68% and 2.47%, respectively. With an increase of excitation wavelength from 280 nm to 500 nm, the emission wavelength ranged from 390 nm to 560 nm. The Raman spectra of lignin showed its peak at 1347 cm^{-1} and 1592 cm^{-1}. The characterisation also revealed the presence of the amorphous and crystalline structures of lignin. The LCDs also had a C=O bond at 1650 and a C–O bond at 1330 cm^{-1}. The LCDs emitted blue light with both one- and two-photon excitation. The emission was caused by irradiation with 320 nm and 800 nm light. The LCDs produced blue, green, and red colours when excited with 320 nm, 400 nm, and 480 nm light. The luminescence lifetime was 6.827 μs. These photoluminescent properties make them ideal for bioimaging. The colour of the LCDs was further dependent on pH and gave maximum intensity when the pH was 3 to 9. The labeling of the LCDs to cells was confirmed by using HeLa cells (Figures 10.4 and 10.5).

Hybrid LCQDs were synthesised using the hydrothermal method. Different ratios of citric acid and ethanediamine were used as precursor molecules for the synthesis.[142] The resultant LCQDs are suitable for bioimaging. The drawbacks of this method are its long synthesis time and energy-consuming process. Shi *et al.* synthesised LCQDs by adding secondary amines to lignin *via* Mannich and Michael reactions followed by carbonisation, milling, and filtration.[143] The resultant LCQDs can detect Fe^{3+} in the 1–100 nm range.[144] The detecting ability further depends on the precursor materials. The LCQDs exhibit excellent bioimaging properties and biocompatibility when taken up

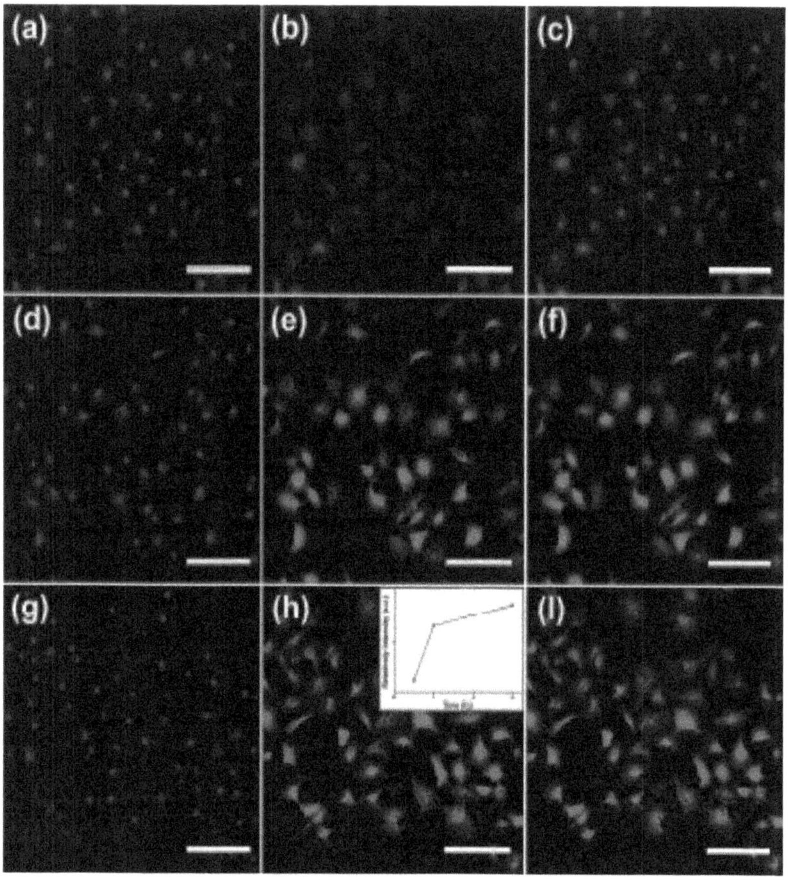

Figure 10.4 Fluorescence images of HeLa cells costained with DAPI and LCDs. Incubation times of (a–c) 0.5 hour, (d–f) 1 hour, and (g–i) 3 hours. The inset graph of (h) indicates the fluorescence intensity. Scale bars: 50 μm. Reproduced from ref. 15 with permission from American Chemical Society, Copyright 2017.

by biological cells. To enhance the fluorescence intensity, often the LCQDs are doped with boron and nitrogen.[134,145] Ding *et al.* synthesised LCQDs by sonicating lignin with nitric acid followed by hydrothermal treatment.[146] The obtained LCQDs had defects and were not suitable for bioimaging. Zhang *et al.* prepared LCQDs with bright green fluorescence which is suitable for cell imaging which are sensitive towards Ag^{+}.[147] Niu *et al.* synthesised LCQDs *via* electrical interaction induced molecular aggregation using ethanol.[148] The LCQDs exhibit fluorescence at 320 nm and 800 nm. The absorption of LCQDs in the near-infrared region makes them suitable for bioimaging.[134]

Lignin-derived graphene quantum dots (LGQDs) were synthesised using *ortho*-aminobenzene sulfonic acid to form LNPs.[149] The synthesis was

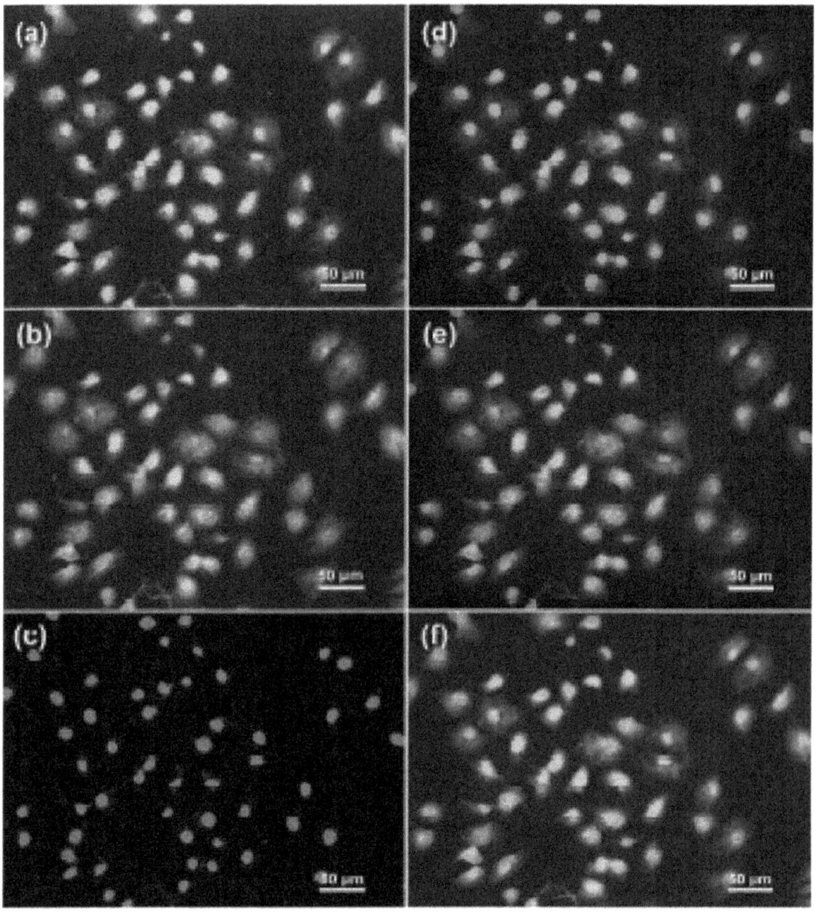

Figure 10.5 Fluorescence images of HeLa cells incubated with DAPI for 10 minutes and LCDs for three hours. (a) One-photon luminescence image, (b) two-photon luminescence image, (c) DAPI image, (d) overlay of the one-photon luminescence image with DAPI image, (e) overlay of the two-photon luminescence image with DAPI image, and (f) merge of all three images. Reproduced from ref. 15 with permission from American Chemical Society, Copyright 2017.

followed by hydrothermal carbonisation by using LGQDs.[134,150] LGQDs had the ability to quench H_2O_2 which was 0.13 nM.[150,151] Lignin-derived reduced fluorescent carbon dots were synthesised using the hydrothermal synthesis method. The chitosan-derived lignin was made using H_2O_2 as an oxidising agent.[152] The reaction time was long, and these carbon dots exhibited poor luminescent properties and were less effective for bioimaging. The lignin-derived reduced carbon dots were synthesised using lignosulfonate as a source of carbon.[16,153] The synthesised reduced fluorescent carbon dots (r-FCDs) were used for nanocarriers of curcumin.[154] The r-FCDs were checked for hemolytic activity and were found to be safe (2.8%) for such applications.[16] The r-FCDs were also investigated for imaging in the human lung adenocarcinoma cell line (A549) as well as colon cancer cell line (SW480)[16] (Figure 10.6). Thus, it is difficult to distinguish the cytoplasm from the nucleus.[16] r-FCDs do not affect the cell viability as evidenced by MTT assay (Figure 10.7). This suggests that r-FCDs are a good bioimaging agent for cancer treatment.

Different types of lignin-hybridised carbon quantum dots are produced by varying monomeric units. Several types of lignin include alkali lignin, kraft lignin, and lignosulfonates. Lignin has a long hydrophobic backbone having carboxyl and phenolic hydroxyl groups. The first carbon quantum dots (CQDs) were synthesised from lignin using H_2O_2 as an oxidising agent through a hydrothermal method.[136] The resultant CQDs were less efficient for bioimaging. Later Ding *et al.* synthesised single-crystalline CQDs from alkali lignin modified with nitric acid *via* a hydrothermal method.[155] This procedure has a prolonged reaction time and consumes high energy. Very few studies report CQDs derived from Al-modified citric acid (CA) and ethylenediamine (EDA) for bioimaging applications.[156] A one-pot hydrothermal approach was used to synthesise lignin-hybridised CQDs *via* lignin–CQD interactions.[157] The synthesised L-CQDs were photoluminescent, water-soluble, less cytotoxic, and biocompatible. The L-CQDs were checked for their cytotoxicity in HeLa cells using MTT assay. The synthesised CQDs could stain the cytoplasm and cell membrane but were unable to label

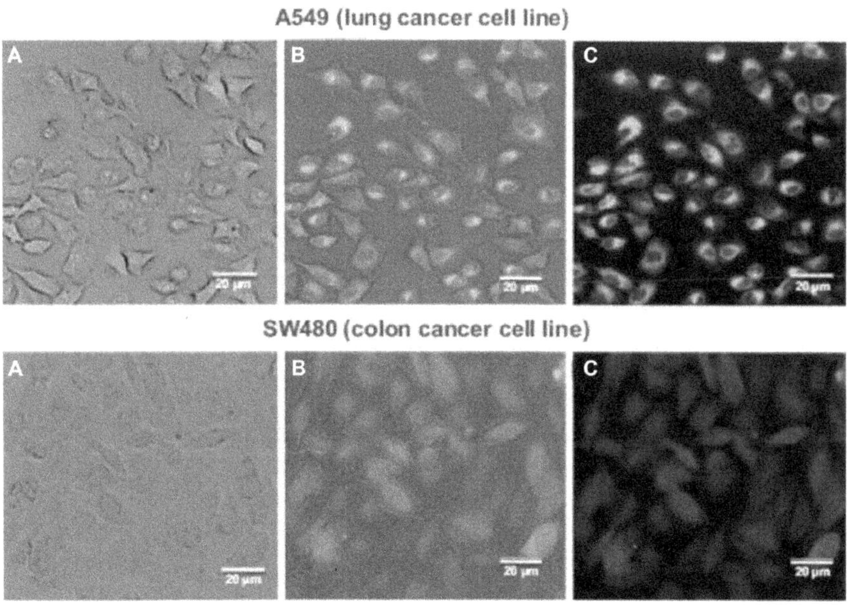

Figure 10.6 Confocal images of cancerous cells labeled with reduced fluorescent carbon dots (r-FCDs). Reproduced from ref. 16 with permission from Elsevier, Copyright 2017.

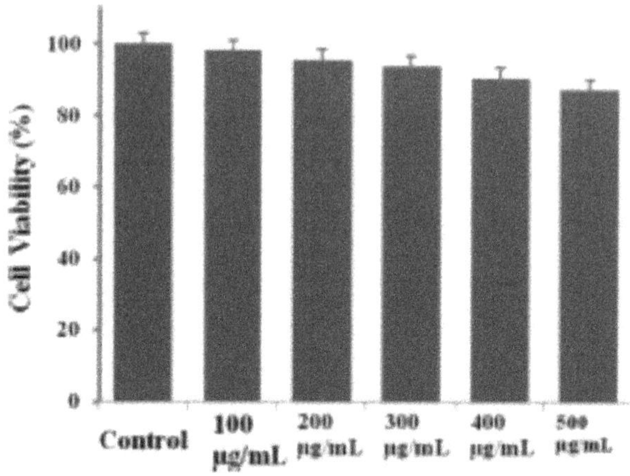

Figure 10.7 Cytotoxicity of r-FCDs as determined by MTT assay. Reproduced from ref. 16 with permission from Elsevier, Copyright 2017.

the nucleus. These CQDs were used for multicolour imaging. They exhibited photoluminescent behaviour in the blue, green, and red channels, suggesting their application in the fields of bioimaging and biolabeling.[158]

Xu *et al.* (2019) also synthesised lignin-hybridised quantum dots (L-CQDs) using different ratios of citric acid and ethanediamine.[159] The synthesised L-CQDs were spherical and 10 nm in size. They were photoluminescent with emission in the range from 454 to 535 nm for excitation wavelengths of 375 to 460 nm. The L-CQDs were less cytotoxic and biocompatible, and had great potential for bioimaging. Amine-rich lignin quantum dots (Al-CQDs)[160] were synthesised from the green precursor of aminated alkali lignin. The Al-CQDs were synthesised by introducing secondary amine groups onto the lignin backbone. The synthesised Al-CQDs had sizes of 4–10 nm and were stable with respect to pH. The synthesised Al-CQDs could be used for cell imaging and they could detect intracellular Fe.[160,161]

10.4.4.1 Lignin for Photodynamic Therapy

Lignin is a polyphenol-rich biopolymer and has antimicrobial activity.[64] Lignin-based photodynamic nanoconjugates were synthesised by several researchers.[111,162] Hydrogels embedded with these nanoconjugates were synthesised by lignin metallic and lignin bimetallic methods. In the metallic method gold is commonly used. In the bimetallic method both gold and silver are used. The photodynamic method was used to achieve pH-triggered controlled release.[111] The hydrogels helped to maintain the nanoconjugates for better ADT. The IC_{50} can distinguish between live and dead cells. These nanomaterials can be used for wound dressings, and antifungal and antiviral coatings.

10.4.4.1.1 Lignin for Photosensitising and Bioimaging. Lignin nanospheres were synthesised using a one-step synthesis process.[76,163] The green synthesis method was used by taking ethanol–water in the solvent–antisolvent method without any toxic chemicals.[76] Using sonication the lignin solution was converted to nanospray in a single step.[164] The entire process took only 10 minutes. At the end of the process, the colour of the lignin changed from brown to pale yellow.[165] By reducing the pressure, the ethanol was removed, and by optimising the temperature, pH, lignin concentration, and volume of lignin, nanospheres were obtained. The synthesised lignin nanoformulation (LNSR) was stable at pH 7 and had a size of around 80 nm. It has C=O and C=C stretching as functional groups within it. The synthesised LNSR is capable of killing microbes and blocking 50% UV-A and 90% UV-B. LNSR has 90% antioxidant activity at 30 $\mu g\,ml^{-1}$ concentration. The presence of the phenolic group is responsible for the photoluminescence properties.[150] The phenol group fluoresces in the presence of UV light.[166] LNSR has a green fluorescence maximum at 554 nm at an excitation wavelength of 480 nm.[167] This LNSR is used for imaging and has treatment potential. The fluorescence properties were investigated in *E. coli* by incubating it with LNSR. LNSR generates singlet oxygen which can be used for PDT.[76,168] The synthesised LNSR can kill *E. coli* (Gram-negative) and *Bacillus megaterium* (Gram-positive) bacteria. The bacteria-killing property of LNSR is due to the generation of ROS in the bacterial membrane.[169,170]

10.4.4.2 Lignin-derived Carbon Nanoparticles for Metal Detection

10.4.4.2.1 Lignin-AuNPs for Detection of Pb^{2+}. NIR-responsive lignin-AuNPs were synthesised by Han *et al.*[171] The synthesised lignin-AuNPs were tested for the adsorption of various metals. Among all, lignin-AuNPs showed a strong reaction towards Pb^{2+}. Lignin-AuNPs showed colour change within 5 minutes after the addition of Pb^{2+} solution (10^{-3} M).[171] The photothermal properties could also be altered by altering the amount of Au. The lead binding was confirmed by SEM, EDS, and FTIR. After Pb^{2+} adsorption, the band at 3380 cm^{-1} (corresponding to the OH stretch) shifted to 3420 cm^{-1}. The band of the benzene ring appeared at 678 cm^{-1} after adsorption of Pb^{2+}.

10.4.4.3 Lignin for Fluorescence Lifetime Imaging Microscopy

Lignin is present in the cell wall as protolignin.[172] The substructure of lignin contributes to the fluorescent properties of lignocellulose.[159,173,174] Fluorescence lifetime imaging microscopy (FLIM) allows us to determine the fluorescence lifetime of fluorophores by using microscopy.[175] The lifetime of a fluorophore is affected by the chemical and physical environments.[176,177] Using different techniques different lifetime spectra were obtained.

FLIM application is used to analyse the calcium ion, pH, and molecular interaction of living cells.[178] Using lignin as a biomarker, immunolabeling of wood fibers and lignin degradation can be studied.[173]

10.4.4.4 LNPs for Cancer Therapy

LNPs can load different drugs and can be modified to target different molecules like pH-sensitive polymers and a targeting agent.[179–182] The modified LNPs could be used to release the drug in a pH-dependent manner. The targeted drug delivery system released the drug and enhanced its interaction with cells in a specific manner. Thus LNPs are used due to their potential for drug delivery. LNPs were used for carrying hydrophobic drugs like BZL and Sorafenib.[180] BZL and Sorafenib showed antiproliferative effects against MDA-MB-231 and MCF-7 and caco-2 cell lines.[180] Later, a pH-responsive polymer was added to LNPs which could release the drug in a pH-responsive manner. LNPs were modified with targeting molecules to increase the cellular uptake by specific cells.[180] CLNPs were conjugated with PEG-PHIS and CPP to make CLNPs-CPP and CLMPs-PEG-PHIS-CPP by using 1-ethyl-3-(3-dimethylaminopropyl)carbodiimide hydrochloride and *N*-hydroxysuccinimide coupling chemistry.[183] Both the LNPs had a positive charge due to the presence of the NH_2 group which was observed at 1640 cm^{-1}. The 1670 cm^{-1} corresponds to C=O and is attributed to amide bond formation.[184]

10.4.4.5 Iron-incorporating Lignin

Iron(III)-complexed LNPs (Fe-LNPs) and Fe_3O_4-infused LNPs (Fe_3O_4-LNPs) were prepared by Lintnene *et al.*[185] For the preparation of Fe-LNPs a solution of iron(III) isopropoxide [$Fe(OiPr)_3$] was added to lignin in THF. With the addition of water, a solution of Fe:lignin in THF was formed. After dialysis, Fe-LNPs were formed. Fe_3O_4-LNPs were synthesised by adding lignin solution and oleic acid-coated Fe_3O_4NPs in equal proportions. Next, this mixture was dialysed for 24 hours to get Fe_3O_4NPs. The oleic acid makes the Fe_3O_4NPs hydrophobic and helps in the loading of lignin. Fe-LNPs and Fe_3O_4-LNPs are round in shape and have a negative charge. Fe_3O_4-LNPs show supermagnetic behaviour which makes them suitable for magnetic resonance study. Fe-LNPs and Fe_3O_4-LNPs at 250 mg mL^{-1} are cytotoxic due to the release of H_2O_2. Both Fe-LNPs and Fe_3O_4-LNPs show more hemolytic capability than the normal LNPs. LNPs of size 220 nm are more easily internalised by tumor cells. BZL-loaded LNPs are more efficient in stopping the growth of cancer cell lines. The antiproliferative effect of BZL is different for different cell lines.

10.5 Conclusion and Future Perspectives

This chapter summarised the application of lignin in the field of visualisation with special emphasis on bioimaging. Due to its inexpensiveness and abundance in nature, lignin is used in various biomedical applications including PDT, antioxidant application, and bioimaging. Lignin is an antioxidant and has an anti-UV effect. It also has antiperoxidative and oxygen radical scavenging properties. All these properties make it more suitable and biocompatible for various cellular imaging studies. Hence several methods were developed to synthesise lignin for imaging. Although many of the synthesised LNPs have been used for labeling, still the physicochemical parameters are not up to the mark for all the synthesised LNPs. LNPs are more prone to agglomeration. This property prevents the synthesis of uniform LNPs. As a consequence, the synthesised LNPs are not of the same size. This property interferes with the application of LNPs in all types of bioimaging. Lignin nanocomposites can carry various metals due to their nanoeffect. Thus they can be used for environmental protection and water treatment and the metals can be visualised easily by using lignin. The toxicological studies also report the toxic properties of lignin. There are lignin degraders or marine protists which can utilise lignin and convert it to carotenoids. LNPs can be improved by using various bioconversion processes. Some of the microbes are used for the bioconversion process. These microbes can be utilised to detoxify the toxicity associated with lignin. The other way to detoxify the toxic properties of lignin is by using nontoxic chemicals in the synthesis process. The synthesis protocols available for LNPs are expensive and use toxic chemicals like DMSP, THF, and DMF. Replacement of these toxic chemicals with nontoxic ones will improve the nontoxic properties of lignin. The application of lignin still faces many problems while using it for bioimaging. The addition of functional groups to the nanoparticles will make lignin appropriate for any desired bioimaging application.

Abbreviations

AcLi	Acetylated lignin – an organosolv lignin
ADT	Antimicrobial photodynamic therapy
AL lignin	Alkali extracted with 16% sugar content
Al-CQDs	Amine-rich lignin quantum dots
C4H	Cinnamate 4-hydroxylase
CA	Citric acid
CC	Cell corner
CLNP	Carboxylated lignin nanoparticle
CML	Compound middle lamella
CQD	Carbon quantum dot
DMF	Dimethylformamide
DMSP	Dimethylsulfoniopropionate
EDA	Ethylenediamine
EDS	Energy-dispersive X-ray spectroscopy
Fe_3O_4-LNPs	Fe_3O_4-infused LNPs
Fe-LNPs	Iron(III)-complexed LNPs
FLIM	Fluorescence lifetime imaging microscopy

FTIR	Fourier transform infrared spectroscopy
H_2O_2	Hydrogen peroxide
L-CQD	Lignin-CQD
LCQDs	Lignin-centred carbon quantum dots
LGQDs	Lignin-derived graphene quantum dots
LNPs	Lignin nanoparticles
LNSR	Lignin nanoformulation
$NaBH_4$	Sodium borohydride
NIR	Near-infrared region
NP	Nanoparticle
PACT	Photodynamic antimicrobial chemotherapy
PDT	Photodynamic therapy
RB	Rose bengal
r-FCD	Reduced fluorescent carbon dot
r-LCQDs	Reduced lignin-centred carbon quantum dots
ROS	Reactive oxygen species
S2	Secondary cell wall
SEM	Scanning electron microscopy
THF	Tetrahydrofuran
THPP	5,10,15,20-Tetrakis(4-hydroxyphenyl)-21*H*,23*H*-Porphyrin
ZHL lignin	Lignin – depolymerised and with 27% sugar content

Acknowledgements

MM lab was supported by the Department of Biotechnology, Ministry of Science and Technology [SERB/EMR/2017/003054, BT/PR21857/NNT/28/1238/2017, and Odisha DBT 3325/ST(BIO)-02/2017], for financial assistance.

References

1. M. Ayyachamy, F. E. Cliffe, J. M. Coyne, J. Collier and M. G. Tuohy, *Biomass Convers. Biorefin.*, 2013, **3**, 255–269.
2. F. G. Calvo-Flores and J. A. Dobado, *ChemSusChem*, 2010, **3**, 1227–1235.
3. J. A. Jennings, *Heterogeneous Base Metal Catalyzed Oxidative Depolymerization of Lignin and Lignin Model Compounds*, 2017.
4. K. E. Achyuthan, A. M. Achyuthan, P. D. Adams, S. M. Dirk, J. C. Harper, B. A. Simmons and A. K. Singh, *Molecules*, 2010, **15**, 8641–8688.
5. Q. Liu, L. Luo and L. Zheng, *Int. J. Mol. Sci.*, 2018, **19**, 335.
6. A. Hagerman and L. G. Butler, *Herbivores: their interactions with secondary plant metabolites*, 1991, vol. 1, pp. 355–388.
7. A.-M. Boudet, *Plant Physiol. Biochem.*, 2000, **38**, 81–96.
8. J. Zhu, U. P. Agarwal, P. N. Ciesielski, M. E. Himmel, R. Gao, Y. Deng, M. Morits and M. Österberg, *Biotechnol. Biofuels*, 2021, **14**, 1–31.
9. Z. Yan, G. Liao, X. Zou, M. Zhao, T. Wu, Y. Chen and G. Fang, *J. Agric. Food Chem.*, 2020, **68**, 8341–8349.

10. M. S. Ganewatta, H. N. Lokupitiya and C. Tang, *Polymers*, 2019, **11**, 1176.
11. M. Rizwan, R. Yahya, A. Hassan, M. Yar, A. D. Azzahari, V. Selvanathan, F. Sonsudin and C. N. Abouloula, *Polymers*, 2017, **9**, 137.
12. S. Jacob, A. B. Nair, J. Shah, N. Sreeharsha, S. Gupta and P. Shinu, *Pharmaceutics*, 2021, **13**, 357.
13. L. Bacakova, J. Pajorova, M. Bacakova, A. Skogberg, P. Kallio, K. Kolarova and V. Svorcik, *Nanomaterials*, 2019, **9**, 164.
14. B. Xue, Y. Yang, Y. Sun, J. Fan, X. Li and Z. Zhang, *Int. J. Biol. Macromol.*, 2019, **122**, 954–961.
15. N. Niu, Z. Ma, F. He, S. Li, J. Li, S. Liu and P. Yang, *Langmuir*, 2017, **33**, 5786–5795.
16. S. Rai, B. K. Singh, P. Bhartiya, A. Singh, H. Kumar, P. Dutta and G. Mehrotra, *J. Lumin.*, 2017, **190**, 492–503.
17. A. A. Myint, W.-K. Rhim, J.-M. Nam, J. Kim and Y.-W. Lee, *J. Ind. Eng. Chem.*, 2018, **66**, 387–395.
18. N. Maldonado-Carmona, T.-S. Ouk, M. J. Calvete, M. M. Pereira, N. Villandier and S. Leroy-Lhez, *Photochem. Photobiol. Sci.*, 2020, **19**, 445–461.
19. M. Schmidt, A. Schwartzberg, P. Perera, A. Weber-Bargioni, A. Carroll, P. Sarkar, E. Bosneaga, J. Urban, J. Song and M. Balakshin, *Planta*, 2009, **230**, 589–597.
20. S. Fujita, D. De Bellis, K. H. Edel, P. Köster, T. G. Andersen, E. Schmid-Siegert, V. Dénervaud Tendon, A. Pfister, P. Marhavý and R. Ursache, *EMBO J.*, 2020, **39**, e103894.
21. F. Baldacci-Cresp, C. Spriet, L. Twyffels, A. S. Blervacq, G. Neutelings, M. Baucher and S. Hawkins, *Plant J.*, 2020, **102**, 1074–1089.
22. N. Zhu, D. Wu and K. Chen, *Plant Methods*, 2018, **14**, 1–16.
23. B. Fergus, A. Procter, J. Scott and D. Goring, *Wood Sci. Technol.*, 1969, **3**, 117–138.
24. U. P. Agarwal, *Planta*, 2006, **224**, 1141–1153.
25. Y. Zeng, B. G. Saar, M. G. Friedrich, F. Chen, Y.-S. Liu, R. A. Dixon, M. E. Himmel, X. S. Xie and S.-Y. Ding, *Bioenergy Res.*, 2010, **3**, 272–277.
26. L. Sun, B. A. Simmons and S. Singh, *Biotechnol. Bioeng.*, 2011, **108**, 286–295.
27. K. J. Moore and H.-J. G. Jung, *J. Range Manage.*, 2001, **54**, 420–430.
28. A. Kärkönen and S. Koutaniemi, *J. Integr. Plant Biol.*, 2010, **52**, 176–185.
29. X. Li and C. Chapple, *Plant Physiol.*, 2010, **154**, 449–452.
30. R. Vanholme, B. Demedts, K. Morreel, J. Ralph and W. Boerjan, *Plant Physiol.*, 2010, **153**, 895–905.
31. Y. Zeng, S. Zhao, S. Yang and S.-Y. Ding, *Curr. Opin. Biotechnol.*, 2014, **27**, 38–45.
32. Y. Tobimatsu and M. Schuetz, *Curr. Opin. Biotechnol.*, 2019, **56**, 75–81.
33. H. Onnerud, L. Zhang, G. Gellerstedt and G. Henriksson, *Plant Cell*, 2002, **14**, 1953–1962.
34. R. A. Dixon and J. Barros, *Open Biol.*, 2019, **9**, 190215.

35. Y. Zeng, S. Zhao, H. Wei, M. P. Tucker, M. E. Himmel, N. S. Mosier, R. Meilan and S.-Y. Ding, *Biotechnol. Biofuels*, 2015, **8**, 1–12.
36. Q. Jin, C. Yan, J. Qiu, N. Zhang, Y. Lin and Y. Cai, *Sci. Hortic.*, 2013, **155**, 123–130.
37. Y. Zeng, M. E. Himmel and S.-Y. Ding, *Biotechnol. Biofuels*, 2017, **10**, 1–16.
38. J.-Y. Gou, L.-J. Wang, S.-P. Chen, W.-L. Hu and X.-Y. Chen, *Cell Res.*, 2007, **17**, 422–434.
39. M. J. Meents, Y. Watanabe and A. L. Samuels, *Ann. Bot.*, 2018, **121**, 1107–1125.
40. L. A. Rogers, C. Dubos, C. Surman, J. Willment, I. F. Cullis, S. D. Mansfield and M. M. Campbell, *New Phytol.*, 2005, **168**, 123–140.
41. S. C. Fry, *New Phytol.*, 2004, **161**, 641–675.
42. R. M. da Costa, S. Pattathil, U. Avci, S. J. Lee, S. P. Hazen, A. Winters, M. G. Hahn and M. Bosch, *New Phytol.*, 2017, **213**, 1710–1725.
43. N. Gierlinger, *Front. Plant Sci.*, 2014, **5**, 306.
44. T. Kanbayashi, Y. Kataoka, A. Ishikawa, M. Matsunaga, M. Kobayashi and M. Kiguchi, *J. Photochem. Photobiol., B*, 2018, **187**, 136–140.
45. M. Kačuráková, N. Wellner, A. Ebringerová, Z. Hromádková, R. Wilson and P. Belton, *Food Hydrocolloids*, 1999, **13**, 35–41.
46. R. Atalla, U. Agarwal and J. Bond, in *Methods in Lignin Chemistry*, Springer, 1992, pp. 162–176.
47. S. Yamauchi, Y. Iijima and S. Doi, *J. Wood Sci.*, 2005, **51**, 498–506.
48. Y. Lu, Y.-C. Lu, H.-Q. Hu, F.-J. Xie, X.-Y. Wei and X. Fan, *J. Spectrosc.*, 2017, 8951658.
49. T. W. Jeffries, in *Physiology of Biodegradative microorganisms*, Springer, 1991, pp. 163–176.
50. J. Pérez, J. Munoz-Dorado, T. De la Rubia and J. Martinez, *Int. Microbiol.*, 2002, **5**, 53–63.
51. H. Liu, J. Sun, S. Y. Leu and S. Chen, *Biofuels, Bioprod. Biorefin.*, 2016, **10**, 648–663.
52. M. Tenkanen, T. Tamminen and B. Hortling, *Appl. Microbiol. Biotechnol.*, 1999, **51**, 241–248.
53. M. R. Tosta, L. L. Prates, D. A. Christensen and P. Yu, *J. Cereal Sci.*, 2019, **89**, 102814.
54. J. M. Robinson, C. E. Burgess, M. A. Bently, C. D. Brasher, B. O. Horne, D. M. Lillard, J. M. Macias, H. D. Mandal, S. C. Mills and K. D. O'Hara, *Biomass Bioenergy*, 2004, **26**, 473–483.
55. D. K. Ro, N. Mah, B. E. Ellis and C. J. Douglas, *Plant Physiol.*, 2001, **126**, 317–329.
56. B. M. Upton and A. M. Kasko, *Chem. Rev.*, 2016, **116**, 2275–2306.
57. J. H. Grabber, *Crop Sci.*, 2005, **45**(3), 820–831.
58. G. Gellerstedt and G. Henriksson, *Monomers, polymers and composites from renewable resources*, 2008, pp. 201–224.
59. C. A. R. Ferro, Lignin nanoparticles for cancer therapy, Doctoral dissertation, 2017.
60. N.-E. El Mansouri and J. Salvadó, *Ind. Crops Prod.*, 2007, **26**, 116–124.

61. A. G. Vishtal and A. Kraslawski, *BioResources*, 2011, **6**, 3547–3568.
62. V. K. Thakur, M. K. Thakur, P. Raghavan and M. R. Kessler, *ACS Sustainable Chem. Eng.*, 2014, **2**, 1072–1092.
63. M. Fache, B. Boutevin and S. Caillol, *ACS Sustainable Chem. Eng.*, 2016, **4**, 35–46.
64. A. Alzagameem, S. E. Klein, M. Bergs, X. T. Do, I. Korte, S. Dohlen, C. Hüwe, J. Kreyenschmidt, B. Kamm and M. Larkins, *Polymers*, 2019, **11**, 670.
65. M. Galkin, *Curr. Opin. Green Sustainable Chem.*, 2021, **28**, 100438.
66. A. Naseem, S. Tabasum, K. M. Zia, M. Zuber, M. Ali and A. Noreen, *Int. J. Biol. Macromol.*, 2016, **93**, 296–313.
67. J. Zhao, W. Xiuwen, J. Hu, Q. Liu, D. Shen and R. Xiao, *Polym. Degrad. Stab.*, 2014, **108**, 133–138.
68. A. Awal and M. Sain, *J. Appl. Polym. Sci.*, 2011, **122**, 956–963.
69. K. B. Sandved, G. T. Prance and A. E. Prance, *Bark: The Formation, Characteristics, and Uses of Bark Around the World*, Timber Press, Portland, OR, 1992.
70. A. Duval and M. Lawoko, *React. Funct. Polym.*, 2014, **85**, 78–96.
71. V. K. Ponnusamy, D. D. Nguyen, J. Dharmaraja, S. Shobana, J. R. Banu, R. G. Saratale, S. W. Chang and G. Kumar, *Bioresour. Technol.*, 2019, **271**, 462–472.
72. B. Du, H. Zhu, X. Wang, L.-P. Xiao, J. Ma, X. Chen, J. Zhou and R.-C. Sun, *Int. J. Biol. Macromol.*, 2021, **174**, 254–262.
73. D. Bajwa, G. Pourhashem, A. H. Ullah and S. Bajwa, *Ind. Crops Prod.*, 2019, **139**, 111526.
74. H. Ľudmila, J. Michal, Š. Andrea and H. Aleš, *Wood Res.*, 2015, **60**, 973–986.
75. S. Beisl, A. Miltner and A. Friedl, *Int. J. Mol. Sci.*, 2017, **18**, 1244.
76. S. Paul, N. S. Thakur, S. Chandna, Y. N. Reddy and J. Bhaumik, *J. Mater. Chem. B*, 2021, **9**, 1592–1603.
77. S. Beisl, A. Miltner and A. Friedl, *Int. J. Mol. Sci.*, 2017, **18**(6), 1244.
78. P. Figueiredo, K. Lintinen, J. T. Hirvonen, M. A. Kostiainen and H. A. Santos, *Prog. Mater. Sci.*, 2018, **93**, 233–269.
79. A. P. Richter, B. Bharti, H. B. Armstrong, J. S. Brown, D. Plemmons, V. N. Paunov, S. D. Stoyanov and O. D. Velev, *Langmuir*, 2016, **32**, 6468–6477.
80. C. Desfrançois, R. Auzély and I. Texier, *Pharmaceuticals*, 2018, **11**, 118.
81. A. Okamoto, T. Asai, Y. Hirai, K. Shimizu, H. Koide, T. Minamino and N. Oku, *Mol. Pharmaceutics*, 2018, **15**, 1495–1504.
82. P. Figueiredo, C. Ferro, M. Kemell, Z. Liu, A. Kiriazis, K. Lintinen, H. F. Florindo, J. Yli-Kauhaluoma, J. Hirvonen and M. A. Kostiainen, *Nanomedicine*, 2017, **12**, 2581–2596.
83. M. Jafari, D. Heidari and P. Ebrahimnejad, *Int. J. Pharm. Res.*, 2016, **15**, 449.
84. D. R. Gang, M. A. Costa, M. Fujita, A. T. Dinkova-Kostova, H.-B. Wang, V. Burlat, W. Martin, S. Sarkanen, L. B. Davin and N. G. Lewis, *Chem. Biol.*, 1999, **6**, 143–151.

85. Z. Li, E. Sutandar, T. Goihl, X. Zhang and X. Pan, *Green Chem.*, 2020, **22**, 7989–8001.
86. H. Zeng, D. Cao, Z. Qiu and C. J. Li, *Angew. Chem.*, 2018, **130**, 3814–3819.
87. P. Rutkowski, *Fuel Process. Technol.*, 2011, **92**, 517–522.
88. M. Witzler, A. Alzagameem, M. Bergs, B. E. Khaldi-Hansen, S. E. Klein, D. Hielscher, B. Kamm, J. Kreyenschmidt, E. Tobiasch and M. Schulze, *Molecules*, 2018, **23**, 1885.
89. S. Laurichesse and L. Avérous, *Prog. Polym. Sci.*, 2014, **39**, 1266–1290.
90. L. Dessbesell, M. Paleologou, M. Leitch, R. Pulkki and C. C. Xu, *Renewable Sustainable Energy Rev.*, 2020, **123**, 109768.
91. W. Zhu, J. Lu and L. Dai, *Part. Part. Syst. Charact.*, 2018, **35**, 1800145.
92. H. Wang, Y. Wang, F. Fu, Y. Qian, Y. Xiao, D. Yang and X. Qiu, *Int. J. Biol. Macromol.*, 2020, **150**, 371–379.
93. S. Bertella and J. S. Luterbacher, *Trends Chem.*, 2020, **2**, 440–453.
94. A. Abdulkhani, J. Hosseinzadeh, A. Ashori, S. Dadashi and Z. Takzare, *Polym. Test.*, 2014, **35**, 73–79.
95. S. Kalia, K. Thakur, A. Celli, M. A. Kiechel and C. L. Schauer, *J. Environ. Chem. Eng.*, 2013, **1**, 97–112.
96. S. Atifi, C. Miao and W. Y. Hamad, *J. Appl. Polym. Sci.*, 2017, **134**(29), 45103.
97. M. F. Cherif, D. Trache, N. Brosse, F. Benaliouche and A. F. Tarchoun, *Waste Biomass Valorization*, 2020, **11**, 6541–6553.
98. G. Marchand, C. Calliste, R. Williams, C. McLure, S. Leroy-Lhez and N. Villandier, *ChemistrySelect*, 2018, **3**, 5512–5516.
99. J. Parkås, M. Paulsson, S. Li, K. Lundquist and U. Westermark, *Nord. Pulp Pap. Res. J.*, 2004, **19**, 146–154.
100. I. Kühnel, B. Saake and R. Lehnen, *React. Funct. Polym.*, 2017, **120**, 83–91.
101. C. C. Gonçalves, T. Bruce, C. D. O. G. Silva, E. X. F. Fillho, E. F. Noronha, M. Carlquist and N. S. Parachin, *Front. Microbiol.*, 2020, **11**, 1081.
102. X. Chen, X. Tian, I. Shin and J. Yoon, *Chem. Soc. Rev.*, 2011, **40**, 4783–4804.
103. P. Raja, A. Bozzi, H. Mansilla and J. Kiwi, *J. Photochem. Photobiol., A*, 2005, **169**, 271–278.
104. M. G. Neumann and A. E. Machado, *J. Photochem. Photobiol., B*, 1989, **3**, 473–481.
105. C. G. Boeriu, D. Bravo, R. J. Gosselink and J. E. van Dam, *Ind. Crops Prod.*, 2004, **20**, 205–218.
106. D. M. Rocca, J. P. Vanegas, K. Fournier, M. C. Becerra, J. C. Scaiano and A. E. Lanterna, *RSC Adv.*, 2018, **8**, 40454–40463.
107. S. Imlimthan, A. Correia, P. Figueiredo, K. Lintinen, V. Balasubramanian, A. J. Airaksinen, M. A. Kostiainen, H. A. Santos and M. Sarparanta, *J. Biomed. Mater. Res. Part A*, 2020, **108**(3), 770–783.
108. D. Gan, W. Xing, L. Jiang, J. Fang, C. Zhao, F. Ren, L. Fang, K. Wang and X. Lu, *Nat. Commun.*, 2019, **10**, 1–10.

109. Y. N. Slavin, K. Ivanova, J. Hoyo, I. Perelshtein, G. Owen, A. Haegert, Y.-Y. Lin, S. LeBihan, A. Gedanken and U. O. Häfeli, *ACS Appl. Mater. Interfaces*, 2021, **13**, 22098–22109.
110. S. Chandna, N. S. Thakur, Y. N. Reddy, R. Kaur and J. Bhaumik, *ACS Biomater. Sci. Eng.*, 2019, **5**, 3212–3227.
111. S. Chandna, N. S. Thakur, R. Kaur and J. Bhaumik, *Biomacromolecules*, 2020, **21**, 3216–3230.
112. E. M. Fernandes, F. C. Lobo and R. L. Reis, *Molecules*, 2021, **26**(6), 1749.
113. R. Yin, T. Dai, P. Avci, A. E. S. Jorge, W. C. de Melo, D. Vecchio, Y.-Y. Huang, A. Gupta and M. R. Hamblin, *Curr. Opin. Pharmacol.*, 2013, **13**, 731–762.
114. G. Jori, C. Fabris, M. Soncin, S. Ferro, O. Coppellotti, D. Dei, L. Fantetti, G. Chiti and G. Roncucci, *Lasers Surg. Med.*, 2006, **38**, 468–481.
115. T. Maisch, R.-M. Szeimies, G. Jori and C. Abels, *Photochem. Photobiol. Sci.*, 2004, **3**, 907–917.
116. T. Dai, Y.-Y. Huang and M. R. Hamblin, *Photodiagn. Photodyn. Ther.*, 2009, **6**, 170–188.
117. E. Polat and K. Kang, *Biomedicines*, 2021, **9**, 584.
118. R. Yin and M. R Hamblin, *Curr. Med. Chem.*, 2015, **22**, 2159–2185.
119. G. Jori, M. Camerin, M. Soncin, L. Guidolin and O. Coppellotti, *Photodynamic inactivation of microbial pathogens: Medical and environmental applications*, 2011, vol. 11.
120. B. Sharma, V. Thakur, G. Kaur and G. R. Chaudhary, *ACS Appl. Bio Mater.*, 2020, **3**, 8515–8524.
121. C. Li, F. Lin, W. Sun, F.-G. Wu, H. Yang, R. Lv, Y.-X. Zhu, H.-R. Jia, C. Wang and G. Gao, *ACS Appl. Mater. Interfaces*, 2018, **10**, 16715–16722.
122. P. Paramanantham, A. P. Antony, S. S. Lal, A. Sharan, A. Syed, M. Ahmed, A. A. Alarfaj, S. Busi, M. Maaza and K. Kaviyarasu, *Sci. Afr.*, 2018, **1**, e00007.
123. W. C. De Melo, P. Avci, M. N. De Oliveira, A. Gupta, D. Vecchio, M. Sadasivam, R. Chandran, Y.-Y. Huang, R. Yin and L. R. Perussi, *Expert Rev. Anti-Infect. Ther.*, 2013, **11**, 669–693.
124. K. Liu, D. Zheng, H. Lei, J. Liu, J. Lei, L. Wang and X. Ma, *ACS Biomater. Sci. Eng.*, 2018, **4**, 1730–1737.
125. W. Hu, J. Zhang, B. Liu, C. Zhang, Q. Zhao, Z. Sun, H. Cao and G. Zhu, *Compos. Commun.*, 2021, **24**, 100616.
126. B. Balasubramaniam, Prateek, S. Ranjan, M. Saraf, P. Kar, S. P. Singh, V. K. Thakur, A. Singh and R. K. Gupta, *ACS Pharmacol. Transl. Sci.*, 2021, **4**, 8–54.
127. Q. Zhao, Y. Zhao, Z. Lu and Y. Tang, *ACS Appl. Mater. Interfaces*, 2019, **11**, 16320–16327.
128. J. Ponomarenko, M. Lauberts, T. Dizhbite, L. Lauberte, V. Jurkjane and G. Telysheva, *Holzforschung*, 2015, **69**, 795–805.
129. W. Yang, J. Owczarek, E. Fortunati, M. Kozanecki, A. Mazzaglia, G. Balestra, J. Kenny, L. Torre and D. Puglia, *Ind. Crops Prod.*, 2016, **94**, 800–811.

130. T. Ho-Yin, S.-C. Cheng, C. S. Yeung, C.-Y. Lau, W.-H. Wong, C. Dong and S.-Y. Leu, *Green Chem.*, 2019, **21**, 1319–1329.
131. C. Frangville, M. Rutkevičius, A. P. Richter, O. D. Velev, S. D. Stoyanov and V. N. Paunov, *Chem. Phys. Chem.*, 2012, **13**, 4235.
132. M. S. Alqahtani, A. Alqahtani, A. Al-Thabit, M. Roni and R. Syed, *J. Mater. Chem. B*, 2019, **7**, 4461–4473.
133. Q. Cao, M. Zhu, J. Chen, Y. Song, Y. Li and J. Zhou, *ACS Appl. Mater. Interfaces*, 2019, **12**, 1210–1221.
134. A. Moreno and M. H. Sipponen, *Mater. Horiz.*, 2020, 7, 2237–2257.
135. Y. Pei, A. Y. Chang, X. Liu, H. Wang, H. Zhang, A. Radadia, Y. Wang, W. W. Yu and S. Wang, *AIChE J.*, 2021, **67**, e17132.
136. W. Chen, C. Hu, Y. Yang, J. Cui and Y. Liu, *Materials*, 2016, **9**, 184.
137. X. Wu, X. Fan, S. Xie, J. Lin, J. Cheng, Q. Zhang, L. Chen and Y. Wang, *Nat. Catal.*, 2018, **1**, 772–780.
138. A. Valizadeh, H. Mikaeili, M. Samiei, S. M. Farkhani, N. Zarghami, A. Akbarzadeh and S. Davaran, *Nanoscale Res. Lett.*, 2012, 7, 1–14.
139. D. L. Li, S. B. Ge, W. X. Peng, Q. D. Wu and J. G. Wu, *Polym. Compos.*, 2017, **38**, 955–965.
140. F. K. Kanodarwala, PhD thesis, The University of New South Wales, 2012.
141. H. Li, Y. Qu and J. Xu, in *Production of biofuels and chemicals with microwave*, Springer, 2015, pp. 61–82.
142. C. Huang, H. Dong, Y. Su, Y. Wu, R. Narron and Q. Yong, *Nanomaterials*, 2019, **9**, 387.
143. S. Shi, W. Qiu, P. Miao, R. Li, X. Lin and Z. Sun, *Nat. Commun.*, 2021, **12**, 1–7.
144. J. Jeevanandam, A. Barhoum, Y. S. Chan, A. Dufresne and M. K. Danquah, *Beilstein J. Nanotechnol.*, 2018, **9**, 1050–1074.
145. A. D. Stiff-Roberts, PhD thesis, University of Michigan, 2004.
146. N. Ding, H. Liu, Y. Sun, X. Tang, T. Lei, F. Xu, X. Zeng and L. Lin, *J. Cleaner Prod.*, 2021, **278**, 123984.
147. A.-P. Zhang, C.-F. Liu, R.-C. Sun and J. Xie, *BioResources*, 2013, **8**, 1604–1614.
148. W. Niu, H. Willett, J. Mueller, X. He, L. Kramer, B. Ma and J. Guo, *Metab. Eng.*, 2020, **59**, 151–161.
149. B. Zhao, M. Borghei, T. Zou, L. Wang, L.-S. Johansson, J. Majoinen, M. H. Sipponen, M. Österberg, B. D. Mattos and O. J. Rojas, *ACS Nano*, 2021, **15**, 6774–6786.
150. R. Wang, G. Xia, W. Zhong, L. Chen, L. Chen, Y. Wang, Y. Min and K. Li, *Green Chem.*, 2019, **21**, 3343–3352.
151. S. Sugiarto, Y. Leow, C. L. Tan, G. Wang and D. Kai, *Bioact. Mater.*, 2021, **8**, 71–94.
152. L. Chen, C.-Y. Tang, N.-Y. Ning, C.-Y. Wang, Q. Fu and Q. Zhang, *Chin. J. Polym. Sci.*, 2009, **27**, 739–746.
153. S. Rai, P. Dutta and G. Mehrotra, *J. Polym. Mater.*, 2017, **34**, 171.
154. S. Rai, B. K. Singh, P. Bhartiya, A. Singh, H. Kumar, P. Dutta, G. Mehrotra and H. K. Singh, *J. Lumin.*, 2017, **190**, 492–503.

155. Z. Ding, F. Li, J. Wen, X. Wang and R. Sun, *Green Chem.*, 2018, **20**, 1383–1390.
156. M. J. Molaei, *RSC Adv.*, 2019, **9**, 6460–6481.
157. H. Wang, Q. Yao, C. Wang, B. Fan, Q. Sun, C. Jin, Y. Xiong and Y. Chen, *Sci. Rep.*, 2016, **6**, 1–10.
158. D. Qu, X. Wang, Y. Bao and Z. Sun, *J. Phys. Mater.*, 2020, **3**, 022003.
159. J. Xu, L. Dai, C. Zhang, Y. Gui, L. Yuan, Y. Lei and B. Fan, *Bioresour. Technol.*, 2020, **305**, 123043.
160. Y. Shi, X. Liu, M. Wang, J. Huang, X. Jiang, J. Pang, F. Xu and X. Zhang, *Int. J. Biol. Macromol.*, 2019, **128**, 537–545.
161. M. O. Caglayan, F. Mindivan and S. Şahin, *Crit. Rev. Anal. Chem.*, 2020, 1–34.
162. R. Kaur, S. K. Bhardwaj, S. Chandna, K.-H. Kim and J. Bhaumik, *J. Cleaner Prod.*, 2021, 128300.
163. R. P. Pandey, K. Rasool, P. A. Rasheed, T. Gomez, M. Pasha, S. A. Mansour, O.-S. Lee and K. A. Mahmoud, *Green Chem.*, 2020, **22**, 678–687.
164. I. A. Gilca, V. I. Popa and C. Crestini, *Ultrason. Sonochem.*, 2015, **23**, 369–375.
165. O. Ajao, J. Jeaidi, M. Benali, A. M. Restrepo, N. El Mehdi and Y. Boumghar, *Molecules*, 2018, **23**, 377.
166. Z. G. Cerovic, G. Samson, F. Morales, N. Tremblay and I. Moya, *Agronomie*, 1999, **19**, 543–578.
167. L. Donaldson and N. Williams, *Plants*, 2018, 7, 10.
168. Y. Song, Y. Li, Y. Zhang, L. Wang and Z. Xie, *J. Mater. Chem. B*, 2019, 7, 7776–7782.
169. Y. Xie, Y. Qian, Z. Li, Z. Liang, W. Liu, D. Yang and X. Qiu, *ACS Sustainable Chem. Eng.*, 2021, **9**, 6479–6488.
170. Y. Wang, H. Wang, Z. Li, D. Yang, X. Qiu, Y. Liu, M. Yan and Q. Li, *J. Colloid Interface Sci.*, 2021, **594**, 316–325.
171. G. Han, X. Wang, J. Hamel, H. Zhu and R. Sun, *Sci. Rep.*, 2016, **6**, 1–8.
172. N. Terashima and K. Fukushima, *Wood Sci. Technol.*, 1988, **22**, 259–270.
173. J. Hafrén and H. M. Oosterveld-Hut, *J. Wood Sci.*, 2009, **55**, 236–239.
174. H. Dong, L. Zheng, P. Yu, Q. Jiang, Y. Wu, C. Huang and B. Yin, *ACS Sustainable Chem. Eng.*, 2019, **8**, 256–266.
175. E. B. van Munster and T. W. Gadella, *Microsc. Tech.*, 2005, 143–175.
176. P. Sarder, D. Maji and S. Achilefu, *Bioconjugate Chem.*, 2015, **26**, 963–974.
177. J. R. Lakowicz, H. Szmacinski, K. Nowaczyk, K. W. Berndt and M. Johnson, *Anal. Biochem.*, 1992, **202**, 316–330.
178. Y. Sun, R. N. Day and A. Periasamy, *Nat. Protoc.*, 2011, **6**, 1324–1340.
179. L. Siddiqui, H. Mishra, P. K. Mishra, Z. Iqbal and S. Talegaonkar, *Med. Hypotheses*, 2018, **121**, 10–14.
180. P. Figueiredo, K. Lintinen, A. Kiriazis, V. Hynninen, Z. Liu, T. Bauleth-Ramos, A. Rahikkala, A. Correia, T. Kohout and B. Sarmento, *Biomaterials*, 2017, **121**, 97–108.

181. C. E. Byrne, C. E. Astete, M. Vaithiyanathan, A. T. Melvin, M. Moradipour, S. E. Rankin, B. L. Knutson, C. M. Sabliov and E. C. Martin, *Nanomedicine*, 2020, **15**, 981–1000.
182. B. Li, X. Xia, J. Chen, D. Xia, R. Xu, X. Zou, H. Wang and C. Liang, *Nanotechnology*, 2020, **32**, 015101.
183. W. Gao and P. Fatehi, *Can. J. Chem. Eng.*, 2019, **97**, 2827–2842.
184. S. Akbari, A. Bahi, A. Farahani, A. S. Milani and F. Ko, *Molecules*, 2021, **26**, 518.
185. K. Lintinen, M. Latikka, M. H. Sipponen, R. H. Ras, M. Österberg and M. A. Kostiainen, *RSC Adv.*, 2016, **6**, 31790–31796.

CHAPTER 11

Lignin-based UV-protective Materials

K. SHIKINAKA*[a] AND Y. OTSUKA[b]

[a] Research Institute for Chemical Process Technology, National Institute of Advanced Industrial Science and Technology, Nigatake, 4-2-1, Miyagino-ku, Sendai 983-8551, Japan; [b] Forestry and Forest Products Research Institute, Matsunosato, 1, Tsukuba 305-8687, Japan
*Email: kaz.shikinaka@aist.go.jp

11.1 Introduction

Protection of living organisms and organic compounds from ultraviolet (UV) radiation in the sunlight, which is categorized into the UVA region (from 315 to 400 nm), the UVB region (from 280 to 315 nm), and the UVC region (from 100 to 280 nm), is important due to its harmfulness to them such as denaturation of proteins and polymers by radicals due to UV absorption as a result of oxidation reactions.[1] Packaging materials for UV absorption that prolong the life of organic materials have also been energetically developed[2] to accomplish some sustainable development goals (*e.g.*, target 17.14).[3] For protection of living organisms and organic materials from UV radiation, various synthesized UV-absorbers (*e.g.*, triazole derivatives[4] and inorganic particles[1,5,6]) and hindered amine light stabilizers (HALSs)[7] have been broadly applied to give UV-absorbing nature for coating reagents, creams, and resins/plastics themselves. Many UV-absorbers and HALSs have been developed in accordance with their compatibility, UV absorbency, anti-volatile nature, and colorability for substances. However, some of the UV-absorbers and HALSs such as triazole derivatives[8] bring some environmental pollution

Biomaterials Science Series No. 16
Lignin-based Materials: Health Care and Medical Applications
Edited by Kuruvilla Joseph, Runcy Wilson, Gejo George and Saritha Appukuttan

Published by the Royal Society of Chemistry, www.rsc.org

and have adverse effects on the health of living organisms. Against this background, the development of novel UV-absorbers and/or HALSs from natural resources is important to save the environment and health of living organisms.

The energy of photons from UV radiation prompts electrons to jump from a bonding or nonbonding orbital into one of the empty antibonding orbitals that changes with the chemical nature and the electronic state. Due to UV radiation electrons potentially jump (1) from nonbonding orbitals to pi (π) antibonding orbitals, (2) from nonbonding orbitals to σ antibonding orbitals, (3) from π bonding orbitals to π antibonding orbitals, and (4) from sigma (σ) bonding orbitals to σ antibonding orbitals.[9] The absorbing wavelengths for types 1, 2, and 3 are from 200 to 800 nm that are located in the UV and visible light regions, so the group that includes lone pairs of electrons or π bonds can absorb the UV light.[9]

Several chromophores in the chemical structure of lignin correspond to its UV-absorption ability.[10] These chromophores[11,12] are (1) double bonds (CH=CH) conjugated with the aromatic ring, (2) chalcone structures, (3) quinone methides and quinones, (4) metal complexes with catechol structures, and (5) free radicals. Lignin absorbs UV light with wavelengths from 250 to 400 nm because of the presence of the bonds from the aromatic groups and lone pair electrons from oxygen atoms. The chromophores of the chemical structure in lignin are (1) aromatic groups by $\pi \rightarrow \pi^*$ electronic transition (absorbing UV light with a wavelength of 200 nm),[13] (2) free and etherified hydroxy groups (absorbing UV light with wavelengths of 240 and 282 nm, respectively),[14] (3) syringyl and guaiacyl groups (absorbing UV light with wavelengths of 270–273 nm and 280–282 nm, respectively),[15] and (4) aromatic groups by $\pi \rightarrow \pi^*$ and $n \rightarrow \pi^*$ transitions (absorbing UV light with a wavelength of 320 nm)[13] (Table 11.1).

The utilization of lignin as a UV-absorbers is significant for breaking away from an oil-refinery-based society to realize a biomass-refinery-based society in the industrial field of UV-protecting reagents.[16] Lignin is obtained as derivatives from plants (hereafter denoted as lignin derivatives) due to its chemical reaction and denaturation such as bond cleavage/generation prior to the isolation process from plants. The ordinary lignin extraction procedures are thermochemical treatment for plants such as pulping methods (*i.e.*, cooking reactions) that require some reaction with acids/bases at high temperatures.[17–19] Numerous physico-chemical procedures have also been

Table 11.1 Chromophores of lignin for UV absorption.

Chromophores	Wavelength of absorbed light/nm
Aromatic ring by $\pi \rightarrow \pi^*$ transition	200
Hydroxy group (free and etherified)	240 and 282
Syringyl group	270–273
Guaiacyl group	280–282
Aromatic ring by $\pi \rightarrow \pi^*$ and $n \rightarrow \pi^*$ transitions	320

studied to isolate polymeric contents from plants (*i.e.*, polysaccharides and lignin) that include the γ-valerolactone-based process,[20] glycerol thermal processing,[21–23] and the sulfite pretreatment of plants to overcome the recalcitrance of lignocelluloses (SPORL).[24] A wet-type ultrafine bead milling with enzymatic saccharification of plants has been proposed to realize simultaneous enzymatic saccharification and comminution (SESC) as a physico-biological technique to obtain polymeric contents in plants. The SESC process enables an isolation of the polymeric contents of plants as a sugar solution (from polysaccharides) and nanoparticulated aromatic polymers (from lignin) without contamination with toxic byproducts and reagents.[25,26]

In this chapter, we review recent progress in lignin-based UV-protective materials that encourages us to apply lignin derivatives as UV-absorbers in the industrial and/or biomedical fields.

11.2 Preparation of Lignin-based UV-protective Materials

To prepare UV-protective materials consisting of lignin derivatives (mainly films), various methods have been proposed by many researchers. Here, we introduce some examples of routes for the preparation of lignin-based UV-protective materials as described in Figure 11.1.

11.2.1 Composites Consisting of Polysaccharides and Lignin

It has been reported that defibered woods such as nanofibered wood[27] and lignocellulose[28,29] exhibit UV-blocking performance. In these reports, defibering wood and preparing lignocellulose have been performed by the

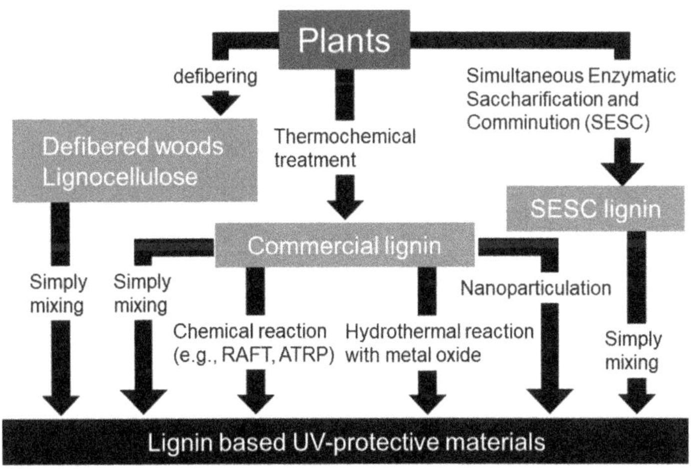

Figure 11.1 Outline for preparation of lignin-based UV-protective composites.

cationization of wood with carbamides,[27] treatment with alkaline hydrogen peroxide,[28] grinding,[28] and *p*-toluenesulfonic acid hydrolysis.[29] These defibered woods endow self-standing films with UV-absorbing nature.

11.2.2 Composites Consisting of Lignin Derivatives

Lignin derivatives isolated by cooking reaction for woods (generally called "commercial lignin") have been studied for their utilization as UV-absorbers. Because lignin derivatives themselves cannot be molded into films and resins due to their non-meltability and non-solubility, the lignin derivatives were composited with organic/inorganic molecules *via* the following procedures to form film and/or resin-shaped materials.

11.2.2.1 Lignin–Polymer Composites Prepared by Simple Mixing

There have been various reports about composites with UV-absorption nature consisting of lignin derivatives and polymers such as polyurethane,[30] cellulose derivatives,[31–33] gelatin,[34] poly(methyl methacrylate),[35] chitosan,[36] and poly(ethylene oxide)[37] that are prepared by solvent casting. To increase the compatibility of lignin derivatives and polymers, modifying lignin derivatives (*e.g.*, attaching graphene[30] and acetylation of lignin derivatives[31]) and/or adding compatibilizing agents[34] are performed as needed.

11.2.2.2 Lignin–Polymer Composites Prepared Through Chemical Reaction

To improve the miscibility of lignin derivatives with other components, their chemical modifications has been performed.

Grafting lignin derivatives is one of the methods to prepare composites consisting of chemically crosslinked lignin derivatives and other components. The ring-opening polymerization of lactide on lignin derivatives using triazabicyclodecene has been reported for the preparation of poly-L-lactic acid (PLLA)-grafted lignin derivatives.[38,39] PLLA-grafted lignin derivatives act as compatibilizing agents of lignin derivatives and PLLA that endow transparent PLLA–lignin composite films with UV-absorption nature.[38] Lignin derivatives have also been grafted by reversible addition–fragmentation chain transfer (RAFT) methods.[40–44] It has been reported that the lignin-based grafted poly(*n*-butyl acrylate-*co*-1-vinylimidazole) prepared by RAFT methods gives UV-absorbing coatings for silica glass.[42] Atom transfer radical polymerization (ATRP) of lignin derivatives also achieves grafting of lignin derivatives.[43] It has been reported that lignin derivative-grafted 2-3-benzoyl-3-hydroxyphenyl acrylate prepared by the ATRP method exhibits UV-blocking performance.[43]

Modifying lignin derivatives with functional groups *via* chemical reaction also induces high miscibility for synthetic polymers that realizes fine dispersion of lignin derivatives in matrix polymers. For example, a modification of a lignin derivative with maleic anhydride makes it compatible with poly(butylene succinate) which endows transparent films with UV-absorbing nature derived from the lignin derivative.[44] Reaction between propargylated lignin derivatives and azide modified cellulose also affords their homogeneous transparent films with UV-absorption properties.[45] Quaternarization of lignin derivatives enables their composition with silica[46] and polyurethane,[47] which gives hybrid materials with UV-absorption properties.

11.2.3 UV-absorption Properties of Composites Consisting of Metal Oxides and Lignin Derivatives

There have been some reports about functional composites consisting of lignin derivatives and metal oxides such as zinc oxide (ZnO) and titanium oxide (TiO_2) that exhibit various properties such as microbicidal performance and photocatalytic activity. Their UV-absorption properties are also studied. These composites can be used as natural sunscreen creams, sustainable UV-protective coatings, and eco-friendly UV-blocking films as described in the following sections.

11.2.3.1 ZnO–Lignin Composites

Combining lignin with ZnO enhances its UV-shielding performance in the full UV spectrum.[48] Composites consisting of ZnO and lignin derivatives (ZnO–lignin composites) have been synthesized *via* hydrothermal reaction between various commercial lignin derivatives and zinc ingredients.[49–52] ZnO–lignin composites are obtained as nanoparticles with a spherical shape[49,50,52] or flowerlike shape.[51] ZnO–lignin composites act as UVAs in their water dispersion[48] and in polymer films such as polyurethane[50,51] and PLLA.[52] Lignin-based carbon/ZnO composites also afford higher UV-absorption materials because the benzene ring structure and C=O bonding together enhance UV light absorption.[53]

11.2.3.2 TiO_2–Lignin Composites

Composites consisting of TiO_2 and lignin derivatives (TiO_2–lignin composites) have been synthesized *via* hydrothermal reaction between various commercial lignin derivatives and titanium ingredients, like ZnO–lignin composites.[54–59] The structure of TiO_2–lignin composites was revealed as lignin-covered titanium core–shell nanoparticles.[54] The UV-blocking nature of TiO_2–lignin composites for their utilization as sunscreens was demonstrated by blending them in creams.[54–56] Additionally, TiO_2–lignin composites also form UV-protective coatings on various polymer films.[57] TiO_2–lignin composites can exhibit UV-absorption nature even in dispersions.[57] Blending TiO_2–lignin composites into polymer films such as poly(propylene carbonate)[58] and waterborne polyurethane[59] gives UV-absorption nature. The strategy shown here is based on a nontoxic, biocompatible lignin shell that neutralizes the free radicals by scavenging them with natural antioxidants before they exit the particles.[60] The TiO_2–lignin composites preserve the scattering and absorption properties of TiO_2 because the particles retain their nanoscale dimensions as preferred by the cosmetic industry.

11.2.4 Micro/Nano-sized Micelles/Particles Consisting of Lignin Derivatives

Lignin derivatives themselves form micro/nano-sized micelles or particles by controlling the polarity of solvents in their dispersion/solution. Here, an aggregation of the hydrophobic skeleton and hydrophilic groups of lignin derivatives in appropriate solvents gives their micro/nano-sized micelles or particles. Additionally, hydrophilic nanoparticles of lignin were recently produced *via* SESC of plants, that is, the ultrafine bead milling of plants simultaneously with enzymatic degradation of polysaccharides in water. These micro/nano-sized micelles and particles of lignin derivatives exhibit UV-absorption nature as described in the following sections.

11.2.4.1 Nano-sized Micelles Consisting of Commercial Lignin Derivatives and Their Composites

Lignin derivatives such as alkali lignin and enzymatic hydrolysis lignin form reverse micelles in a mixture of organic solvents (*e.g.*, dioxane/cyclohexane[61]). Dialyzing sodium salt from lignin sodium hydroxide solution also affords lignin derivative nanomicelles.[62] These micelles are stable upon drying and blending with other components. Even after physical milling with melted polymer, these micelles retain their shape in the polymer composites.[61]

The composites consisting of lignin reverse micelles and high density polyethylene exhibit UV-absorbing nature with good mechanical performance.[61] A solution of lignin nanomicelles and poly(vinyl alcohol) gives UV-protective films with water vapour barrier performance by solution casting.[62]

11.2.4.2 Micro/Nano-sized Particles Consisting of Commercial Lignin Derivatives and Their Composites Consisting of Organic Polymers

Lignin nanoparticles are obtained by adding antisolvent to solutions of lignin derivatives such as enzymatic hydrolysis lignin, acetic acid lignin, and alkali lignin. The examples of solvent combination are tetrahydrofuran (THF)/deionized water,[63] acetic acid/deionized water,[64] and methanol/THF/dimethyl sulfoxide/dimethyl formamide/deionized water.[65] As needed, stabilizers such as polyvinylpyrrolidone and sodium dodecyl sulfate[65] are added. The diameter of lignin nanoparticles ranges from several tens of nanometers to sub-micrometers and is controlled by tuning the volume ratio of solvents.[65] Lignin nanoparticles retain their shape after aging for several tens of days.[65] From kraft lignin,[66] enzymatic hydrolysis lignin,[67] organosolv lignin,[68] and alkali lignin,[69] nano/micro-particulated lignin has been obtained.

Lignin nanoparticles can be used as modifiers for creams and various polymers. For example, solution casting of a mixture consisting of nanoparticles of lignin and poly(vinyl alcohol) gives transparent composite films with excellent UV-absorption properties, which make them promising candidates for packaging of medicine and food.[63,65] Creams blended with lignin nanoparticles exhibit high sun protection factors (SPFs) that enable them to act as sunscreens and help in preventing skin from staining.[64] Furthermore, lignin nanoparticles can be used as functional crosslinkers for polyurethane[70] while retaining their UV-absorbing nature.

11.2.4.3 Hydrophilic Lignin Nanoparticles Obtained by SESC of Plants and Their Composites Consisting of Organic Polymers or Clay Mineral

Recently, hydrophilic nanoparticles of lignin were produced *via* SESC of plants, that is, the ultrafine bead milling of plants simultaneously with enzymatic degradation of polysaccharides in water.[25,26,71,72] In principle, this process makes an extraction of polymeric contents of plants including lignin under extremely mild conditions (*i.e.*, 50 °C, $4<pH<6$, and 1 psi) possible without utilization of toxic reagents. The polysaccharides in plants can be retrieved as mono/oligo-sugars through SESC treatments. The sugars obtained by SESC give ethanol[73] and methane[74] *via* their fermentation.

The lignin derivative extracted by SESC (denoted "SESC lignin") has nanoparticles with diameters of several tens of nm dispersed in water as shown in Figure 11.2.

Dried SESC lignin is a transparent brown-colored film and it can be ground into a fine powder.[25] SESC lignin is finely mixed with other polymer contents (*e.g.*, synthetic polymers).[75] The physical properties of SESC lignin are ideal for its utilization as a functional material (*i.e.*, antioxidant nanoparticles[76] and a hard segment of shape-memory polymers).[77] Another noteworthy characteristic of SESC lignin is its heatproof properties for synthetic polymers;[75,78–80] *e.g.*, the addition of several percentages of SESC lignin causes a dramatic increase in the thermal decomposition temperature of poly(ethylene carbonate) (PEC), poly(propylene carbonate), and poly(vinyl alcohol). Applying SESC lignin to improve the heat tolerance of synthetic

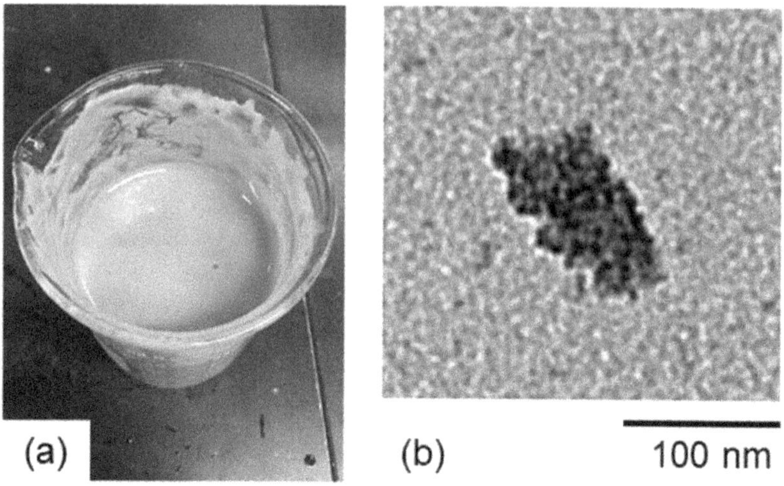

Figure 11.2 (a) Photograph of SESC lignin dispersed in water as slurry. (b) Transmission electron microscopy (TEM) image of dried SESC lignin.

polymers never requires some toxic reagents; *i.e.*, SESC lignin is a heat-proofing additive consisting of only plant-based contents. Furthermore, it was recently reported that addition of a small amount of SESC lignin causes the large improvement of ionic conductivity on poly(ethylene oxide) and PEC.[81]

SESC lignin also acts as a UV-absorber. Blending several weight percentages of SESC lignin into poly(vinyl alcohol) affords transparent UV-absorption films (Figure 11.3).[80]

Because of the hydrogen bonding between the hydroxyl groups of SESC lignin and that of poly(vinyl alcohol), SESC lignin disperses in a polymer matrix as nanoparticles that induce high transparency and good mechanical performance of SESC lignin–poly(vinyl alcohol) composite films; *i.e.*, an addition of SESC lignin never spoils the mechanical properties of poly(vinyl alcohol) films.[80]

Dried SESC lignin contains brittle and inhomogeneous small pieces that cannot be molded as products such as films and resins.[25] The Claist®[82] production technique encourages us to prepare a soft film with self-standing nature consisting of SESC lignin and clay minerals[83] as shown in Figure 11.4.[84]

This composite film consisting of clay minerals and SESC lignin showed UV-protecting properties due to the UV-absorption nature of SESC lignin as shown in Figure 11.5. The characteristics as the inorganic molecules (clay minerals) and as the aromatic polymer (SESC lignin) provide the non-flammability of the films. In the films, clay minerals and SESC lignin exhibited phase separation (Figure 11.4). There are many nano-sized holes in the SESC lignin phase that gives their moisture and gas transmittance nature.

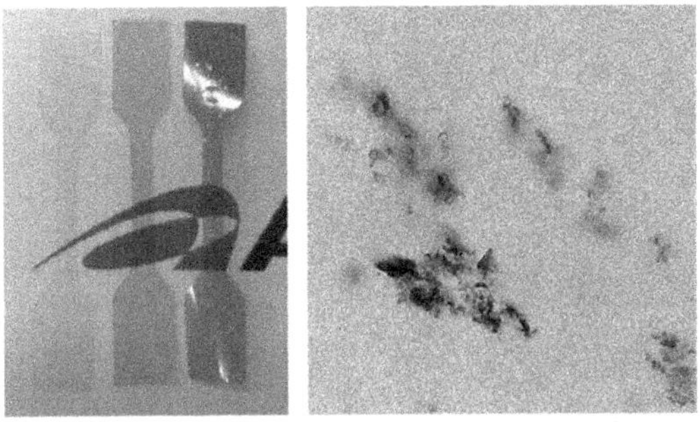

Figure 11.3 (a) Photograph of poly(vinyl alcohol)–SESC lignin composite films: (left) neat poly(vinyl alcohol) film, (middle) poly(vinyl alcohol) film with 5% w/w of SESC lignin, and (right) poly(vinyl alcohol) film with 50% w/w of SESC lignin. (b) Cross-sectional TEM image of a poly(vinyl alcohol) film with 5% w/w of SESC lignin.

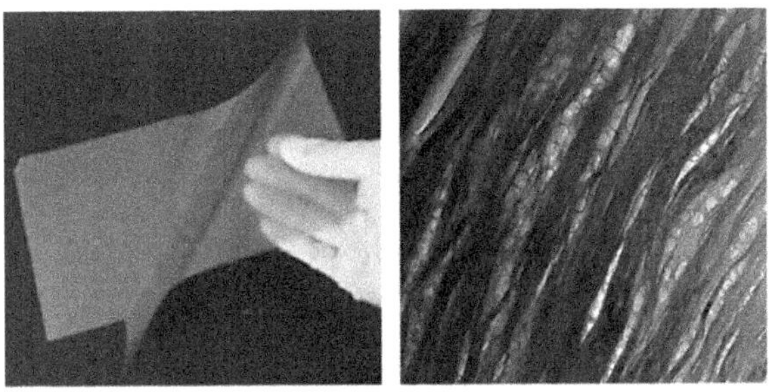

Figure 11.4 (a) Photograph of a typical clay mineral–SESC lignin composite film with approx. A4-paper size (Li^+ montmorillonite : SESC lignin = 8 : 2 w/w%). (b) Cross-sectional TEM image of the clay mineral–SESC lignin composite film.

Furthermore, the physical characteristics of the films (*e.g.*, light and gas transmittance) could be tuned by controlling their structure. For example, the helium and oxygen transmittance of the films were tuned in the range of several mol m^{-2} s^{-1} Pa^{-1} by changing the species of clay minerals. Especially, the oxygen transmittance ratio of the films consisting of synthetic clay minerals, SESC lignin, and poly(vinyl alcohol) was extremely lower than that of poly(ethylene terephthalate) films (2.0×10^{-17} mol m^{-2} s^{-1} Pa^{-1}).[85]

In a recent study, a novel technique for permanent whitening of SESC lignin *via* solvent-controlled encapsulation was proposed.[86] Even after being permanently whitened, the SESC lignin retained its specific UV-absorption nature, which originated from its chemical components such as guaiacyl and syringyl groups. This result encourages us to apply lignin derivatives as

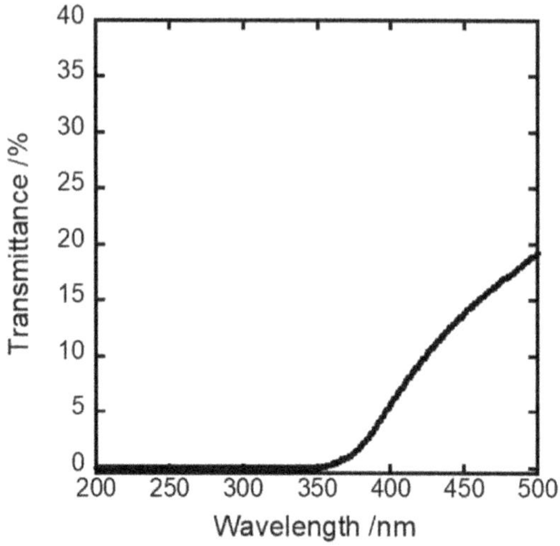

Figure 11.5 Typical UV–vis spectrum of a clay mineral–SESC lignin composite film (Li^+ montmorillonite : SESC lignin = 8 : 2 w/w%).

UV-absorbers with high design flexibility that gives a broad-range usage of lignin to realize conversion from an oil-based refinery to a biomass-based refinery.

11.3 Summary and Future Perspectives

As shown in this chapter, lignin derivatives can be used as UV-absorbers. Blending lignin derivatives with other components such as small molecules, organic polymers, metal oxides, and clay minerals affords lignin-based UV-protective materials. The morphology of lignin derivatives strongly affects their functions like UV-absorption performance. Generally, simple mixing, chemical reaction with organic/inorganic molecules, and nanoparticulation are introduced for the utilization of lignin derivatives as UV-absorbers. Lignin-based UVAs enable us to develop green and sustainable UV-absorbers that are totally free from toxic reagents. The UV-absorber nature of lignin derivatives encourages us to apply them as novel biomass-based polymers for achievement of SDGs.[87] Recently, the role of monomer units in lignin for UV absorption has been revealed by an extraction and estimation of lignin low molecular fraction.[88] A mechanism of UV absorbance by lignin derivatives has also been reviewed in detail.[89] Further understanding about the relationship between the chemical structure and UV-absorption nature of lignin derivatives accelerates the utilization of lignin derivatives as UV-absorbers and HALSs.

The singular UV extinction characteristic of lignin derivatives described in this chapter results in an aggregation-induced emission (AIE) of lignin derivatives that exhibits blue lignin fluorescence, *i.e.*, the novel potential of lignin derivatives for AIE-active materials based on unconventional chromophores.[90–93] Thus, the UV-absorption nature of lignin derivatives encourages us to design lignin derivatives not only as UV-absorbers but also as photoluminescence compounds.

Acknowledgements

The studies described in Section 11.2.4.3 were supported by grants from JST ALCA (Grant Number JPMJAL1601) and JST-Mirai R&D (Grant Number JPMJMI19E8) and assisted by various collaborators.

References

1. C. C. Han, F. Wang, C. Gao, P. Liu, Y. F. Ding, S. M. Zhang and M. S. Yang, *J. Mater. Chem. C*, 2015, **3**, 5065.
2. T. J. Simmons, S. H. Lee, J. Miao, M. Miyauchi, T. J. Park, S. S. Bale, R. Pangule, J. Bult, J. G. Martin, J. S. Dordick and R. J. Linhardt, *Wood Sci. Technol.*, 2011, **45**, 719.
3. SDG indicators, https://unstats.un.org/sdgs/indicators/indicators-list/.
4. D. M. Wiles and D. J. Carlsson, *Polym. Degrad. Stab.*, 1980, **3**, 61.
5. B. Tigges, M. Möller and O. Weichold, *J. Colloid Interface Sci.*, 2010, **345**, 41.
6. X. L. Wang, S. X. Zhou and L. M. Wu, *J. Mater. Chem. C*, 2013, **1**, 7547.
7. C. W. Klampfl and M. Himmelsbach, *Anal. Chim. Acta*, 2016, **933**, 10.
8. D. A. Pillard, J. S. Copnell, D. L. Dufresne and M. T. Hernandez, *Water Res.*, 2001, **35**, 557.
9. H. Förster, *Molecular Sieves*, 2004, vol. 4, p. 337.
10. K. V. Sarkanen and C. H. Ludwig, in *Lignin: Occurrence, Formation, Structure and Reactions*, ed. K. V. Sarkanen and C. H. Ludwig, Wiley Interscience, New York, 1971.
11. J. Wang, Y. Deng, Y. Qian, X. Qiu, Y. Ren and D. Yang, *Green Chem.*, 2016, **18**(3), 695.
12. S. I. Falkehag, J. Marton and E. Adler, *Adv. Chem. Ser.*, 1966, **59**, 75.
13. T. You and F. Xu, *Applications of Molecular Spectroscopic Methods to the Elucidation of Lignin Structure*, InTechOpen Rijeka, 2016, p. 235.
14. D. N.-S. Hon and N. Shiraishi, *Wood and Cellulosic Chemistry*, Marcel Dekker, New York, 2nd edn, 2001.
15. S. C. Lee, T. M. T. Tran, J. W. Choi and K. Won, *Int. J. Biol. Macromol.*, 2019, **122**, 549.
16. H. Wang, Y. Pu, A. Ragauskas and B. Yang, *Bioresour. Technol.*, 2019, **271**, 449.
17. E. Adler, *Wood Sci. Technol.*, 1977, **11**, 169.
18. S. Aziz and K. V. Sarkanen, *Tappi J.*, 1989, **72**, 169.
19. K. Shikinaka, N. Fujii, S. Egashira, Y. Murakami, M. Nakamura, Y. Otsuka, S. Ohara and K. Shigehara, *Green Chem.*, 2010, **12**, 1914.
20. J. S. Luterbacher, J. M. Rand, D. M. Alonso, J. Han, J. T. Youngquist, C. T. Maravelias, B. F. Pfleger and J. A. Dumesic, *Science*, 2014, **343**, 277.

21. W. Zhang, J. R. Barone and S. Renneckar, *ACS Sustainable Chem. Eng.*, 2015, **3**(3), 413.
22. W. Zhang, N. Sathitsuksanoh, J. R. Barone and S. Renneckar, *Bioresour. Technol.*, 2016, **199**, 148.
23. W. Zhang, N. Sathitsuksanoh, B. A. Simmons, C. E. Frazier, J. R. Barone and S. Renneckar, *RSC Adv.*, 2016, **6**(36), 30234.
24. J. Y. Zhu, M. S. Chandra, F. Gu, R. Gleisner, R. Reiner, J. Sessions, G. Marrs, J. Gao and D. Anderson, *Bioresour. Technol.*, 2015, **179**, 390.
25. K. Shikinaka, Y. Otsuka, R. R. Navarro, M. Nakamura, T. Shimokawa, M. Nojiri, R. Tanigawa and K. Shigehara, *Green Chem.*, 2016, **18**, 5962.
26. R. R. Navarro, Y. Otsuka, M. Nojiri, S. Ishizuka, M. Nakamura, K. Shikinaka, K. Matsuo, K. Sasaki, K. Sasaki, K. Kimbara, Y. Nakashimada and J. Kato, *BMC Biotechnol.*, 2018, **18**, 79.
27. J. A. Sirviö, M. Y. Ismail, K. Zhang, M. V. Tejesvi and A. Ämmälä, *J. Mater. Chem. A*, 2020, **8**, 7935.
28. Y. Jiang, Z. Wang, X. Liu, Q. Yang, Q. Huang, L. Wang, Y. Dai, C. Qin and S. Wang, *ACS Sustainable Chem. Eng.*, 2020, **8**, 17508.
29. H. Bian, L. Chen, M. Dong, L. Wang, R. Wang, X. Zhou, C. Wu, X. Wang, X. Ji and H. Dai, *Int. J. Biol. Macromol.*, 2021, **166**, 1578.
30. S. Ismail, S. Shahabadi, J. Kong and X. Lu, *ACS Sustainable Chem. Eng.*, 2017, **5**, 3148.
31. M. Parit, P. Saha, V. A. Davis and Z. Jiang, *ACS Omega*, 2018, **3**, 10679.
32. M. T. Haqiqi, W. Bankeeree, P. Lotrakul, P. Pattananuwat, H. Punnapayak, R. Ramadhan, T. Kobayashi, R. Amirta and S. Prasongsuk, *ACS Omega*, 2021, **6**, 9653.
33. A. Hambardzumyan, L. Foulon, B. Chabbert and V. Aguié-Béghin, *Biomacromolecules*, 2012, **13**, 4081.
34. M. J. Mehta and A. Kumar, *Chem. – Eur. J.*, 2019, **25**, 1269.
35. F. Avelino, D. R. de Oliveira, S. E. Mazzetto and D. Lomonaco, *Int. J. Biol. Macromol.*, 2019, **125**, 171.
36. N. Izaguirre, O. Gordobil, E. Robles and J. Labidi, *Int. J. Biol. Macromol.*, 2020, **155**, 447.
37. R. Wu, S. Wang, Y. Leng and Q. Li, *J. Appl. Polym. Sci.*, 2020, **137**, 48593.
38. Y.-L. Chung, J. V. Olsson, R. J. Li, C. W. Frank, R. M. Waymouth, S. L. Billington and E. S. Sattely, *ACS Sustainable Chem. Eng.*, 2013, **1**, 1231.
39. S. Y. Park, J.-Y. Kim, H. J. Youn and J. W. Choi, *Int. J. Biol. Macromol.*, 2019, **138**, 1029.
40. G. Gao, M. A. Karaaslan, J. F. Kadla and F. Ko, *Green Chem.*, 2014, **16**, 3890.
41. D. Kai, K. Y. Zhang, L. Jiang, H. Z. Wong, Z. B. Li, Z. Zhang and X. J. Loh, *ACS Sustainable Chem. Eng.*, 2017, **5**, 6016.
42. W. Wang, F. Wang, C. Zhang, J. Tang, X. Zeng and X. Wan, *Chem. Eng. J.*, 2021, **404**, 126358.

43. Y. Wu, Y. Qian, A. Zhang, H. Lou, D. Yang and X. Qiu, *Ind. Eng. Chem. Res.*, 2020, **59**, 17057.
44. Y. Zhang, S. Zhou, X. Fang, X. Zhou, J. Wang, F. Bai and S. Peng, *Eur. Polym. J.*, 2019, **116**, 265.
45. H. Sadeghifar, R. Venditti, J. Jur, R. E. Gorga and J. J. Pawlak, *ACS Sustainable Chem. Eng.*, 2017, **5**, 625.
46. W. Xiong, X. Qiu, D. Yang, R. Zhong, Y. Qian, Y. Li and H. Wang, *Chem. Eng. J.*, 2017, **326**, 803.
47. X. Li, X. Chen, S. Zhang, Y. Yin and C. Wang, *Eur. Polym. J.*, 2021, **159**, 110763.
48. J. M. Gutiérrez-Hernández, A. Escalante, R. N. Murillo-Vázquez, E. Delgado, F. J. González and G. Toríz, *J. Photochem. Photobiol., B*, 2016, **163**, 156.
49. R. Kaur, N. S. Thakur, S. Chandna and J. Bhaumik, *J. Mater. Chem. B*, 2020, **8**, 260.
50. H. Wang, X. Qiu, W. Liu, F. Fu and D. Yang, *Ind. Eng. Chem. Res.*, 2017, **56**, 11133.
51. H. Wang, W. Lin, X. Qiu, F. Fu, R. Zhong, W. Liu and D. Yang, *ACS Sustainable Chem. Eng.*, 2018, **6**, 3696.
52. F. Luzi, I. Tortorella, A. Di Michele, F. Dominici, C. Argentati, F. Morena, L. Torre, D. Puglia and S. Martino, *Nanomaterials*, 2020, **10**, 2176.
53. H. Wang, X. Qiu, W. Liu and D. Yang, *Appl. Surf. Sci.*, 2017, **426**, 206.
54. Y. Li, D. Yang, S. Lu, X. Qiu, Y. Qian and P. W. Li, *ACS Sustainable Chem. Eng.*, 2019, 7, 6234.
55. J. Yu, L. Li, Y. Qian, H. Lou, D. Yang and X. Qiu, *Ind. Eng. Chem. Res.*, 2018, **57**, 15740.
56. D. Guo, J. Zhang, L. Sha, B. Liu, X. Zhang, X. Zhang and G. Xue, *BioResources*, 2020, **15**(4), 7374.
57. R. Kaur, N. S. Thakur, S. Chandna and J. Bhaumik, *ACS Sustainable Chem. Eng.*, 2021, **9**, 11223.
58. W. Wu, T. Liu, X. Deng, Q. Sun, X. Cao, Y. Feng, B. Wang, V. A. L. Roy and R. K. Y. Li, *Int. J. Biol. Macromol.*, 2019, **126**, 1030.
59. H. Wang, Y. Wang, F. Fu, Y. Qian, Y. Xiao, D. Yang and X. Qiu, *Int. J. Biol. Macromol.*, 2020, **150**, 371.
60. M. Morsella, N. d'Alessandro, A. E. Lanterna and J. C. Scaiano, *ACS Omega*, 2016, **1**(3), 464.
61. Y. Qian, X. Qiu, X. Zhong, D. Zhang, Y. Deng, D. Yang and S. Zhu, *Ind. Eng. Chem. Res.*, 2015, **54**, 12025.
62. X. Zhang, W. Liu, W. Liu and X. Qiu, *Int. J. Biol. Macromol.*, 2020, **142**, 551.
63. F. Xiong, Y. Wu, G. Li, Y. Han and F. Chu, *Ind. Eng. Chem. Res.*, 2018, **57**, 1207.
64. H. Zhang, X. Liu, S. Fu and Y. Chen, *Ind. Eng. Chem. Res.*, 2019, **58**, 13858.
65. T. Ju, Z. Zhang, Y. Li, X. Miao and J. Ji, *RSC Adv.*, 2019, **9**, 24915.

66. M. Ma, L. Dai, J. Xu, Z. Liu and Y. Ni, *Green Chem.*, 2020, **22**, 2011.
67. T. Pang, G. Wang, H. Sun, L. Wang, Q. Liu, W. Sui, A. M. Parvez and C. Si, *ACS Sustainable Chem. Eng.*, 2020, **8**, 9174.
68. Z.-H. Liu, N. Hao, S. Shinde, Y. Pu, X. Kang, A. J. Ragauskas and J. S. Yuan, *Green Chem.*, 2019, **21**, 245.
69. P. Jiang, Q. Li, C. Gao, J. Lu, Y. Cheng, S. Zhai, Q. An and H. Wang, *Bioresour. Technol.*, 2019, **289**, 121640.
70. G. Qi, W. Yang, D. Puglia, H. Wang, P. Xu, W. Dong, T. Zheng and P. Ma, *Mater. Des.*, 2020, **196**, 109150.
71. K. Shikinaka, Y. Otsuka, M. Nakamura, E. Masai and Y. Katayama, *J. Oleo Sci.*, 2018, **67**(9), 1059.
72. K. Shikinaka, M. Nakamura, R. R. Navarro and Y. Otsuka, *Trends Glycosci. Glycotechnol.*, 2020, **32**(186), E63.
73. Y. Otsuka, M. Nojiri, N. Kusumoto, R. R. Navarro, K. Hashida and N. Matsui, *RSC Adv.*, 2020, **10**, 39753.
74. R. R. Navarro, Y. Otsuka, K. Matsuo, K. Sasaki, K. Sasaki, T. Hori, H. Habe, M. Nakamura, Y. Nakashimada, K. Kimbara and J. Kato, *Bioresour. Technol.*, 2020, **300**, 122622.
75. K. Shikinaka, H. Sotome, Y. Kubota, Y. Tominaga, M. Nakamura, R. R. Navarro and Y. Otsuka, *J. Mater. Chem. A*, 2018, **6**, 837.
76. K. Shikinaka, M. Nakamura, R. R. Navarro and Y. Otsuka, *ChemistryOpen*, 2018, 7, 709.
77. K. Shikinaka, Y. Funatsu, Y. Kubota, Y. Tominaga, M. Nakamura, R. R. Navarro and Y. Otsuka, *Soft Matter*, 2018, **14**, 9227.
78. H. Sotome, K. Shikinaka, A. Tsukidate, Y. Tominaga, M. Nakamura and Y. Otsuka, *Polym. Degrad. Stab.*, 2020, **179**, 109273.
79. K. Shikinaka, A. Tsukidate, Y. Tominaga, H. Inoue and Y. Otsuka, *Polym. J.*, 2021, **53**(7), 841.
80. K. Shikinaka, M. Nakamura and Y. Otsuka, *Polymer*, 2020, **190**, 122254.
81. Z. Liu, K. Shikinaka, Y. Otsuka and Y. Tominaga, *Chem. Commun.*, 2022, **58**, 4504.
82. T. Ebina and F. Mizutani, *Adv. Mater.*, 2007, **19**, 2450.
83. M. Stöter, S. Rosenfeldt and J. Breu, *Annu. Rev. Mater. Res.*, 2015, **45**, 129.
84. K. Shikinaka, M. Nakamura, R. R. Navarro and Y. Otsuka, *Green Chem.*, 2019, **21**, 498.
85. K. Shikinaka, A. Suzuki and Y. Otsuka, *RSC Adv.*, 2021, **11**, 23385.
86. K. Shikinaka and Y. Otsuka, *Green Chem.*, 2022, **24**, 3243.
87. R. E. Neale, P. W. Barnes, T. M. Robson, P. J. Neale, C. E. Williamson, R. G. Zepp, S. R. Wilson, S. Madronich, A. L. Andrady, A. M. Heikkilä, G. H. Bernhard, A. F. Bals, P. J. Aucamp, A. T. Banaszak, J. F. Bornman, L. S. Bruckman, S. N. Byrne, B. Foereld, D.-P. Häder, L. M. Hollestein, W.-C. Hou, S. Hylander, M. A. K. Jansen, A. R. Klekociuk, J. B. Liley, J. Longstreth, R. M. Lucas, J. Martinez-Abaigar, K. McNeill, C. M. Olsen, K. K. Pandey, L. E. Rhodes, S. A. Robinson, K. C. Rose, T. Schikowski, K. R. Solomon, B. Sulzberger, J. E. Ukpebor, Q.-W. Wang,

S.-Å. Wängberg, C. C. White, S. Yazar, A. R. Young, P. J. Young, L. Zhu and M. Zhu, *Photochem. Photobiol. Sci.*, 2021, **20**, 1.

88. K. Mikame, Y. Ohashi, Y. Naito, H. Nishimura, M. Katahira, S. Sugawara, K. Koike and T. Watanabe, *ACS Sustainable Chem. Eng.*, 2021, **9**, 16651.
89. Y. Zhang and M. Naebe, *ACS Sustainable Chem. Eng.*, 2021, **9**, 1427.
90. Y. Xue, X. Qiu, Y. Wu, Y. Qian, M. Zhou, Y. Deng and Y. Li, *Polym. Chem.*, 2016, 7, 3502.
91. Z. Ma, C. Liu, N. Niu, Z. Chen, S. Li, S. Liu and J. Li, *ACS Sustainable Chem. Eng.*, 2018, **6**, 3169.
92. Y. Xue, X. Qiu and X. Ouyang, *Int. J. Biol. Macromol.*, 2020, **154**, 981.
93. M. Takada, Y. Okazaki, H. Kawamoto and T. Sagawa, *ACS Omega*, 2022, 7, 5096.

Subject Index